SPRINGER PROCEEDINGS IN PHYSICS 110

SPRINGER PROCEEDINGS IN PHYSICS

90 **Computer Simulation Studies in Condensed-Matter Physics XV**
Editors: D.P. Landau, S.P. Lewis, and H.-B. Schüttler

91 **The Dense Interstellar Medium in Galaxies**
Editors: S. Pfalzner, C. Kramer, C. Straubmeier, and A. Heithausen

92 **Beyond the Standard Model 2003**
Editor: H.V. Klapdor-Kleingrothaus

93 **ISSMGE**
Experimental Studies
Editor: T. Schanz

94 **ISSMGE**
Numerical and Theoretical Approaches
Editor: T. Schanz

95 **Computer Simulation Studies in Condensed-Matter Physics XVI**
Editors: D.P. Landau, S.P. Lewis, and H.-B. Schüttler

96 **Electromagnetics in a Complex World**
Editors: I.M. Pinto, V. Galdi, and L.B. Felsen

97 **Fields, Networks, Computational Methods and Systems in Modern Electrodynamics**
A Tribute to Leopold B. Felsen
Editors: P. Russer and M. Mongiardo

98 **Particle Physics and the Universe**
Proceedings of the 9th Adriatic Meeting, Sept. 2003, Dubrovnik
Editors: J. Trampetić and J. Wess

99 **Cosmic Explosions**
On the 10th Anniversary of SN1993J (IAU Colloquium 192)
Editors: J. M. Marcaide and K. W. Weiler

100 **Lasers in the Conservation of Artworks**
LACONA V Proceedings, Osnabrück, Germany, Sept. 15–18, 2003
Editors: K. Dickmann, C. Fotakis, and J.F. Asmus

101 **Progress in Turbulence**
Editors: J. Peinke, A. Kittel, S. Barth, and M. Oberlack

102 **Adaptive Optics for Industry and Medicine**
Proceedings of the 4th International Workshop
Editor: U. Wittrock

103 **Computer Simulation Studies in Condensed-Matter Physics XVII**
Editors: D.P. Landau, S.P. Lewis, and H.-B. Schüttler

104 **Complex Computing-Networks**
Brain-like and Wave-oriented Electrodynamic Algorithms
Editors: I.C. Göknar and L. Sevgi

105 **Computer Simulation Studies in Condensed-Matter Physics XVIII**
Editors: D.P. Landau, S.P. Lewis, and H.-B. Schüttler

106 **Modern Trends in Geomechanics**
Editors: W. Wu and H.S. Yu

107 **Microscopy of Semiconducting Materials**
Proceedings of the 14th Conference, April 11–14, 2005, Oxford, UK
Editors: A.G. Cullis and J.L. Hutchison

108 **Hadron Collider Physics 2005**
Proceedings of the 1st Hadron Collider Physics Symposium, Les Diablerets, Switzerland, July 4–9, 2005
Editors: M. Campanelli, A. Clark, and X. Wu

109 **Progress in Turbulence 2**
Proceedings of the iTi Conference in Turbulence 2005
Editors: M. Oberlack et al.

110 **Nonequilibrium Carrier Dynamics in Semiconductors**
Proceedings of the 14th International Conference, July 25-29, 2005, Chicago, USA
Editors: M. Saraniti, U. Ravaioli

Volumes 64–89 are listed at the end of the book.

M. Saraniti U. Ravaioli
(Eds.)

Nonequilibrium Carrier Dynamics in Semiconductors

Proceedings of the
14th International Conference,
July 25–29, 2005, Chicago, USA

With 223 Figures

Springer

Professor M. Saraniti
Department of Electrical
and Computer Engineering
Illinois Institute of Technology
Suite 103, Siegel Hall
3301 South Dearborn Street
Chicago, IL 60616, USA

Professor U. Ravaioli
Institute for Advanced Science and Technolgy
University of Illinois
405 Nroth Mathes Avenue
Urbana, IL 61801, USA

Published in association with Canopus Publishing Limited, Bristol, UK

ISSN 0930-8989

ISBN-10 3-540-36587-7 Springer Berlin Heidelberg New York

ISBN-13 978-3-540-36587-7 Springer Berlin Heidelberg New York

Library of Congress Control Number: 2006929190

Springer is a part of Springer Science+Business Media.
springer.com

Printed in the UK

Cover concept: eStudio Calamar Steinen
Cover production: *design & production* GmbH, Heidelberg
Printing: Short Run Express, Exeter, UK

Printed on acid-free paper SPIN: 11575108 54/3141/mh 5 4 3 2 1 0

Preface

This volume contains invited and contributed papers of the 14th International Conference on Nonequilibrium Carrier Dynamics in Semiconductors (HCIS-14) held July 24-29, 2005 in Chicago, Illinois.

The conference featured five invited and 62 contributed talks, as well as 49 posters and an international contingent of more than 80 scientists. Following the tradition of the conference, the topics discussed identified the most promising developments of nonlinear transport studies. Among these, interesting contributions were offered on mesoscopic systems, coherence in charge transport, ultrafast phenomena and TeraHertz devices. Two sessions were devoted to high field transport in nitrides, while the discussion on spintronics and thermoelectric phenomena clearly indicated the importance of these topics for the next generations of devices. Finally, a session was devoted to molecular electronics and two to bioelectronics, stressing the interest of the community in the study of charge transport in complex macromolecular systems.

On behalf of the Program and International Advisory Committees, we thank the participants, who made the conference a successful and pleasant experience and the generous support of DARPA, IBM, the Beckman Institute of the University of Illinois, and the Illinois Institute of Technology in Chicago. We are also indebted to Ms. Sara Starkey and Ms. Carol Osmer for their invaluable contribution to the conference organization and administration.

Marco Saraniti
Umberto Ravaioli

Contents

Preface v

Electron transport in curved low dimensional electron systems
N Shaji, H Qin, I Knezevic, C Deneke, O G Schmidt, M A Eriksson and R H Blick 1

Fabrication and characterization of InAs mesoscopic devices
M Koyama, M Furukawa, H Ishii, M Nakai, T Maemoto, S Saas and M Inoue 7

Nonlinear effects on quantum interference in electron billiards
C A Marlow, R P Taylor, M Fairbanks and H Linke 11

Prediction of entanglement detection by I-V characteristics
T Zibold, P Vogl and A Bertoni 15

Simulation of entanglement creation for carrier-impurity scattering in a 2D system
P Bordone and A Bertoni 19

Super-Poissonian current fluctuations in tunnelling through coupled quantum dots
G Kießlich, A Wacker and E Schöll 23

Ultrafast formation of coupled phonon-plasmon modes in InP observed with femtosecond terahertz spectroscopy
C Kübler, R Huber, S Tübel, F Köhler, M C Amann and A Leitenstorfer 29

Optical coherent control of polariton modes in ZnSe single-quantum wells
I Kudyk, L Wischmeier, T Voss, I Rückmann and J Gutowski 33

Optical properties of coupled quantum disk-waveguide structure
M Yamaguchi, H Tanaka, M Yokoi, H Takagi and N Sawaki 37

Picosecond spin-preserving carrier capture in InGaAs/GaAs quantum dots
S Trumm, M Wesseli, H Krenner, D Schuh, M Bichler, J J Finley and M Betz 41

Influence of surfaces on the pure dephasing of quantum dots
T Kuhn, B Krummheuer and V M Axt 45

Exploiting the non-Markovian nature of carrier-phonon dynamics: multi-pulse control of decoherence in quantum dots
P Machnikowski, V M Axt, T Kuhn and L Jacak 49

Numerical study of weak localization effects in disordered cavities
L Bonci, M Macucci, G Iannaccone and M G Pala 55

Carrier scattering by optical phonons, two-phonon processes in photon absorption, and spontaneous polarization in wurtzites
M Dutta, G J Brown, D Ramadurai, D Geerpuram, J Yang, B Kohanpour, C Chen and M A Stroscio 59

Terahertz plasma oscillations in nanotransistors
W Knap and J Łusakowski 63

High-intensity THz radiation from a large interdigitated array photo-conductive emitter
S Winnerl, A Dreyhaupt, F Peter, D Stehr, M Helm and T Dekorsy 73

Broadband terahertz emission from ion-implanted semiconductors
J Lloyd-Hughes, E Castro-Camus, M D Fraser, H H Tan, C Jagadish and M B Johnston 77

THz collective real-space oscillations of ballistic electrons in wide parabolic potential wells: an exotic transport regime
M Betz, S Trumm, M Eckardt, A Schwanhäußer, S Malzer, F Sotier, A Leitenstorfer, T Müller, K Unterrainer and G H Döhler 81

Effect of injector doping on non-equilibrium electron dynamics in mid-infrared GaAs/AlGaAs quantum cascade lasers
V D Jovanović, D Indjin, N Vukmirović, Z Ikonić, P Harrison, E H Linfield, H Page, X Marcadet, C Sirtori, C Worrall, H Beere and D A Ritchie 85

Experimental investigation of hot carriers in THz and mid-IR quantum cascade lasers
G Scamarcio, V Spagnolo, M S Vitiello and C Di Franco 89

Time- and spectrally-resolved THz photoconductivity in quantum hall devices
C Stellmach, Y B Vasilyev, R Bonk, A Hirsch, N G Kalugin, G Hein, C R Becker and G Nachtwei 95

Transport properties and terahertz emission in narrow minigap GaAs-GaAlAs superlattices
A A Andronov, E P Dodin, A Y Klimov, V V Rogov, Y. N. Nozdrin, D I Zinchenko, A A Marmalyuk and A A Padalitsa 99

Investigation of antenna-coupled MOM diodes for infrared sensor applications
B Rakos, H Yang, J A Bean, G H Bernstein, P Fay, Á I Csurgay and W Porod 105

Transport and noise in ultrafast unipolar nanodiodes and nanotransistors
T González, A M Song, B G Vasallo, D Pardo and J Mateos 109

Monte Carlo study of coupled SO phonon-plasmon scattering in Si MOSFETs with high κ- dielectric gate stacks: hot electron and disorder effects
J R Barker, J R Watling, A Brown, S Roy, P Zeitzoff, G Bersuker and A Asenov 115

Implementation of separable scattering mechanisms in three-dimensional quantum mechanical simulations of devices
M J Gilbert, R Akis and D K Ferry 121

A 2D-NEGF quantum transport study of unintentional charges in a double gate nanotransistor
A Martinez, J R Barker, A Svizhenko, M Bescond, M P Anantram, A R Brown and A Asenov 125

Wigner function RTD simulations with DMS barriers
H L Grubin 129

High field transport in GaN and AlGaN/GaN heterojunction field effect transistors
S Yamakawa, J Branlard, M Saraniti and S M Goodnick 133

Impact ionization and high-field electron transport in GaN
A Kuligk, N Fitzer and R Redmer 139

Studies of high field transport in a high-quality InN film by ultrafast Raman spectroscopy
K T Tsen, D K Ferry, H Lu and W J Schaff 143

Monte Carlo investigation of dynamic transport in nitrides
L Reggiani, P Shiktorov, E Starikov, V Gruzinskis, L Varani, J C Vaissiere and J P Nougier 147

High-field transport in nitride channels: a hot-phonon bottleneck
A Matulionis, L F Eastman and J Liberis 151

Quantum transport and spin polarization in strongly biased semiconductor superlattices with Rashba spin-orbit coupling
P Kleinert and V V Bryksin 155

Temperature dependent transport in spin valve transistor structures
R Heer, J Smoliner, J Bornemeier and H Brückl 159

Spin filtering effects in a quantum point contact
R Akis and D K Ferry 163

Exchange effects in the Wigner-function approach
E Cancellieri, P Bordone and C Jacoboni 167

Few-particle quantum transmitting boundary method: scattering resonances through a charged 1D quantum dot
A Bertoni and G Goldoni 171

The *R-Σ* approach to tunnelling in nanoscale devices
M Rudan, A Marchi, R Brunetti, S Reggiani and E Gnani 175

Monte Carlo simulation of solid-state thermionic energy conversion devices based on non-planar heterostructure interfaces
Z Bian and A Shakouri 179

Simulations of inelastic tunnelling in molecular bridges
A Gagliardi, G C Solomon, A Pecchia, A Di Carlo, T Frauenheim, J R Reimers and N S Hush 183

Phonon effects in nanotubes: phase space reduction and electron conductance
A Raichura, M Dutta and M A Stroscio 187

Carbon nanotubes films for sensing applications: from piezoresistive sensor to gas sensing
M Lucci, P Regoliosi, F Brunetti, A Reale, A Di Carlo, E Tamburri, A Fiori, S Orlanducci, M L Terranova and P Lugli 191

Electro-thermal transport in silicon and carbon nanotube devices
E Pop, D Mann, J Rowlette, K Goodson and H Dai 195

Silicon-based ion channel platforms
S J Wilk, L Petrossian, M Goryll, J M Tang, R S Eisenberg, M Saraniti, S M Goodnick and T J Thornton 201

Implicit water simulations of non-equilibrium charge transport in ion channels
U Ravaioli, T A van der Straaten and G Kathawala 205

An investigation of the dependence of ionic conduction on the dielectric properties of porin
S J Aboud, D Marreiro and M Saraniti 211

Physical mechanisms for ion-current levelling off in the KcsA channel through combined Monte Carlo/molecular dynamics simulations
E Piccinini, F Affinito, R Brunetti, C Jacoboni and M Rudan 217

Simulations of the gramicidin A channel by using the TR-PNP model
S Hu and K Hess 221

Phonon emission and absorption by holes in the HOMO bands of duplex DNA
T Yamanaka, M Dutta, T Rajh and M A Stroscio 225

An impedance network model for the electrical properties of a single-protein nanodevice
V Akimov, E Alfinito, C Pennetta, L Reggiani, J Minic, T Gorojankina, E Pajot-Augy and R Salesse 229

Field effect transistor constructed of novel structure with short-period $(GaAs)_n/(AlAs)_m$ superlattice
V T Trofimov, M V Valeiko, N A Volchkov, A I Toropov, K S Zhuravlev, E V Kiseleva, S V Obolenskii, M A Kitaev and V A Kozlov 233

Predominance of geminate process of exciton formation in AlGaAs layers at low excitation
E V Kozhemyakina, A V Efanov, K S Zhuravlev, J Fuerst and H Pascher 237

Electron-distribution function for the Boltzmann equation in semiconductors
O Muscato 241

Giant increase of electron saturated drift velocity in a MODFET channel
V G Mokerov, J Pozela, K Pozela and V Juciene 245

Technological crossroads: silicon or III-V for future generation nanotransistors
M J Gilbert and D K Ferry 249

Optical phonon modes and electron-phonon interaction in a spheroidal quantum dot
M Ishida, M Yamaguchi, and N Sawaki 253

Terahertz negative differential conductivity in heterostructures due to population inversion and bunching of ballistic electrons
V A Kozlov, A V Nikolaev, and V A Verbus 257

Carrier dynamics of single ZnO nanowires
L Wischmeier, C Bekeny, and T Voss 261

Traditional hot-electron MOS devices for novel optoelectronic applications
T Dekorsy, J Sun, W Skorupa, M Helm, L Rebohle and T Gebel 265

Investigation of self-heating effects in individual SOI devices and device-device interactions
M Arifuzzaman and D Vasileska 269

Measurements of the electrical excitation of QH-devices in the real time domain
G Vasile, C Stellmach, G Hein and G Nachtwei 273

Impact ionization and avalanche multiplication in AlGaAs: a time-resolved study
M Betz, S Trumm, M Eckardt, A Schwanhäußer, F Sotier, A Leitenstorfer, M Hanson, D Driscoll, A C Gossard, S Malzer and G H Döhler 277

Fermi-Dirac statistics in Monte Carlo simulations of InGaAs MOSFETs
K Kalna, L Yang and A Asenov 281

Monte Carlo study of the suppression of diffusion noise
L Varani, E Starikov, P Shiktorov, V Gruzhinskis, C Palermo, J C Vaissière and J P Nougier 287

TeraHertz emission from nanometric HEMTs analyzed by noise spectra
J-F Millithaler, L Varani, C Palermo, J Mateos, T González, S Perez, D Pardo, W Knap, J Lusakowski, N Dyakonova, S Bollaert and A Cappy 291

Electron transport in novel Sb-based quantum cascade lasers
V Spagnolo, M S Vitiello, G Scamarcio, D G Revin and J W Cockburn 295

Quantum phonon-limited high-field electron transport in semiconductors
G Ferrari, E Cancellieri, P Bordone and C Jacoboni 301

Transit time and velocity distribution functions in decananometer gate-length SOI MOSFETs
M J Martín and R Rengel 305

Collision of fano resonances in a molecular ring
E R Hedin, A M Satanin and Y S Joe 309

Simulation of domain formation in p-Si/SiGe quantum cascade structures
Z Ikonic, P Harrison and R W Kelsall 313

Calculation of optical gain and electron relaxation rates in single- and double-phonon resonant quantum cascade lasers in a magnetic field
J Radovanović, A Mirčetić, V Milanović, Z Ikonić, D Indjin, P Harrison and R W Kelsall 317

Curvature-dependent conductance resonances in quantum cavities
G J Meyer, R H Blick and I Knezevic 321

Mid-infrared optical absorption in germanium under intense laser fields
H Furuse, Y Nakata, H Kubo and N Mori 325

Interface related radiative recombination on a type-II broken-gap single GaInAsSb/InAs heterojunction
K A Korolev, K D Moiseev, V A Berezovets, M P Mikhailova, Y P Yakovlev, R V Parfeniev, C J Meinning and B D McCombe 329

Drift and diffusion in superlattices within the Wannier-Stark approach
M Rosini and L Reggiani 333

Ballistic transport in arbitrary oriented nanowire MOSFETs
M Bescond, N Cavassilas, L Raymond and A Asenov 337

Scanning tunnelling microscopy of ultrathin silicon-on-insulator
P P Zhang, E Tevaarwerk, B N Park, D E Savage, G Celler, I Knezevic, P G Evans, M A Eriksson and M G Lagally 341

Effect of regular and irregular potential perturbations in mesoscopic cavities
P Marconcini and M Macucci 345

Simulation of electronic/ionic mixed conduction in solid ionic memories
H I Kwon, U Ravaioli and J D Lee 349

Full-band modeling of magnetic semiconductors
S Beysserie, I Remond, S Goodnick and M Saraniti 353

Cellular Monte Carlo modeling of $Al_xIn_{1-x}Sb$/InSb quantum well transistors
J Branlard, N Faralli, T Dutta-Roy, S M Goodnick, D K Ferry, S J Aboud and M Saraniti 359

Non-parabolic model for the solution of 2-D quantum transverse states applied to narrow conduction channel simulation
Z Yang, A Godoy, U Ravaioli and F Gámiz 365

Self-consistent quantum transport theory of carrier capture in heterostructures
T Kubis, A Trellakis and P Vogl 369

Electron transport in curved low dimensional electron systems

N. Shaji[1*], H.Qin[1], I. Knezevic[1], C. Deneke[2], O.G. Schmidt[2], M. A. Eriksson[3] and R.H. Blick[1].

[1]Laboratory for Molecular-Scale Engineering, Electrical and Computer Engineering, University of Wisconsin-Madison, 1415 Engineering Drive, Madison, WI 53706, USA.

[2]Max-Plank-Institut für Festkörperforschung, Heisenbergstr. 1, D-70569 Stuttgart, Germany.

[3]Department of Physics, University of Wisconsin-Madison, 1150 University Avenue, Madison, WI 53706-1390.

Summary. To investigate geometric potentials in low dimensional electron systems, we have conducted first studies on topography dependant electron transport in complete tubes, using built in strain between lattice mismatched semiconductors. Initial studies reveal two regimes of electron transport which are probed by a varying perpendicular magnetic field. At low magnetic field, an increased zero field peak in magneto resistance followed by a negative magneto resistance is observed due to increase in electron scattering along curved regions. At high magnetic field, we find a linear increase in resistance of the curved region as compared to planar regions.

1. Introduction

Investigating electron transport in suspended low-dimensional electron systems is a new approach which allows us to study dissipation phenomena such as the interaction of single electrons with discrete phonon modes directly [1]. The next step is to suspend the electronic system and to change the topology and study curved and rolled up electron systems.

It has been shown theoretically that the confinement potentials of low dimensional systems with a mechanical degree of freedom can depend on their geometry [2, 3, 4]. Such non-planar systems combined with precision band engineering can be used to mechanically tune the required geometric confinement potential. This additional tuning of low dimensional systems

* Present Address: 2439 Engineering Hall, 1415 Engineering Drive, Madison, Wisconsin, 53706. Tel.: +1-608-213-0836; Fax: +1- 608-262-1952; e-mail: nakul@cae.wisc.edu.

through mechanical relief gives non-planar systems an advantage over its planar counterpart.

Starting from Ref [2], we see that similar to electron confinement via electrostatic gates in planar systems, geometrically confined potentials in non-planar systems can be modeled to first order as simple square well potentials. The binding energies of such non-planar systems are found to be inversely proportional to the square of curvature radius of the non-planar system. In addition such confinement potentials cause a phase shift in the electronic wave function propagating phase coherently through the system, corresponding to Berry's phase [5].

Confining a 2DEG in such curved and rolled geometries marks the first step to obtaining a non-planar low dimensional electron system. First studies performed by peeling a planar hall bar off the supporting substrate and attaching it to curved geometries have shown that the magneto-resistance oscillations in millimeter sized bent electron gases depend on the dispersion of Landau levels and a cosine variation of linear resistance [6]. To obtain tubular geometries with smaller diameters, we make use of the built-in strain in heterostructures. When lattice mismatched semiconductors are grown layer by layer epitaxially, a strain is built in as the epitaxial layer tries to align its lattice with that of the substrate. Release of this strain by removing the sacrificial layer below the strained bilayer causes the bilayer to bend forming tubular geometries as shown in Fig. 1(a) [7, 8]. Recent experiments on tubes formed from such strained 2DEG structures [9] have shown a wash out of magneto-resistance oscillations with tube formation.

2. Experiment

The heterostructure we report on consists of a transport layer formed by 10 nm GaAs cap layer followed by 10 nm $Al_{0.33}Ga_{0.67}As$, 2 nm GaAs (silicon delta doped), 20 nm $Al_{0.33}Ga_{0.67}As$, 20 nm GaAs quantum well (2DEG). The strained bilayers following the transport layer consists of 20 nm $Al_{0.33}Ga_{0.67}As$, 14 nm $In_{0.2}Ga_{0.8}As$ (strained) and 10 nm AlAs (sacrificial layer) over a GaAs substrate. Since $In_{0.2}Ga_{0.8}As$ has a larger lattice constant, the layers curve up to form the tube when the strain is released by removing the sacrificial layer of AlAs.

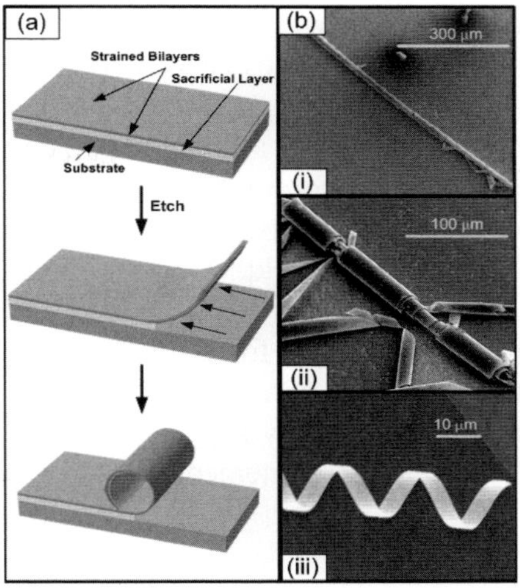

Fig. 1. (a) The built-in strain between mismatched semiconductors is released by removing the sacrificial layer below it. Tube rolls upwards as InGaAs has larger lattice constant than AlGaAs. (b) Fabricated tubes (i) 720 μm long single turn tube. (ii) Multi turn tube. (iii) Spiral coils.

Fig. 1(b) shows the various types of tubes fabricated from this strained heterostructure. Single turn tubes as long as 720 μm were fabricated when the width of the mesa was equal to πD, where D is the tube diameter (i). To fabricate multi turn tubes the initial mesa was patterned to have a width much larger than πD, to perform multiple rotations when the strain is released (ii). When the initial width of the mesa was smaller than πD, the bilayer was unable to complete a single rotation, and instead would share its strain with nearby elements performing angular rotations which resulted in the formation of helical coils (iii).

A hall bar was fabricated to characterize electron transport through the sample. The mesa was a 150 μm square with leads 600 μm long and 60 μm in width. AuGe/Ni/AuGe ohmic contacts were annealed at 420 degrees celsius. The sample at 2K showed conduction only in the presence of light. With an applied back gate bias and a varying magnetic field applied perpendicular to the crystal surface, the sample showed a superposition of oscillations from parallel conducting channels. This is due to the photoconduction from 14nm InGaAs and through electron transport in the 2DEG. Similar magneto oscillations in thin slabs of InGaAs have been reported [10]. The extracted carrier sheet density for the 2DEG is $n_s = 3.8 \times 10^{14}$ m^{-2} and the mobility is 680 cm^2/Vs. Due to this low mobility, the electron transport through the sample was non-ballistic.

The sacrificial layer of AlAs was removed by dipping the mesa in 1% HF. Upon releasing the strain the leads curved up to form tubes. Due to the length of the leads being much larger than the tube diameter and being pinned down at one end by contacts, all the tubes did not survive this process, which limited us to taking two-point measurements.

The two point measurements on the planar sample shows an increase in resistance with applied magnetic field. A closer look at the low magnetic field region shows a giant magneto resistance at zero magnetic field and a negative magneto resistance region for fields less than 0.7T. Such peaks and negative resistance region in parallel conducting sample have been reported before [11] and a fit to theory suggests an interplay of weak localization in both 2DEG and bulk confinement.

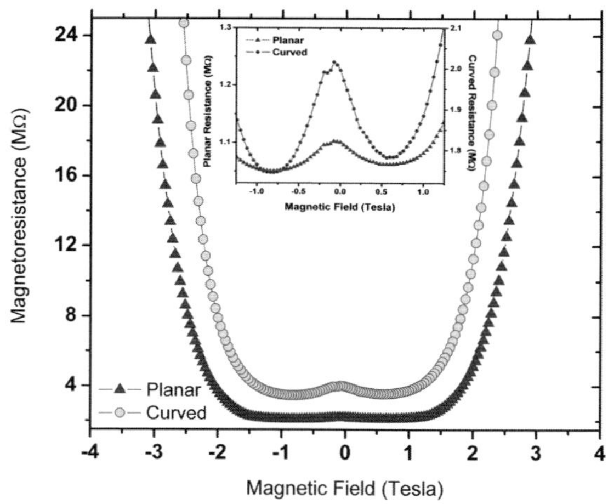

Fig. 2. Comparison of magneto-resistance variation in both planar and rolled up mesa. The inset shows the low magnetic field regime where a zero field peak and a negative magneto resistance region is seen in both planar and rolled mesa.

Upon tube formation, we see two regimes of electron transport (Fig 2). The inset of Fig. 2 shows the low magnetic field region where we see an unexpected increase in zero-field peak resistance from 52 kΩ to 260 kΩ indicating an additional scattering mechanism that is dominant in curved regions. A likely candidate for this scattering in curved region is surface scattering as we release a new InGaAs surface which was initially attached to the sacrificial layer. This release causes the formation of dangling bonds which now can scatter the electrons in curved regions.

At higher magnetic fields, there is a linear change in magneto-resistance as if there is a linear change in effective electronic width of the sample. A possible explanation is that at high fields, the resistance of the curved region is higher than in the planar region, and the linear change in overall *planar width* of the mesa is reflected in the linear increase in overall resistance. Proposals for this increase in resistance at curved regions include a change in local piezoelectric potential and confinement of carriers to locally bound states due to the geometry. More measurements are needed to confirm the exact nature reason of this behaviour.

We have shown clear topography induced changes in electron transport through a parallel conducting two dimensional electron systems. With better material engineering, a higher mobility 2DEG can be confined in these non-planar systems to probe pure ballistic electron transport. Incorporating larger strain in such systems would help realize tubular low dimensional geometries with smaller diameters useful for probing geometric potentials and achieving topographical quantum systems.

We thank ARO and NSF MRSEC for financial support.

References

1. Weig, E. M. et al.: 'Single-Electron-Phonon Interaction in a Suspended Quantum Dot Phonon Cavity', *Phys. Rev. Lett.*, **92**, 046804, 2004.
2. Chaplik, A. V. and Blick, R. H.: 'On geometric potentials in quantum-electromechanical circuits', *New J. Phys.*, **6**, 33, 2004.
3. da Costa, C. T.: 'Quantum mechanics of a constrained particle', *Phys. Rev. A.*, **23**, 1982–1987, 1981.
4. Foden, C. L. et al.: 'Quantum magnetic confinement in a curved two-dimensional electron gas', *J. Phys.: Condens. Matter.*, **6**, L127-L134, 1994.
5. Berry, M. V.: 'Quantal phase factors accompanying adiabatic changes', *Superlattices and Microstructures*, **33**, 347–35, 2003.
6. Lorke, A. et al.: 'Curved two-dimensional electron gases', *Phys. Rev. Lett.*, **92**, 046804, 2004.
7. Prinz, V. Ya. et al.: 'Free-standing and overgrown InGaAs/GaAs nanotubes, nanohelices and their arrays ', *Physica E.*, **6**, 828, 2000.
8. Schmidt, O. G. and Eberl, K.: 'Thin solid films roll up into nanotubes', *Nature*, **410**, 168, 2001.
9. Mendach, S. et al.: 'Preparation of curved two-dimensional electron systems in InGaAs/GaAs-microtubes', *Physica E.*, **23**, 274–279, 2004.
10. McElhinney, M. et al.: 'Quantum transport measurements on Si δ- and slab-doped In0.53Ga0.47As grown by molecular beam epitaxy ', *Journal of Crystal Growth*, **150**, 266, 1995.
11. Mace, D. R. et al.: 'Negative magnetoresistance in a parallel-conducting InGaAs structure', *J. Phys.: Condens. Matter.*, **4**, L487-L494, 1992.

Fabrication and Characterization of InAs Mesoscopic Devices

M. Koyama, M. Furukawa, H. Ishii, M. Nakai, T. Maemoto, S. Sasa, and M. Inoue

New Materials Research Center, Osaka Institute of Technology
5-16-1 Ohmiya, Asahi-ku, Osaka 535-8585, Japan

Summary. The transport properties of symmetry-broken InAs mesoscopic devices are reported. We fabricated InAs mesoscopic structures with a triangular anti-dot structure to serve as a ballistic rectifier. In this structure, rectification effects relying on the ballistic transport were observed at room temperature and 77K. These results show the superiority of InAs/AlGaSb heterostructures for the realization of ballistic mesoscopic devices.

1 Introduction

InAs-based heterostructures have various advantages, such as small effective mass and strong electron quantum confinement, for the realization of quantum effect devices. In addition, due to the long phase coherence time, InAs-based heterostructures offer the possibility to observe ballistic electron transport properties at relatively high temperatures compared to GaAs/AlGaAs heterostructures. Therefore, these material systems are suitable for the development of mesoscopic devices which rely on ballistic electron transport. As one of the typical sample of such applications, ballistic rectifiers are actively being researched these days. Based on the research of Song et al., it was thought that ballistic rectification is based on quasi-classical transport properties [1][2]. Fleischmann et al. reported the microscopic theoretical model with Landauer-Büttiker approach [3]-[5]. In this paper, we report on an InAs mesoscopic structure for a high temperature operation as a ballistic rectifier.

2 Fabrication

The epitaxial layer of the InAs/AlGaSb heterostructure was grown by molecular beam epitaxy on a semi-insulating GaAs(100) substrate. In order to improve the crystal quality of the InAs channel layer, an undoped 1.5 μm thick AlSb layer was grown as a buffer layer to accommodate the lattice mismatch of about 7% between GaAs and InAs. The heterostructure consists of an AlSb buffer layer, AlSb/GaSb superlattices, a 200 nm AlGaSb bottom barrier, an 8 nm AlSb barrier layer, a 15 nm InAs channel layer, a 15 nm AlGaSb upper barrier layer, and finally a 10 nm GaSb cap layer. Hall-effect measurements by the van der Pauw method showed electron mobility of 20,000 cm^2/Vs, sheet carrier density of 1.8×10^{12} cm^{-2} at 300K, and 140,000 cm^2/Vs, 1.0×10^{12} cm^{-2} at 77K, respectively.

An atomic force microscope image of the central part of the device is shown in the Fig. 1. Definition of the antidot and probes of device was achieved by electron beam lithography with ZEP-520A resist and wet chemical etching. The etchant was phosphoric acid based (H_3PO_4 : H_2O_2 : H_2O = 1 : 1: 100). Next, the Hall bridge was fabricated by photolithography. In order to eliminate the leakage current from the buffer layer, all regions except for the Hall bridge mesa were covered with SiO_2 insulator. Non-alloyed ohmic metals, In (20 nm)/Au (120 nm), were then deposited directly onto the InAs channel layer by thermal evaporation and were defined by lift-off. Metal pads for bonding were formed at the same time. The size of this device is as small as the open quantum dot structures in which we have observed ballistic transport properties and electron wave interference effects at 4.2K [6].

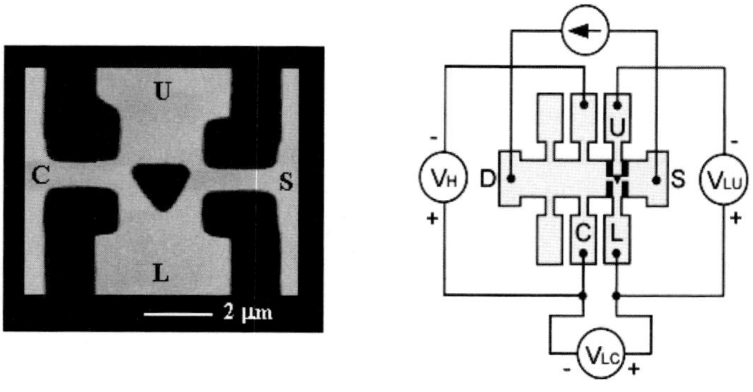

(a) AFM image (b) Measurement Configuration

Fig. 1 Atomic force microscope image of the central part of ballistic rectifier (a). The circuit diagram of the measurement system (b). DC current was applied between Drain and Source.

3 Experimental Results and Discussion

We measured the *I-V* characteristics at room temperature and 77K. Fig. 2 shows the output voltage (V_{LU}) as a function of input current (I_{DS}). For both temperatures, the output voltage V_{LU} shows negative polarity despite of the I_{DS} polarity. Therefore, a rectification effect was observed for both temperatures. As shown in the Fig. 2 (a), the stronger effect was observed at 77K probably due to the increased mean free path of about 2 μm. We believe that these characteristics reflect the ballistic transport in InAs. Although negative V_{LU} was observed on reversal of the drain-source current, the magnitude of V_{LU} shows asymmetry with respect to $I_{DS} = 0$. This result implies the asymmetry of the triangular anti-dot and the distances between the dot and source or drain wires. The distance between the reservoir and source or drain wire may also affect the characteristics. (From Fig. 1 (a), the left wire is slightly misaligned upward with respect to the triangular anti-dot.) Therefore, it is likely that the injection ratio of electrons toward probe U from the left wire is larger than that from right wire.
Figure 2 (b) shows the logarithmic plot of the I_{DS}-V characteristics for $I_{DS} > 0$ measured at 77K. The V_{LU} shows nonlinear characteristics due to the rectification effect while V_{LC} follows Ohm's low. Compared to GaAs/AlGaAs, the nonlinear effect in InAs persists for I_{DS} well above 100 μA indicating the superiority of InAs device. V_H is the reference voltage

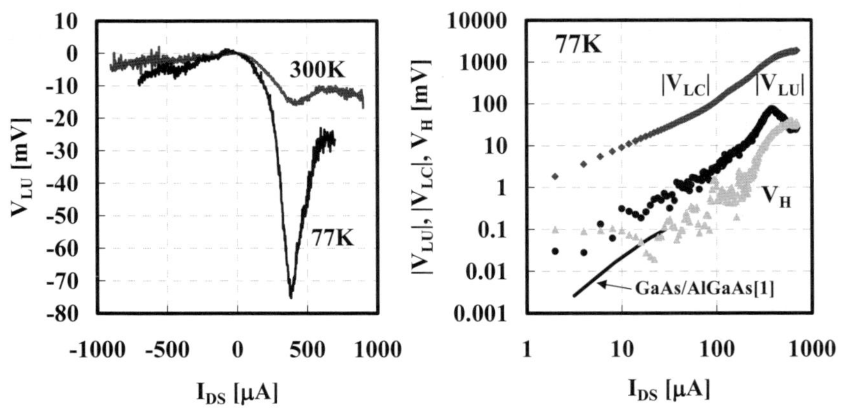

(a) The output voltage versus input direct current

(b) Logarithmic plot of the *I-V* characteristics

Fig. 2 The output voltage (V_{LU}) as a function of input direct current (I_{DS}) measured at room temperature and 77K (a). Logarithmic plot of the *I-V* characteristics measured at 77K.

measured across the channel and outside the left wire (Fig. 1 (b)). Comparing V_{LU} and V_H, the polarity of each voltage is opposite. Therefore, V_{LU} was not affected by the reference voltage V_H. However, for both temperatures, the rectification effects drastically decreased for $I_{DS} > 400$ μA. It is likely that the anti-collimation effect increased by increasing the applied voltage resulting in the increase in the propagation toward probe U.

4 Conclusion

We fabricated and characterized an InAs mesoscopic ballistic rectifier with a triangular anti-dot structure. Clear rectification characteristics were observed for both room temperature and 77K and persisted up to higher current level over 100 μA compared to GaAs/AlGaAs. These results show a potential for higher temperature operation ballistic rectifier by using InAs/AlGaSb heterostructure.

References

2. 1. Song, A. M. et al.: 'Non linear Electron Transport in an Asymmetric Microjunction: A Ballistic Rectifier', *Phys. Rev. Lett.*, **80**, 3831-3834, 1998.
2. Song, A. M.: 'Formalism of nonlinear transport in mesoscopic conductors', *Phys. Rev. B*, **59**, 9806-9809, 1999.
3. Fleischmann, R. and Geisel, T.: 'Mesoscopic Rectifiers Based on Ballistic Transport', *Phys. Rev. Lett.*, **89**, 016804, 2002.
4. Büttiker, M. and Sánchez, D.: 'Comment on "Mesoscopic Rectifiers Based on Ballistic Transport" ', *Phys. Rev. Lett.*, **90**, 119701-1, 2002.
5. Geisel, T. and Fleischmann, R.: 'Geisel and Fleischmann Reply', *Phys. Rev. Lett.*, **90**, 119702-1, 2002.
6. Maemoto, T. et al.: 'Magneto transport in an InAs/AlGaSb quantum wire with a weak periodic potential', *Physica B*, **272**, 110-113, 1999.

Nonlinear Effects on Quantum Interference in Electron Billiards

C. A. Marlow, R. P. Taylor, M. Fairbanks, and H. Linke

Physics Department, University of Oregon, Eugene OR 97403-1274, USA

Summary. Magnetoconductance fluctuations are used to study the effect of an applied bias on an electron billiard. At lower bias, nonlinear effects can be well described by electron heating alone, while at higher bias ($V > 2$mV, ~5% of the electron Fermi energy) non-equilibrium effects become significant. At high bias, we also observe that the spectral content of the MCF is sensitive to the nonequilibrium effects. Spectral behavior is consistent with a fractal scaling of the conductance fluctuations with magnetic field, resulting in the first observation of fractal conductance fluctuations outside of the linear regime of transport.

1 Introduction

In this work, we use electron quantum interference effects to study the effect of an applied bias on electron transport. The electron billiards used to study these effects were defined by e-beam lithography and wet etching of the two-dimensional electron gas (2DEG) formed in the GaInAs quantum well in the GaInAs/InP heterostructure (see Fig. 1(a)). A square (Figs. 1(b)) and rectangular (Fig. 1(c)) were studied with areas, after depletion, of 0.8 μm^2 and 3.4 μm^2, and Fermi energies of 35 meV and 38 meV, respectively. In both cases, the phase coherence length and mean free path were greater than the device dimensions resulting in phase-coherent, ballistic transport. Quantum interference effects lead to fluctuations in billiard conductance as a function of a perpendicular applied magnetic field, B. These magnetoconductance fluctuations, MCF, are a sensitive reproducible probe of the electron dynamics within the billiard[1] and will be used here to monitor the effect of an applied bias on electron transport.

In the presence of an applied bias, electrons are injected into the billiard with excess energy. If the electrons have time to thermalize before leaving the billiard, they relax through electron-electron scattering and the excess energy is distributed amongst the electrons in the billiard causing an increase in overall electron temperature inside the billiard. Previous experiments have found that for small applied bias voltages, in the μV range, the primary effect of the bias is electron heating.[2,3] The electron heating can be characterized by an effective temperature, $T_e(V)$, written:[2]

$$T_e(V) = \frac{T_L}{2} + \frac{1}{2}\sqrt{T_L^2 + e^{-\gamma\tau_\varepsilon(V)} \frac{3}{2\pi^2}\left(\frac{eV}{k}\right)^2} \tag{1}$$

where T_L is the temperature of the lattice, $\exp(-\gamma\tau_\varepsilon(V))$ is the fraction of electrons that thermalize before escaping the billiard, γ is the escape rate,[4] and $\tau_\varepsilon(V)$ is the electron-electron interaction time. At the temperatures used here phase-breaking is dominated by electron-electron scattering, so the experimentally measured phase breaking length, τ_ϕ, will be used for τ_ε when calculating $T_e(V)$. τ_ϕ was determined from the measured MCF using a well-established method that analyzes the correlation field of the fluctuations as a function of magnetic field.[5]

We use this heating model to study the importance of nonequillibrium effects in the mV range. MCF measurements were taken as a function of T and V and directly compared using Eq.1 to translate V to $T_e(V)$; any departure between the two behaviors we interpret as nonequilibrium effects.

2 Experimental Results

The two-terminal magnetoconductance through the billiards was measured as a function of a perpendicular B using a standard low frequency ac lock-in technique. In order to apply a bias across the billiard, a tuneable dc bias V was added to a small ac signal (rms amplitude 20 μV on order of the thermal energy $kT \approx 20$ μeV). Measurements were made at a range of temperatures with $V = 0$ mV and also for a range of dc biases (up to 3 mV) at $T = 230$ mK.

Figure 1(d) shows the MCF for the square billiard measured for a range of T (black curves) and V (gray curves). The bias values have been related to the associated temperature using Eq. 1. At low bias, the fluctuations taken at a bias are similar to those at the corresponding temperature $T_e(V)$, consistent with previous observations in GaAs/AlGaAs billiards[3] where agreement was seen in the μV range. At higher bias, however, a departure is seen between the MCF measured at V and those at the related T, indicating that at higher bias, the effect of the bias on the fluctuations is not just electron heating.

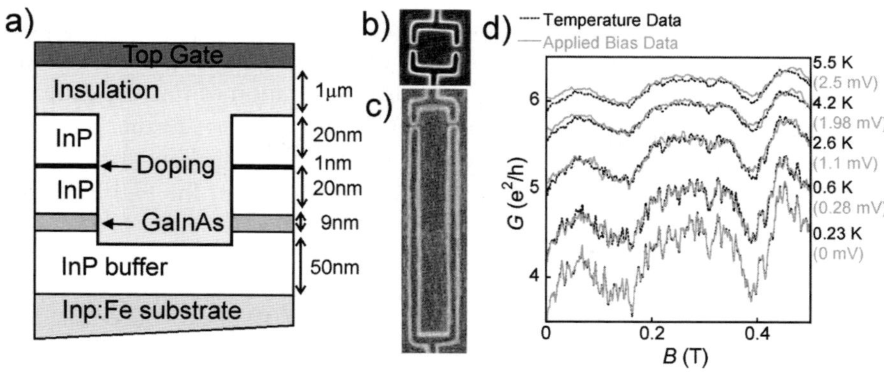

Fig. 1. a) Schematic representation of GaInAs/InP billiard system, scanning electron micrographs of the b) square and c) rectangular billiard and d) MCF for the square billiard, measured for a range of temperatures at zero bias (black curves) and range of bias at $T = 230$ mK (gray curves). Traces are offset for clarity.

We also investigate the effect of the bias on the spectral content of the fluctuations. The power spectrum, $S(f)$, of the MCF for the square billiard at $T = 0.6$ K and $V = 0$ mV is shown in Fig. 2(a). All MCF for the billiards presented here show $1/f^{\alpha}$ scaling, where $f = 1/\Delta B$ and α is the spectral exponent which characterizes the entire spectral content of the fluctuations. The α values observed indicate a fractal scaling[6] of the fluctuations consistent with previously observed fractal conductance fluctuations, FCF, in similar systems.[7]

Figures 2(b) and 2(c) show the dependence of the spectral exponent, α, on both T and T_e (V) for the square and rectangular billiard respectively. We see that α increases with increasing T, consistent with previous results.[7] The same increase is seen with increasing T_e (V). It appears, however, that at T_e (V) ~ 4 K, (corresponding to ~ 2 mV) the evolution of α with V departs significantly from that of α (T), confirming again that the bias is having an additional effect besides heating.

3 Discussion

The deviations of the characteristics of the MCF indicate that the effect of the high bias is nontrivial; the bias does not just increase electron energy, but instead changes electron dynamics within the billiard. Future work needs to experimentally investigate the precise role of nonequillibrium electrons in the generation of MCF.

The fractal nature of MCF has been observed to be is robust to changes in many system parameters.[7] The effect of the nonequillibrium electrons on the spectral exponent is unexpected. Future exploration of this dependence may provide insight into the origins of FCF. In addition, further analysis of FCF in the nonlinear region may help uncover the mechanisms responsible for the symmetry breaking of the spectral content of the FCF with respect to reversal of magnetic field seen previously.[8]

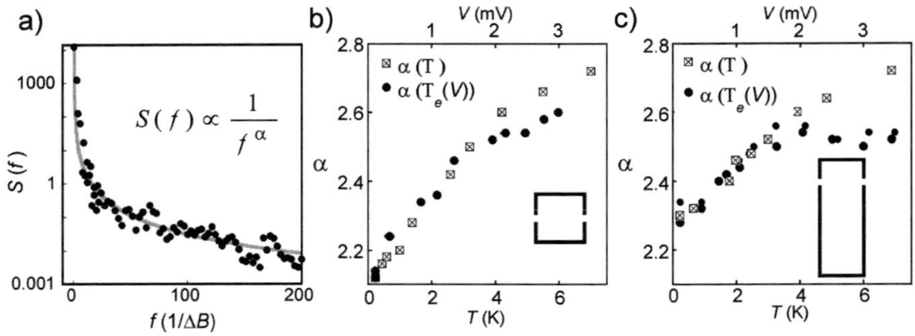

Fig. 2. a) Power spectra of MCF measured on square billiard (T = 0.6 K, V = 0 mV), spectral exponent α as a function of T and T_e (V) for b) square and c) rectangular billiard.

References

1. Beenakker, C. W. J. and van Houten, H.: 'Quantum transport in semiconductor structures', *Solid State Physics*, edited by H Ehrenreich and D Turnbull, Academic Press, **44**, 1991.
2. Linke, H. et al.: 'Non-equilibrium electrons in ballistic quantum dot', Phys. Stat. Sol., **204**, 318, 1997.
3. Switkes, M. et al.: 'High bias transport and magnetometer design in open quantum dots',Appl. Phy. Lett., **72**, 471, 1998.
4. Jensen, R. V.: 'Chaotic scattering, unstable periodic orbits, and fluctuations in quantum transport',Chaos, **1**, 101, 1991.
5. Bird, J. P. et al.:'Phase breaking in ballistic quantum dots: a correlation field analysis', Surf. Sci., **361/362**, 730, 1996.
6. Barnsley, M. F. et al.: *The Science of Fractal Images*, Springer-Verlag, 1988.
7. Micolich, A. P. et al.:'Three key questions on fractal conductance fluctuations: Dynamics, quantization, and coherence', Phys. Rev. B **70**, 085302, 2004; Marlow, C. A. et al.: submitted, 2005.
8. Marlow, C. A. et al.: to be submitted, 2005.

Prediction of Entanglement Detection by I-V Characteristics

T. Zibold[1], P. Vogl[1], A. Bertoni[2]

[1]Walter Schottky Institute, Technische Universität München, 85748 Garching, Germany
[2]National Research Center on "nanoStructures and bioSystems at Surfaces" (S3), INFM-CNR, 41100 Modena, Italy

Summary. We present a theoretical analysis of a ballistic GaAs/AlGaAs quantum device that allows the straightforward control and detection of the entanglement between an open, stationary double quantum wire system and a singly occupied electrostatically defined double quantum dot. The read-out involves the measurement of only the DC I-V characteristics.

1 Introduction

While there is no shortage in theoretical proposals for the creation of entangled qubits in semiconductors (see, e.g., [1,2,3,4]), the experimental realization of their write-in and read-out processes has proven to be extremely difficult, mostly because of the short coherence times and strong interactions between elementary excitations in solids [5]. Charge-based qubits have been proposed either based on double quantum dots (DQD) or electron wave packets propagating ballistically within two adjacent quantum wires (QWR) [6]. Recently, it has been pointed out that qubits or, more generally, quqits (quaternary state bits) can also be realized on the basis of stationary scattering states in open devices [7]. While these stationary wave functions cannot be normalized, the transmission and reflection play the role of the computational basis. The unitarity of ballistic scattering guarantees a well-defined Hilbert space. To the best of our knowledge, however, no concrete write-in and read-out processes for entangled quqits or qubits associated with stationary scattering states have been proposed so far.

We present a quantitative theoretical analysis of a concrete quantum transport device that allows the straightforward control and detection of the entanglement between an open, stationary QWR qubit based on a solid-state Mach-Zehnder interferometer and a DQD qubit. The read-out involves the measurement of only the DC $I-V$ characteristics, no higher order current correlations are required to detect entanglement.

2 Method and Results

Figure 1 shows a schematic top view of the proposed device consisting of two stacked GaAs/AlGaAs two-dimensional electron gases (2DEG). The Mach-Zehnder interferometer in dark gray is composed of two adjacent QWR that are coupled by two tunneling windows [8]. The latter act as rotation gates (R). Along the upper QWR, there is an adjustable barrier in between the rotation gates that acts as a tunable phase gate V_G. An electrostatically defined DQD (light gray) is located in a second 2DEG underneath the QWR layer. The tunnel coupling between the two quantum dots as well as their energy spectra are assumed to be controlled by external gates that are not shown [9]. By applying a bias voltage V equal to 50 μV between the upper left and the remaining three contacts, current flows predominantly from the upper left to the upper right contact (J_0) and from the upper left to the lower right contact (J_1), respectively, depending on the phase gate voltage V_G.

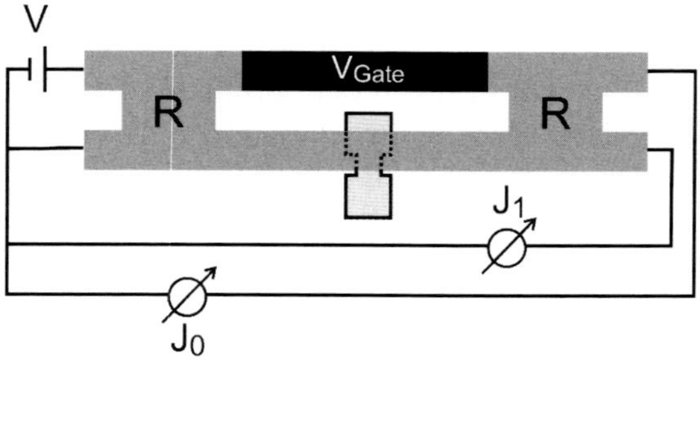

Fig. 1. Schematic top view of the proposed quantum transport device. The device is realized by two stacked GaAs/AlGaAs 2DEGs. The top 2DEG is depleted by external gates to form a Mach-Zehnder interferometer (dark gray). In the bottom 2DEG two coupled quantum dots (light gray) are located. For sake of clarity, the figure is not drawn on scale.

We have calculated the ballistic current through this 3D device, using a single-band effective mass description for the electronic Hamiltonian and by including the Coulomb interaction between the electrons in the QWR and the DQD non-perturbatively. To this end, we have extended the CBR method [10] to deal with Green's functions of two-particle Hamiltonians for distinguishable particles. After determination of the single-particle scattering states of the open QWR system and the eigenstates of the closed DQD system, the two-particle Hamiltonian is diagonalized in the basis of the product states, including the Coulomb interaction between the two subsystems. The single-particle basis states associated with the QWR are determined realistically from the retarded Green's function of the open device using our device simulator nextnano^3. To simplify the two-electron problem, however, we have mapped the DQD subsystem onto a two-level tunneling Hamiltonian that is characterized by a tunnel coupling t in the range between 0 and 20 μeV and a bare splitting of $\Delta = 10$ μeV between the two levels [9]. The charge distribution of the electron in the DQD is approximated by two localized charge distributions that are weighted by the projections of the two-particle eigenstates onto the DQD basis states.

In the present calculations, we take both GaAs 2DEGs to be 10 nm thick; the vertical distance between the QWR and the DQD layer is set to 80 nm. Each of the quantum wires is 55 nm wide, 1000 nm long, and the lateral distance between them amounts to 20 nm. The rotation gates have a length of 85 nm. One of the quantum dots is located exactly beneath the center of the barrier region in between the 2 quantum wires. The lateral distance between the 2 quantum dots is chosen to be 60 nm. The Fermi energy has been set to 1.6 meV in the lowest subband. This causes the tunneling windows to act as almost perfect rotation gates for low bias voltage.

At first, we study the $I-V$ characteristics of the device for a situation where the tunneling between the quantum dots is inhibited ($t=0$). For $V_G = 0$, the rotation gates are dimensioned in such a way that interference causes the current between the upper left and right contact to be zero ($J_0 = 0$) and to attain a maximal value between the left top contact and the lower right contact ($J_1 = J_{\max}$). Application of a finite voltage V_G to the phase gate results in DC currents J_0 and J_1 that have intermediate values between 0 and $J_{\max}$ and are phase-shifted by π relative to one another.

We now allow the dot electron to tunnel between the quantum dots. This leads to an entanglement of the QWR and the DQD. This entanglement puts the QWR subsystem into a mixed state which causes the interference through the upper and the lower wire to be suppressed. The degree of suppression can be characterized by the visibility v which we define by $v = \left[(J_1 - J_0)/(J_1 + J_0)\right]_{V_G=0}$. This quantity can be shown to be related to the von-Neumann entropy by $S = -\sum_{i=+,-} p_i \ln p_i$, where $p_\pm = (1 \pm v)/2$.

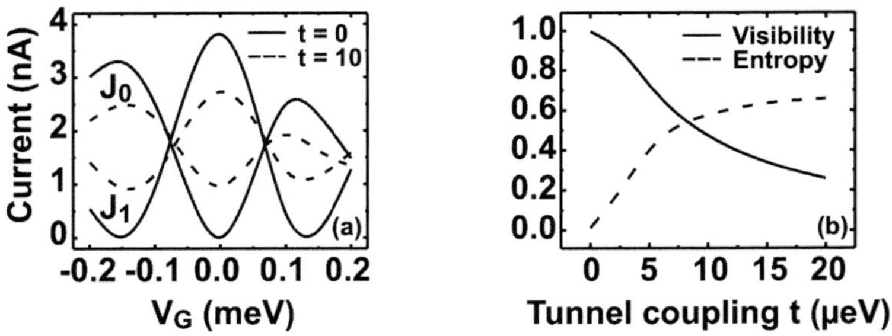

Fig. 2. (a) Currents J_0 and J_1 as a function of the gate voltage of the phase gate for two different tunneling couplings t in μeV. (b) Visibility (full line) and corresponding von-Neumann entropy (dashed line) as a function of the tunnel coupling.

In Fig. 2 (a) and (b), we show the results of our 3D 2-particle Green's function calculation of the ballistic current through the entangled QWR-DQD system. Fig. 2(a) shows the currents J_0 and J_1 as a function of V_G for two different values of the tunnel coupling. In both cases, the currents J_0 and J_1 show an oscillatory pattern. However, in the case of non-vanishing tunnel coupling, the visibility of the interference gets strongly suppressed. Fig. 2(b) shows the visibility and the corresponding von-Neumann entropy at $V_G = 0$ as a function of the tunnel coupling which quantifies the degree of entanglement of the QWR and the DQD subsystems.

References

1. Loss, D. and Sukhorukov E. V.: *Phys. Rev. Lett.*, **84**, 1035, 1998.
2. Reina J. H., Quiroga, L., Johnson, N. F.: *Phys. Rev. A*, **62**, 012305, 2000.
3. Ionicioiu, R., Zanardi, P., and Rossi, F.: *Phys. Rev. A*, **63**, 050101, 2001.
4. Cerletti V., Coish, W. A. Gywat O., Loss, D., *Nanotechnology*, **16**, R27, 2005.
5. Krenner H. J., Stufler S., Sabathil M., Clark, E. C., Ester, P., Bichler M., Abstreiter G., Finley J. J., and Zrenner A.: e-print cond-mat/0505731.
6. Bertoni, A., Bordone, P., Brunetti, R., Jacoboni C., and Reggiani S., *Phys. Rev. Lett.* **84**, 5912 (2000).
7. Akguc, G. B., Reichl, L. E., Shaji, A., and Snyder, M. G.: *Phys. Rev. A*, **69**, 042303, 2004.
8. Sabathil, M., Mamaluy, D., and Vogl, P.: *Semicond. Sci. Technol.*, **19**, S137, 2004.
9. van der Wiel, W. G., De Franceschi, S., Elzerman, J. M., Fujisawa T., Tarucha, S., and Kouwenhoven, L. P.: *Rev. Mod. Phys.*, **75**, 1, 2003.
10. Mamaluy, D., Sabathil, M., and Vogl P.: *J. Appl. Phys.*, **93**, 4628, 2003.

Simulation of Entanglement Creation for Carrier-Impurity Scattering in a 2D System

P. Bordone and A. Bertoni

National Research Center S3, INFM-CNR , and Dipartimento di Fisica, Università di Modena e Reggio Emilia, via Campi 213/A, I-41100, Modena, Italy.

Summary. We present a time dependent numerical analysis of the entanglement created between an electron freely propagating in a 2D system and a charged particle bound to a specific site by a harmonic potential. The latter can be considered as a simplified model of a shallow impurity. The dynamics of the carrier initially bound in the harmonic potential is coupled to that of the incoming electron through a screened Coulomb interaction. The entanglement is found to depend significantly on the energy of the freely propagating particle, on the confining energy of the harmonic potential and on the sign of the charge bound by the harmonic potential. This approach allows a quantitative estimate of the decoherence undergone by a propagating carrier due to a single unelastic scattering.

1 Introduction

The decoherence of a quantum system is ascribed to its entanglement with another system considered as the environment [1], thus the quantitative evaluation of entanglement formation dynamics can shed light on the transition between quantum and classical behavior of a carrier that undergoes a scattering event. It is then crucial to quantify the amount of decoherence suffered by a carrier when it interacts, unelastically, with phonons, impurities or other carriers. In each of these cases the carrier gets entangled with the scatterer thus making impossible to continue to describe it by means of a single particle wavefunction. As a consequence if the final quantum state of the scatterer, considered as part of the environment, is unknown, the carrier loses its quantum coherence.

We present a time dependent numerical analysis of the scattering process that creates entanglement between an electron freely propagating in a 2D system and a charged particle bound to a specific site by a harmonic potential. The latter can be considered as a simplified model of a shallow impurity

2 The Physical Model

We consider an electron (subsystem A), moving in a 2D system and described at the initial time t_0 by a minimum uncertainity Gaussian wave packet, interacting through a Coulomb potential with a second charge (subsystem B) confined in a 2D harmonic potential [2]. The second particle is assumed to be in the ground state of the harmonic potential. At time t_0 the particles are set far enough that their Coulomb interaction is negligible. Therefore the two-particle wavefunction can be assumed separable. To solve the dynamics in terms of two independent Schrödinger equations we move to the center of mass coordinates: $\mathbf{r}=\mathbf{r}_A-\mathbf{r}_B$; $\mathbf{R}=(\mathbf{r}_A+\mathbf{r}_B)/2$. Then the initial condition, that represents the pure separable two-particle state, reads

$$\Psi(\mathbf{r},\mathbf{R},t_0)=\frac{1}{2\pi\sigma_A^2}\exp\left\{-\frac{(\mathbf{R}-\mathbf{R}_{00})^2}{2\sigma_A^2}\right\}\exp(-i\mathbf{k}_0\cdot\mathbf{R})\exp\left\{-\frac{(\mathbf{r}-\mathbf{r}_{00})^2}{8\sigma_A^2}\right\}\exp\left(-i\frac{\mathbf{k}_0}{2}\cdot\mathbf{r}\right) \quad (1)$$

where σ^2_A, $<\mathbf{r}_A>$ and $\mathbf{k}_0=<\mathbf{k}_A>$ are the variance, the mean position and the wave vector of the free electron respectively, $\mathbf{r}_0$ is the center position of the harmonic potential, $\mathbf{r}_{00}=<\mathbf{r}_A>-\mathbf{r}_0$ and $\mathbf{R}_{00}=(<\mathbf{r}_A>+\mathbf{r}_0)/2$. The two particles are assumed distinguishable, so the symmetrization of the wavefunction does not come into play. The time dependent Schrödinger equations, describing the coherent evolution of the biparticle system (see Fig. 1) are solved numerically, using the initial condition given in Eq. (1). The coupling effect of the Coulomb interaction during the evolution can be seen from a two-particle or from a single-particle perspective. In the first case the two carriers A and B get entangled and the pure state described by the wave function $\psi(\mathbf{r}_A,\mathbf{r}_B)$ becomes not separable in the two subspaces of the single particles. In the second case, i.e. tracing over the degrees of freedom of one particle (thus considered as the environment), the state of the other has to be described in terms of a density matrix representing a mixed state. In this perspective the Coulomb interaction induces decoherence: the amount of entanglement between the two subsystems quantifies the amount of decoherence of one of the subsystems, when the other is considered as the environment [1].

The entanglement between the particles is evaluated numerically at several time steps of the quantum evolution as the von Neumann entropy of the reduced density matrix [1]:

$$\varepsilon(t)=-Tr_A\left[\rho_A(t)\log_d\left(\rho_A(t)\right)\right] \quad (2)$$

with

$$\rho_A(t) = Tr_B\left(\rho(\mathbf{r}_A, \mathbf{r}_B, \mathbf{r}'_A, \mathbf{r}'_B)\right) = \sum_{\mathbf{r}_B} \psi(\mathbf{r}_A, \mathbf{r}_B, t)\psi^*(\mathbf{r}'_A, \mathbf{r}_B, t) \quad (3)$$

where d is the dimension of the single-particle Hilbert space and corresponds, in the present case, to the number of points in which $\mathbf{r}_A$ is discretized [3]. Equation (2) is symmetric in the coordinates of subsystems A and B. Therefore the decoherence produced by the scattering on the two particles is the same.

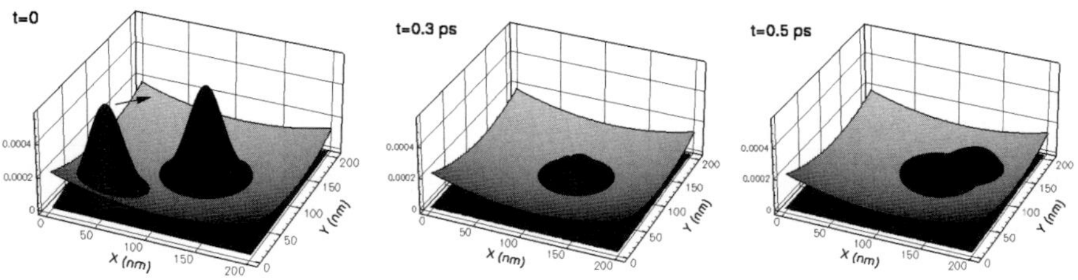

Fig. 1. Particle density at 3 different time steps for an initial kinetic energy of the travelling electron of 10 meV. The grey surface represents the injected density electron , while the black surface represents the probability density of the charge bound by a harmonic potential (sketched by the lite-grey surface. (see text).

3 Results

As stated before, at the initial time the two-particle wavefunction can be considered separable and the entanglement is zero. As the particles get closer their quantum correlation builds up and reaches a stationary value once scattering is completed. Numerical results have been obtained adopting GaAs material parameter, but can be considered representative of a more general behavior. Fig. 1 shows the particles densities at three different times. It should be noticed that the density for a single particle is obtained by integrating over the other particle's position and, as a consequence, the graphs shown do not represent wavefunctions, except for the initial time, when the wavefunction is separable. In Fig. 2(a) the time evolution of the entanglement for three different values of the incoming electron initial energy is reported (the harmonic potential is $\hbar\omega$=2 meV). The simulation runs until the scattered carrier gets far away from the impurity and the entanglement reaches a stationary value. Increasing the incoming-electron energy makes the Coulomb interactions less effective, thus leading to a reduction in the amount of entanglement created. Fig. 2(b) shows the entanglement as a function of time for three different values of the harmonic potential, for an initial energy of the travelling electron of 10 meV (solid, dashed and dotted curves). The increase of the harmonic potential makes it more difficult for the incoming electron to excite the impurity, thus leading to a decrease in the final entanglement. In other words the the degree of decoherence suffered by the travelling particle is reduced. If the bound particle has a positive charge (dashed-dotted curve in Fig. 2(b)) the decoherence suffered by the incoming electron is found to be significantly larger with respect to the previous case. This latter effect is in qualitative agreement with the results obtained by one of the authors for the case of scattering between free carriers [3].

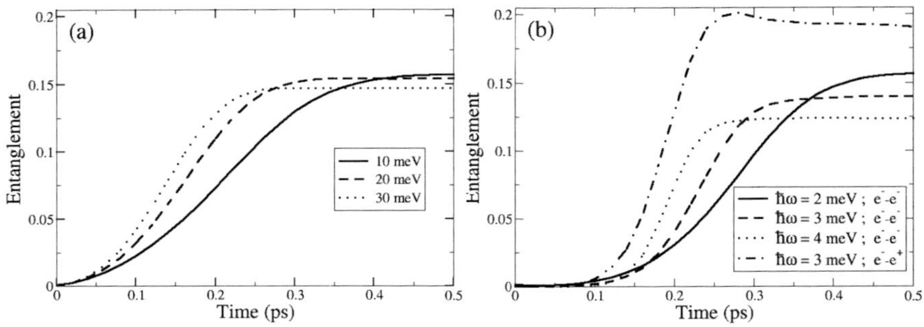

Fig. 2. (a) Time evolution of the entanglement for 3 values of the incoming electron initial energy (ħω=2 meV). (b) Time evolution of the entanglement for 3 different values of the harmonic potential, for an initial energy of the incoming electron of 10 meV (solid, dashed and dotted curves). The dashed-dotted curve represents the case in which the bound particle has a positive charge.

This work has been partially supported by the U.S. Office of Naval Research (contract No. N00014-03-1-0289 / N00014-98-1-0777)

References

1. see, for example, Giulini, D. *et al.* , : *Decoherence and the appearence of a classical world in quantum theory*, Springer, Berlin, 1996; and Vedral M., et. al.: 'Quantifying Entanglement', *Phys. Rev. Lett.*, **78**, 2275-2279, 1997
2. A similar model has been used for 1D systems, see for example, Caldeira A.O., Legget A.J.: ‘Influence of damping on quantum interference: An exactly soluble model’, *Phys. Rev. A*, **31**, 1059-1066, 1985.
3. Bertoni, A.: 'Simulation of electron decoherence induced by carrier-carrier scattering', J. Comput. Electron., **2**, 291-295, 2003.

Super-Poissonian Current Fluctuations in Tunneling Through Coupled Quantum Dots

G. Kießlich (a), A. Wacker (b) and E. Schöll (a)

(a) Institut für Theoretische Physik, Technische Universität Berlin, Hardenbergstr. 36, D-10623 Berlin, Germany

(b) Department of Physics, University of Lund, Box 118, SE-22100 Lund, Sweden

Summary. The shot noise behavior of sequential tunneling through two vertically coupled quantum dots is studied by means of a master equation approach. In particular, we propose a mechanism to observe super-Poissonian shot noise in such a system. The crucial ingredient is the Coulomb interaction between electrons in individual quantum dots.

The measurement of fluctuations in the current through resonant tunneling devices is a powerful tool to obtain information which is not solely accessible from conductance measurements [1]. For instance, the tunnel barrier geometry or the impact of Coulomb interaction on the transport through single quantum dots (QDs) can be extracted by studying their shot noise properties [2,3]. Here, we consider the tunneling through two vertically coupled QDs. Such a system was experimentally considered with respect to the average tunneling current e.g. in Ref.[4]. Additionally, the shot noise for tunneling through such a self-organized QD stack was recently measured in [5]. In this experiment some indications for super-Poissonian noise in the bias range of current resonances were reported which we theoretically address in this work. We attribute this observation to the effect of Coulomb interaction between carriers in individual QDs. Positive correlations as a consequence of Coulomb interaction were also experimentally found in resonant tunneling diodes [6] or in a single-electron transistor setup [7]. For parallel QDs the emergence of super-Poissonian noise was theoretically examined in Refs.[3,8].

In Fig. 1a the coupled QD system is sketched: QD1 is connected to the (e)mitter contact with rate Γ_e and QD2 is coupled to the (c)ollector contact with rate Γ_c. They are mutually coupled with Γ_i. These rates enter the model which is based on the master equation description elaborately introduced in [9]. Therein the coupling between the QDs is treated within Fermi's golden rule: $\Gamma_i = 4|\Omega|^2/(\Gamma_e+\Gamma_c)$, where Ω denotes the matrix element for tunneling between the QDs. Note that in this description the inter-QD coupling also depends on the couplings to the reservoirs. Even though in this framework any effects due to quantum coherence in the tunneling process are neglected, the average current surprisingly provides full agreement with the coherent result [9]. QD1/QD2 contains one spin-degenerate single-particle state with energy $\varepsilon_1/\varepsilon_2$, respectively. The interaction of electrons inside QD1 and QD2 is described with constant charging energies U_1 and U_2 , respectively. Carriers in different QDs interact via U_{inter} which is assumed to be zero throughout this paper.

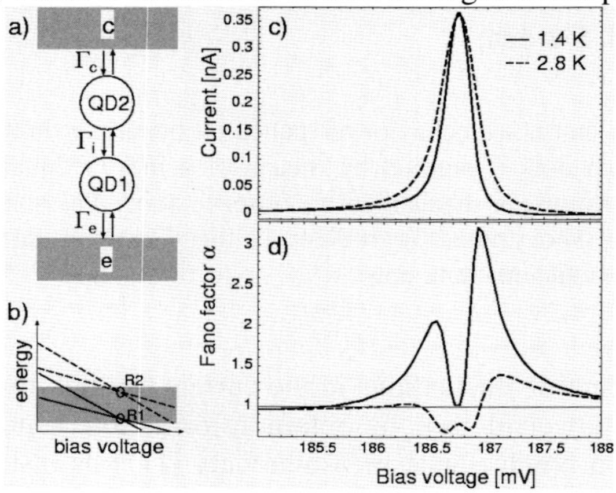

Fig. 1. a) Sketch of the coupled quantum dot system with (e)mitter and (c)ollector contact. b) Resonances of single-particle states R1 and resonances of two-particle states R2. c) Calculated current vs. bias voltage. d) Fano factor vs. bias voltage. Parameters: Γ_e=40μeV, Γ_c=1μeV, Ω=15μeV, $\varepsilon_1(0)$=48meV, $\varepsilon_2(0)$=104.025meV, η_1=0.25, η_2=0.55, $U_1=U_2$=10meV, U_{inter}=0, μ_e=11.7meV, T=1.4K (solid line), T=2.8K (dashed line).

The average current I and the zero-frequency noise S are calculated along the lines of Ref.[3]. The Fano factor as a measure for deviations from uncorrelated tunneling (Poissonian noise) is defined as $\alpha=S/(2eI)$. Then, $\alpha<1$ indicates sub-Poissonian noise corresponding to negative temporal correlations and $\alpha>1$ refers to super-Poissonian noise corresponding to positive temporal correlations or bunching of tunneling events.

The single-particle energies of the QDs depend linearly on the applied bias voltage V as $\varepsilon_{1/2}(V)=\varepsilon_{1/2}-e\eta_{1/2}V$ with the leverage factors $\eta_{1/2}$ for QD1/QD2, respectively. Fig. 1b sketches the energy vs. bias voltage dependence of the single-particle levels (solid lines) and the two-particle states (dashed lines). The shaded region indicates the occupied states in the emitter contact, the chemical potential of collector states is assumed to be energetically much lower. Then, a current is flowing through the QD system if a resonance of states in different QDs exhibits in this energy range. The specific situation depicted in Fig. 1b leads to super-Poissonian noise in the tunneling current: the charging energies of both QDs are assumed to be equal $U_1=U_2$, and the resonances of single-particle states R1 and of the doubly-occupied states R2 are available for emitter electrons.

For this regime the current-voltage characteristic and the Fano factor vs. bias voltage are shown in Fig. 1c and d, respectively. The parameters were estimated with respect to the experiment [5]. Two transport channels R1 and R2 contribute in one current peak. Since R2 lies slightly below the chemical potential in the emitter μ_e the current peak broadens with increasing temperature. In the bias range of the current peak the Fano factor shows an interesting behavior for a temperature T=1.4 K: super-Poissonian noise at the edges of the current peak and almost Poissonian noise at the current peak maximum. For increasing temperature the on-resonance Fano factor becomes sub-Poissonian and the super-Poissonian noise at the edges of the current peak vanishes. This striking temperature dependence is due to the thermal occupancy of the two-particle states (R2).

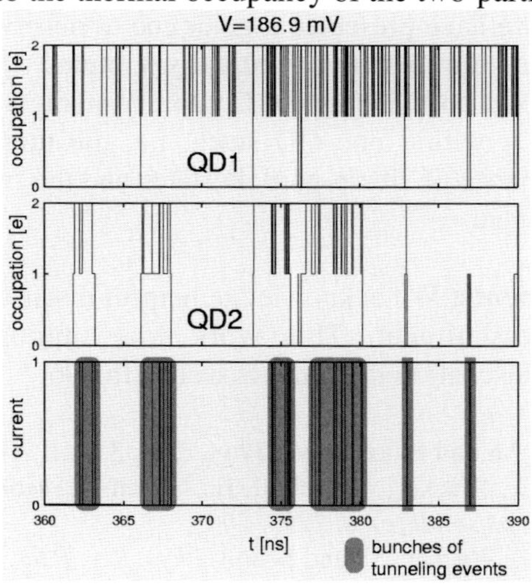

Fig. 2. Monte-Carlo simulation of the occupations in QD1 and QD2 and the electron jumps into the collector (current) for a bias voltage V=186.9mV. Parameters are the same as in Fig. 1.

How can one understand this behavior? For this purpose we look at the time evolution of the occupations in QD1 and QD2 and the current given by the jumps of electrons into the collector. To obtain a realisation for the stochastic process we apply a Monte Carlo simulation with the same parameters leading to the master equation results in Fig. 1c and d (for details of the simulation see [10]). A section of the realisation is shown in Fig. 2 for a bias voltage V=186.9 mV (at the voltage of the right Fano factor maximum in Fig. 1d). The upper graph corresponds to the occupation in QD1, the middle graph shows the occupation of QD2, and the lower graph contains the jumps of electrons into the collector. It can be seen that QD1 is mostly occupied with one electron. Therefore QD1 can easily be occupied with two electrons. Crucial for the occurrence of a tunneling current is the occupation of the single-particle state in QD2. The probability that one electron can enter the single-particle level in QD2 is highest when the levels are aligned which occurs for V=186.75 mV. This becomes apparent in the time series of the QD2 occupation (not shown here) and consequently in the respective current: the tunneling events are statistically distributed and the noise is Poissonian. In contrast, for a slight misalignment of the levels the probability for entering QD2 with one electron decreases. Such events are less frequent now. But, whenever one electron enters QD2 the R2 channel is opened which results in a bunching of tunneling events (shaded regions in Fig. 2). Note that the interaction between carriers in different QDs reduces this bunching effect.

To conclude, we have presented a guideline to observe super-Poissonian noise in tunneling through two vertically coupled QDs. The proposed mechanism is based on the effect of Coulomb interaction between carriers in individual QDs within one QD stack, i.e. the tunneling of electrons through the resonance of single-particle states and the two-particle states at the same bias voltage.

Acknowledgments: We acknowledge helpful discussions with F. Hohls regarding the experiment. This work was supported by Deutsche Forschungsgemeinschaft in the framework of Sfb 296.

1. Blanter, Y. M. and Büttiker, M., *Phys. Rep.*, **336**, *1, 2000.*
2. Kießlich, G., Wacker, A., Schöll, E., Nauen, A., Hohls, F., and Haug, R. J., *phys. status solidi (c),* ***0****, 1293, 2003.*
3. Kießlich, G., Wacker, A., and Schöll, *Phys. Rev. B,****68****, 125320, 2003.*
4. Borgstrom, M., Bryllert, T., Sass, T., Gustafson, B., Wernersson, L.-E., Seifert, W., and Samuelson, L., *Appl. Phys. Lett., 78, 3232, 2001.*
5. Barthold, P., *Diploma thesis, Universität Hannover, 2004.*
6. Iannaccone, P., Lombardi, G., Macucci, M., and Pellegrini, B., *Phys. Rev. Lett.,* ***80****, 1054, 1998.*
7. Safonov, S. S., Savchenko, A. K., Bagrets, D. A., Jouravlev, O. N., Nazarov, Y. V., Linfield, E. H., and Ritchie, D. A., *Phys. Rev. Lett.,* ***91****, 136801, 2003.*

8. Kießlich, G., Sprekeler, H., Wacker, A., and Schöll, E., *Semicond. Sci. Technol.,* ***19****, S37, 2004.* Thielmann, A., Hettler, M. H., König, J., and Schön, G., *Phys. Rev. B,* ***71****, 045341, 2005.*
9. Sprekeler, H., Kießlich, G., Wacker, A., and Schöll, E., *Phys. Rev. B,* ***69****, 125328, 2004.*
10. Kießlich, G., Wacker, A., and Schöll, E., *phys. status solidi (b),* ***234****, 215, 2002.*

Ultrafast Formation of Coupled Phonon-Plasmon Modes in InP Observed with Femtosecond Terahertz Spectroscopy

C. Kübler[1], R. Huber[2], S. Tübel[2], F. Köhler[3], M.-C. Amann[3], and A. Leitenstorfer[1]

1 Fachbereich Physik, Universität Konstanz, D-78457 Konstanz, Germany
2 Physik-Department E11, TU München, D-85748 Garching, Germany
3 Walter-Schottky-Institut, TU München, D-85748 Garching, Germany

Summary. We study the ultrafast transition of an optical phonon resonance to a coupled phonon-plasmon system. The buildup of coherent beats of the emerging hybrid modes after 10 fs photoexcitation of *i*-InP is monitored in the time domain via ultrabroadband THz spectroscopy. The anticrossing is mapped out as a function of time and density. In agreement with quantum kinetic calculations, the buildup time is found to scale with the oscillation cycle of the upper branch of the coupled resonance.

1 Introduction

Longitudinal optical (LO) lattice vibrations in polar semiconductors couple to charge density waves of a free carrier plasma via long-range electrostatic forces, forming LO phonon-plasmon coupled (LOPC) modes[1]. The mixed resonances are of fundamental importance for the transport properties of highly doped or photoexcited polar materials since they mediate both carrier-lattice and carrier-carrier interactions. If a dense electron-hole plasma is generated by an ultrashort laser pulse on a time scale shorter than a typical oscillation cycle of the lattice and the photogenerated plasma, it is predicted by quantum kinetic studies that the LOPC modes are not established instantaneously[2,3]. In fact, the femtosecond formation of carrier-carrier correlations has been observed by means of ultrabroadband THz spectroscopy[4]. The role of the polar lattice remained unresolved up to now.

We study a 230-nm-thin epitaxial layer of intrinsic InP contacted to a diamond substrate. A dense carrier plasma is injected via interband absorption of a 10 fs laser pulse with a central photon energy of 1.55 eV. In the first set of experiments, we excite $N = 2 \times 10^{18}$ electron-hole (*e-h*) pairs per cm^3 (Section 2). The polarization response of this nonequilibrium system is probed in real time by two-dimensional ultrabroadband THz spectroscopy developed in references 4 and 5.

To investigate the density dependence of the phonon-plasmon dynamics in more detail, we repeat the NIR pump–THz probe experiment at different carrier densities (Section 3). Employing a 12-fs Ti:sapphire amplifier system at a repetition rate of 4 MHz[6], we vary N over almost 2 orders of magnitude.

2 Ultrafast Formation of Coupled Modes

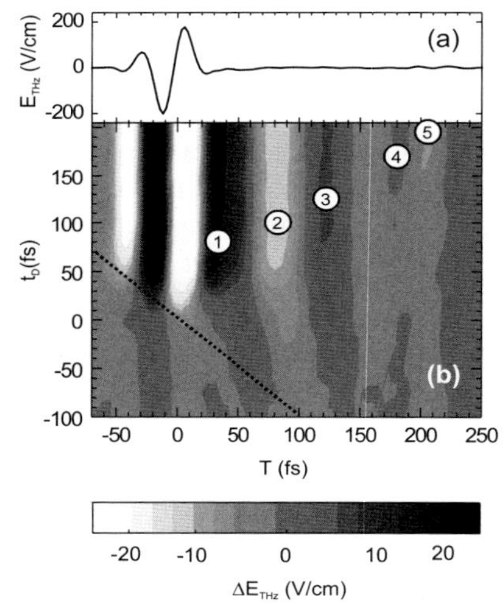

Fig. 1. Two-dimensional multi-THz data of the photoinduced phonon-plasmon system in *i*-InP at room temperature: **(a)** Real-time evolution of the electric probe field E_{THz} as a function of *T*. **(b)** The pump induced changes ΔE_{THz} of the transmitted electric field are recorded as a function of the sampling time *T* and the delay t_D after photoexcitation of $N = 2 \times 10^{18}$ *e-h* pairs per cm^3. The dotted line indicates the temporal position of the maximum of the pump pulse.

Figure 1 (a) shows ETHz as transmitted through the unexcited sample. The THz electric field change ΔE_{THz} induced by carrier injection is depicted as a gray scale map versus electro-optic sampling delay *T* (horizontal) and pump-probe delay t_D (vertical) in Fig. 1(b). The excited system responds instantaneously to the single-cycle probe pulse at early delay times, $-20 \leq t_D \leq 20$ fs. With increasing values of t_D, a trailing oscillation emerges. Up to five retarded half cycles appear for $t_D \geq 150$ fs (labels ''1'' to ''5'') at delay times *T* where the probe field E_{THz} is essentially zero without excitation. It is important to notice that the half cycles are not equidistant. This finding may be attributed to the beats between two newly emerging oscillation modes, as will become clear from a quantitative interpretation of the time domain data. For this purpose, we extract the femtosecond dynamics of the transverse dielectric function in the long wavelength limit, where the

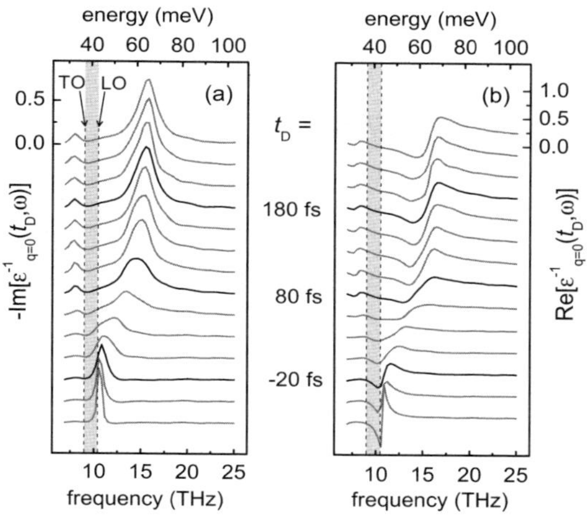

Fig. 2. Dynamics of the dielectric function of InP at a time t_D after photoexcitation ($N=2\times10^{18}cm^{-3}$):
(a) $-Im(1/\varepsilon_{q=0})$ indicates the energy loss of a carrier by interaction with the many-particle system, whereas **(b)** $Re(1/\varepsilon_{q=0})$ describes the renormalization of electric fields via dynamical screening.

transverse and longitudinal parts of the dielectric tensor are degenerate. To this end, the THz time scale is transformed into the frequency domain and the layer structure of the sample is accounted for by a standard matrix formalism.

At a density of 2×10^{18} cm^{-3}, many-particle correlations are found to form within 125 fs, coupling the carrier plasma and the LO lattice mode. The interaction of the two coupled resonances leads to a softening of the phonon-like L_- mode to below the TO phonon frequency due to dynamic screening. The topology of the bifurcation of the LO resonance is seen nicely in Fig. 2(a): While the LO phonon converts into the L_+ mode via intermediate broadening and shift to higher frequency, a sharp L_- mode gains oscillator strength at a constant frequency position on the same time scale.

Figure 3(a) shows the measured spectral positions of both peaks of $-Im(1/\varepsilon_{q=0})$ after well defined coupled oscillation modes have emerged. The two resonance branches exhibit anticrossing as a function of excitation density, as expected from the fully coupled phonon-plasmon system. This behavior is well reproduced by the classical Drude theory[1]. However, this model provides no description of the femtosecond formation dynamics of phonon-plasmon coupling. The time scale, on which hybrid resonances form, is found to depend on the excitation density. For a quantitative analysis we define a characteristic LOPC buildup time τ by the delay at which the L_+ pole reaches 80% of its final height.[i] The dynamics proceeds distinctly faster in denser systems. This tendency and quantitatively consistent values of τ are reproduced by nonequilibrium Green function the-

[i] In the experiment, reliable values of τ require an exceptionally high signal-to-noise ratio. Sufficient data quality could be obtained at elevated densities.

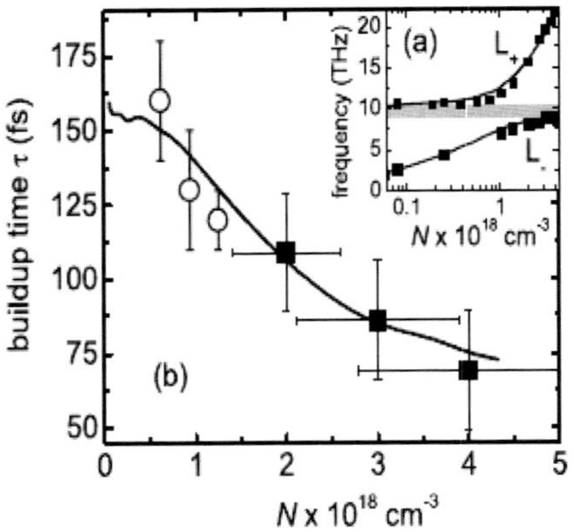

Fig. 3. Density dependence of the phonon-plasmon dynamics: **(a)** Spectral positions of the L_- and L_+ modes for various excitation densities (squares), and calculated (solid curves) via the Drude model. **(b)** Measured values of the buildup time τ for phonon-plasmon coupling (squares). The hollow circles show the result of a quantum kinetic calculation. The solid curve depicts a multiple of the inverse of the measured L_+ frequency ($1.6 \times 2\pi/\omega_+$).

ory[2,7]. Both experiment and theory suggest that the buildup time scales with the inverse L_+ frequency, i.e. the fastest collective response time of the many-body system.

3 Summary

We have observed the ultrafast transition of a dielectric polar lattice to a fully coupled and conducting phonon-plasmon system. The formation of coherent beats between both branches of the anticrossing LOPC resonances is directly resolved in the polarization response of *i*-InP after 10 fs photoexcitation. Our results support predictions of a quantum kinetic theory[3] and are qualitatively reproduced by simulations based on the non-equilibrium Greens functions formalism[7]. Many-particle correlations and screening effects leading to the formation of collective hybrid resonances are found to emerge approximately within the inverse eigenfrequency of the upper LOPC mode.

References

1. A. Mooradian et al., *Phys. Rev. Lett.* **16**, 999, 1966
2. Q.T. Vu et al., Phys. Rev. Lett **85**, 3508, 2000
3. Q.T. Vu et al., *Phys. Rev. B.* **62**, 7179, 2000.
4. R. Huber et al., *Nature* **414** ; 286 , 2001.
5. R.A. Kaindl et al, *Nature* **423**, 734, 2003
6. R.Huber et al., *Opt. Lett.* **28, 2118**, 2003
7. R. Huber et al., *Phys. Rev. Lett.* **94**, 027401, 2005.

Optical Coherent Control of Polariton Modes in ZnSe Single-Quantum Wells

I. Kudyk, L. Wischmeier, T. Voss, I. Rückmann, and J. Gutowski

Institute of Solid State Physics, University of Bremen, P.O. Box 330440, D-28334 Bremen

Summary. We present results obtained in real-time resolved linear pulse-transmission and in four-wave-mixing experiments which make use of optical coherent control to selectively enhance or suppress the excitation of polariton resonances in a semiconductor quantum well. By use of a pair of phase-locked pulses the coherent control of polariton modes and their quantum beats are studied. The observed transients are discribed very well by a phenomenological model.

The optical properties of ZnSe single-quantum-well structures with a rather large active-layer thickness in the range of some tens of nanometers are dominated by several polariton modes and their interactions. The polariton is a quasi-particle generated due to coupling between the light field and the excitonic polarization. Polariton propagation in shallow-confinement heterostructures has been investigated in [1] for linear transmission spectra. The interference between two propagating polariton modes, observed as quantum beats superimposed to the decay of the coherent polarization, has been studied in time-resolved four-wave-mixing (FWM) experiments [2]. Here, we present results obtained in linear pulse-transmission and FWM experiments using optical coherent control to selectively enhance or suppress the excitation of heavy-hole (hh) exciton-polariton resonances in a semiconductor quantum well. The center-of-mass quantization of the heavy-hole exciton polariton in a thin ZnSe layer according to $k_m = m\pi / d$, (where d is the well thickness, m integer) results in different polariton resonances which are labeled hh_m. The coherent control of polariton modes is studied in a 25 nm ZnSe single quantum well (SQW) which is embedded in two 1 μm ZnSSe barriers grown by molecular-beam epitaxy on GaAs substrate (A. Ueta, D. Hommel, University of Bremen). To permit measurements in transmission geometry the substrate was removed by chemical etching. A frequency-doubled Ti-sapphire laser with a pulse width (FWHM) of 120 fs at a repition rate of 82 MHz was used. The spectral position of the laser pulses ($\lambda \approx 440$ nm) was tuned such that four

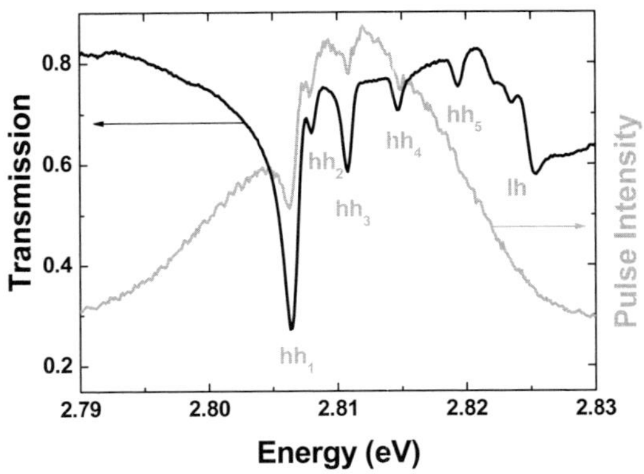

Fig. 1. Linear transmission spectrum of the 25 nm ZnSe-SQW (black). Spectral profile of laser pulse after having passed through the sample (grey). The heavy-hole exciton polariton renonances hh_1 to hh_4 are simultaniously excited.

heavy-hole exciton-polariton resonances labeled hh_1 to hh_4 are simultaniously excited. In Fig. 1 the spectral profile of the transmitted pulse is shown in comparison to the linear transmission spectrum as obtained with a Xe lamp is shown. In the linear transmission spectrum pronounced heavy-hole polariton modes and an absorption line due to the light-hole (lh) exciton are seen. The behaviour of the linear transmission is further influenced by the ZnSSe cladding layers which act as an outer Fabry-Perot resonator.

The real-time resolved measurements are carried out by use of an up-conversion technique (the signal is superimposed to the infrared reference pulse in a sum-frequency generating crystal with t_{upc} the delay time between both pulses). For the coherent control a pair of phase-locked ultra-short laser pulses is used. The pulse pair is provided by an actively stabilized Michelson interferometer. The time delay between the two pulses is $t_{int}= t_{int}^{0} + \Delta t_{int}$ where t_{int}^{0} denotes the basic temporal separation of the pulse pair and Δt_{int} a fine tuning on an attosecond time scale. For all presented measurements the sample was kept in a cryostat at a temperature of 4 K. Without coherent control (only one pulse) a complex beating structure of the signal can be observed due to the influence of several beat periods (inset of Fig. 2). The different beats are caused by the interactions between the dominant ground mode hh_1 and higher modes. In Fig. 2 spectrally integrated real-time resolved transients obtained in linear pulse-transmission experiments are shown which demonstrate the separate control of the polariton modes. The polaritonic polarization generated by the first pulse of the two phase-locked pulses can be amplified or diminished during the coherence time by the second pulse depending on the relative phase between the pulses [3]. Both transients are detected for different time delays Δt_{int}. The transients clearly show a different beat structure which is superim-

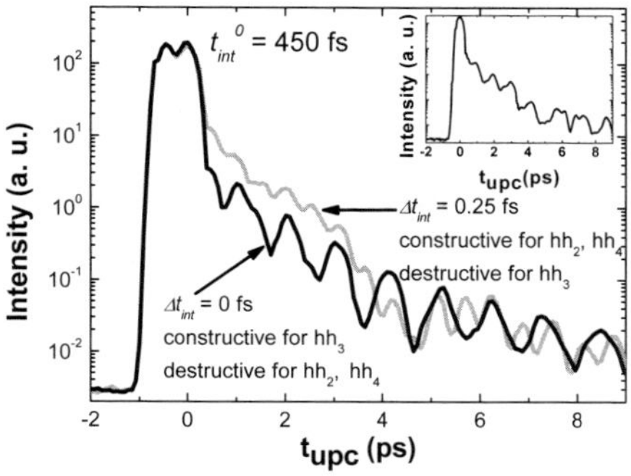

Fig. 2. Real-time resolved transients of two phase-locked pulses after propagation through a 25 nm ZnSe single-quantum well. Inset: Transient of one pulse (no coherent control).

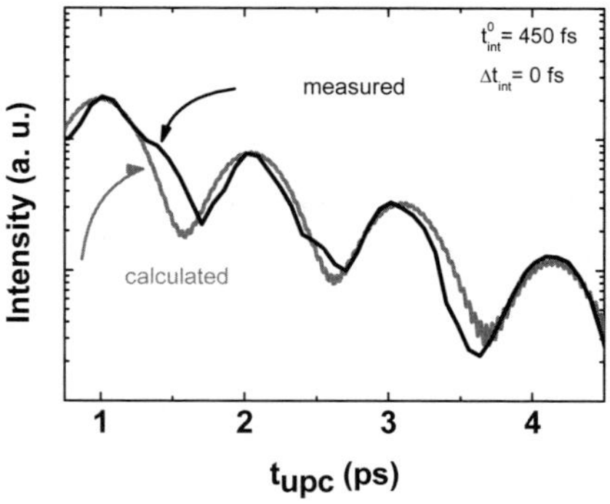

Fig. 3. Real-time measured (black) and calculated (grey) transients at a relative delay time of the the two phase-locked pulses of $t_{int}^{\,0}$ = 450 fs and Δt_{int} = 0 fs.

posed to the exponential dephasing of the coherent polarization. In the case of Δt_{int} = 0 fs the beating between the hh_1 and hh_3 resonances dominates the shape of the signal since for hh_3 constructive interference applies and an intermediate interference situation is given for hh_1. For Δt_{int} = 0.25 fs the beat structure can be identified to result from the interplay of the hh_1 (intermediate case) with the hh_2 and hh_4 polariton modes (constructive interference, each). The hh_1 resonance is always involed due to the large oscillator strength of this mode and the limited experimentally achievable contrast between constructive and destructive cases. In Fig. 3 the measured and calculated transients for $t_{int}^{\,0}$ = 450 fs and Δt_{int} = 0 fs are shown. The coherent polarization decay of the superimposed modes can be described for low excitation intensities as a linear superposition of the four involved polariton modes ($P = \exp(-t/\tau)\ \Sigma A_i \cos(\omega_i t)$). The energy positions and intensities of the hh_1 and hh_3 modes are taken from the linear transmission spectrum (Fig. 1). The good agreement between measured and calculated real-time transients allows to obtain the polarization decay time. It is found that the more modes interact the faster the coherent polarization of the polariton decays. This may be due to the different radiative decay times of the polariton modes [4].

Results obtained for the coherent control of polariton resonances in the nonlinear optical regime are shown in Fig. 4. The FWM signals emitted in direction $2k_1$-k_2 was measured spectrally resolved. The pulse pair was incident on the sample from direction k_1, and an additional single pulse was applied from direction k_2. The k_2 pulse arrives 500 fs after the first k_1 pulse. In Fig. 4a the FWM signal as generated by one pulse only from the

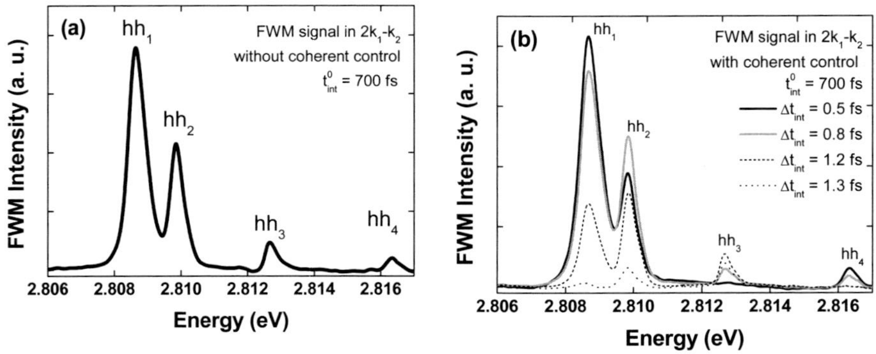

Fig. 4. (a) Spectrally resolved FWM signal in the direction $2k_1$-k_2 without the second phase-locked pulse. (b) Both phase-locked pulses were used. Spectra of the FWM signal at different delay times Δt_{int} are shown.

direction k_1 shows the modes hh_1 to hh_4. In Fig. 4b the two-pulse FWM spectra for different delay times Δt_{int} exhibit a change in the signal intensity of the polariton modes with changing delay time. The maximum intensities of the modes occurs at different delay times, respectively. This variation of the signal intensity is again due to constructive and destructive interference of the respective polarizations. The first k_1 pulse and the k_2 pulse generate the nonlinear polariton polarization which is coherently controlled by the second phase-locked pulse from direction k_1 arriving last at the sample.

Acknowledgements: The authors want to thank A. Ueta and D. Hommel (University of Bremen) for providing the sample.

References

1. Schumacher, S. et al.: 'Polariton propagation in shallow-confinement heterostructures: Microscopic theory and experiment showing the breakdown of the dead-layer concept', Phys. Rev. B **70**, 235340, 2004
2. Pantke, K.-H.: 'Nonlinear quantum beats of propagating polaritons', Phys. Rev. Lett. **70**, 327, 1993.
3. Heberle, A. P.: 'Ultrafast Coherent Control and Destruction of Excitons in Quantum Wells', Phys. Rev. Lett. **75**, 2598, 1995
4. Kudyk, I. et al.: 'Coherent control of polariton modes in real-time resolved pulse-transmission experiments', phys. stat. sol. (b), submitted

Optical Properties of Coupled Quantum Disk-Waveguide Structure

M. Yamaguchi, H. Tanaka, M. Yokoi, H. Takagi, and N. Sawaki

Dept. of Electrical Engineering and Computer Science, Nagoya University

Summary. We prepared a GaAs quantum well waveguide near which a GaAs quantum disk (QD) was located, and evaluated the transmission change in the waveguide due to the presence of the quantum disk. In case of TM polarized wave, the absorption due to the light-hole exciton in the QD was observed in accordance with the selection rule. The TE mode, on the other hand, exhibited two spectral peaks whose wavelengths coincide with the heavy- and light-hole excitons in the QD.

1 Introduction

A quantum well waveguide plays an important role in integrated optics and has been adopted to a device to handle multi-wavelengths. If one can read out the electronic states in a quantum device with the travelling wave through a waveguide, parallel signal processing might be realized by means of a multi-wavelength-waveguide. In previous papers [1, 2], we proposed a method to read out the electronic state in a quantum disk by the resonant coupling between the evanescent part of a travelling wave and the electronic state in the disk. In this paper, we will discuss the coupling phenomena in terms of the TE and TM polarized modes.

2 Experiments and Discussion

A coupled QD-waveguide structure was prepared by selective etching of a GaAs/AlGaAs double QW wafer grown by molecular beam epitaxy [1]. Figure 1 (a) shows the schematic structure of the sample. The core of the waveguide was made of a 5nm GaAs QW embedded in $Al_{0.3}Ga_{0.7}As$ (17 nm) / GaAs (3 nm) superlattices (SLs) which was clad by $Al_{0.4}Ga_{0.6}As$ (17 nm) / GaAs (3 nm) SLs (lower clad) and air (upper clad). The reason why we adopted AlGaAs/GaAs SLs as the core and the clad instead of an AlGaAs layer is to avoid the degradation of the optical properties due to rough heterointerfaces. In this structure, the exciton is confined in the GaAs QW layer while the polariton is extended in the whole core layers. On this waveguide, a QD having a 7.3 nm thick GaAs QW layer was placed. As a control sample, a sample without the QD was provided. In these QW waveguides the optical spectra of the travelling exciton-polariton were studied by analyzing the emission intensity collected at the end of the waveguide as shown in Fig. 1 (b). We utilized the photoluminescence (PL) from the photo-excited QW in the center of the core layers as the source for the propagating wave. We could investigate the travelling exciton-polariton in the wide wavelength range because the luminescence from the QW has a broad spectral peak.

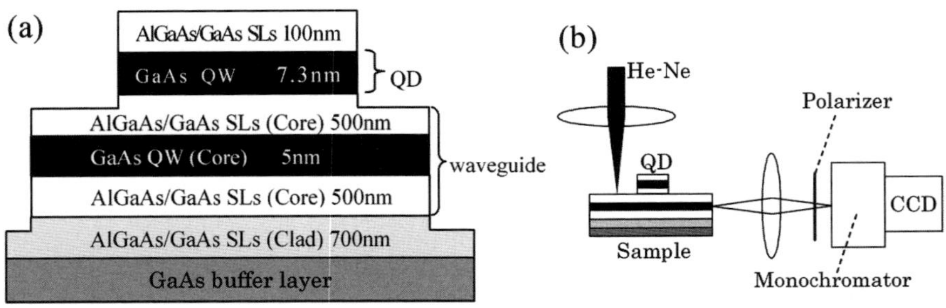

Fig. 1. (a) Schematic sample structure and (b) the measurement system.

In this measurement, the relatively large square disk (10 μm X 20 μm) was fabricated in order to observe the influence of the QD on the travelling wave easily. We predict the absolute effect of the QD on the transmission to be very small [2]. Hence, we reference the measurements on the sample with QD to measurements on the control sample without QD and obtain the transmission ratio by dividing the two emission spectra.

Figure 2 shows the typical spectral changes of TE and TM polarized waves due to the QD at 77 K, respectively. The PL spectrum from the QD surface is also shown in the figure.

For the TE mode, two dips were observed at 786 nm and 794 nm. This suggests that the absorption occurs by the resonant coupling of the exciton-polariton propagating light in the waveguide with the QD electronic state. As shown in Fig. 2, these two dips at 786 nm and 794 nm were in close agreement with the PL peaks due to the heavy hole-exciton (hh-e) and the light hole-exciton (lh-e) in the QD, respectively.

We simulated the characteristics of propagating polariton by solving Maxwell's equation numerically in order to investigate these absorptions

[2]. In this calculation, we assumed that the QD size along the waveguide was not 20 μm but 1 μm because we ascertained only the phenomenon of the absorption. This size was comparable to the electron mean free path in the nominally undoped QW. The calculated results indicated that the resonant coupling was performed via the evanescent part of the travelling wave. The coupling strength was represented as the decrease of the transmission probability of the propagating light. The decrease was estimated about 1 % of the transmitted light intensity at the resonant wavelength due to the heavy-hole in the 1 μm QD. In the measurement, the absorption rate was about 3 % which was larger than the estimated result. While this difference might be attributed to large QD size, the absorption rate was not so large in spite of the quite large QD size. Because the damping constant used for the experiment might be larger than that used by calculation which was data of high purity GaAs bulk. In the large QD, however, we demonstrated the observation of the resonant coupling due to hh-e and lh-e by the evanescent part of the TE polarized travelling wave.

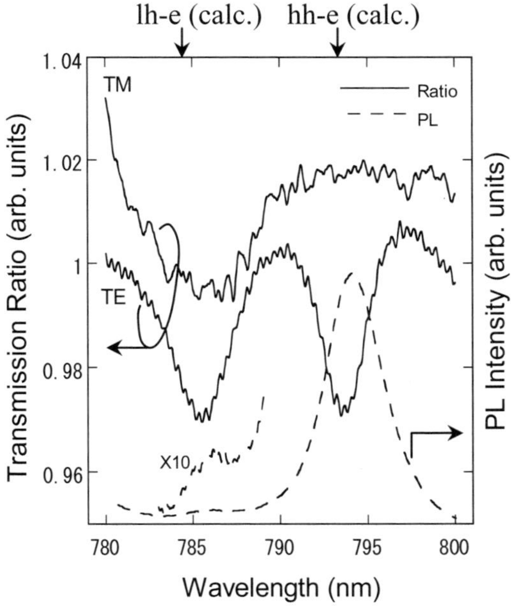

Fig. 2. The transmission ratio of the TE and TM polarized propagating light and PL spectrum of the QD at 77K. TM spectrum is shifted + 1 % for clarity. The calculated energies of hh-e and lh-e are also indicated.

In case of the TM polarized wave, on the other hand, we observed the signal due to the lh-e but no signal was observed for the hh-e. This is attributed to the selection rule in the polarization [3]. While the spectrum in the

TM mode was broad, the absorption was about 3 % as well as the TE mode. The reason of the broadness is not clear. However, the distinguished difference in the absorption spectrum between TE and TM polarized waves was obtained.

In summary we have studied the optical spectra in a coupled quantum disk-waveguide system, where the coupling was performed via the evanescent part of the travelling wave. In the TM polarized wave the absorption due to the lh-e in the QD was obtained in accordance with the selection rule of the polarization. On the other hand, in case of TE polarized wave, two peaks due to the hh-e and the lh-e were observed. In spite of the weak evanescent wave, we succeeded in observing the QD energy states clearly. Therefore, this technique was proven to be effective to read out the electronic state in the dot.

Acknowledgements

This work was partly supported by the “Tatematsu foundation” and “Research Foundation for Opto-Science and Technology”. The use of facilities of Nagoya University Venture Business Laboratory (NU-VBL) is acknowledged.

References

1. Tanaka, H. Takagi, H. Yamaguchi, M. and Sawaki, N.: *Proc. of Int. Conf. on Elec. Eng. 2004*, **3-2,** 653-656, 2004.
2. Takagi, H. Tanaka, H. Yamaguchi, M. and Sawaki, N.: *J. Phys. D*, **38**, 1551-1555, 2005.
3. Gontijo, I. Tessier, G. Livingstone, M. Galbraith, I. and Walker, A.C.: *J. Appl. Phys.*, **80**, 4027-4032, 1996.

Picosecond Spin-Preserving Carrier Capture in InGaAs/GaAs Quantum Dots

S. Trumm,[1] M. Wesseli,[1] H. Krenner,[2] D. Schuh,[2] M. Bichler,[2] J. J. Finley,[2] and M. Betz[1]

[1] Physik-Department E 11, Technische Universität München, D-85747 Garching, Germany
[2] Walter-Schottky-Institut and Physik-Department E24, Technische Universität München, D-85747 Garching, Germany

Summary. Carrier capture into self-organized InGaAs/GaAs quantum dots is studied in a femtosecond transmission experiment. Resonantly generating carriers in the wetting layer, we analyze the population of both the band edge of the wetting layer and the quantum dot excited states. We find a carrier capture time of 3 ps that is independent of the carrier density providing that it remains small compared to the number of available electronic states. Moreover, we find that the capture process is predominantly spin-preserving in nature. These results suggest that phonon mediated scattering governs the quantum dot filling.

Semiconductor quantum dots (QDs) are attractive both as model systems to study the physics of zero-dimensional structures and for their device applications. In particular, the discrete energy level structure of QDs offers significant advantages for modern semiconductor laser technology. In QD based devices, nonequilibrium carrier dynamics plays a central role in determining the performance limitations. As a result, especially intra-QD relaxation processes have been subject to a considerable research effort [1-7]. However, the important initial step of the population transfer from the two dimensional wetting layer (WL) into the zero-dimensional structures has not been isolated experimentally and remains subject of lively discussions [8-10].

In this contribution, we directly analyze the transfer of photoinjected carriers from the WL to fully quantized QD levels. The experiment relies on the transient bleaching of the WL band edge and the QD interband transitions in a femtosecond transmission experiment. For low excitation densities, we find a capture time of 3 ps indicating a very efficient QD filling. Most interestingly, the capture time does not significantly depend on the density of carriers photoinjected into the WL. Moreover, exploiting the selection rules for circularly polarized light, the carrier capture of electrons is shown to be spin-preserving.

The nanostructures investigated in this study are self-organized InGaAs QDs grown on a GaAs. The ground state transitions of the inhomogeneously broadened ensemble are centered at 1.25 eV and the interband energetic spacing of the transitions from different QD shells is approximately 40 meV, i.e. close to the optical phonon energies of GaAs The scheme of our experimental approach is displayed in the inset of Fig. 1(a). A pump pulse with a duration of 100 fs is tuned to the absorption continuum of the WL, thus resonantly creating electron-hole pairs. After relaxation towards the band edge, the carriers are transfered into fully quantized QD levels. Correspondingly, a transient optical bleaching signal is established reflecting the occupation of the electronic states. These transmission changes of the QD monolayer are detected with a 20 fs broadband probe pulse, that is derived from the second branch of a two color-Ti:sapphire laser, is spectrally dispersed after transmission through the specimen and detected with a photodiode.

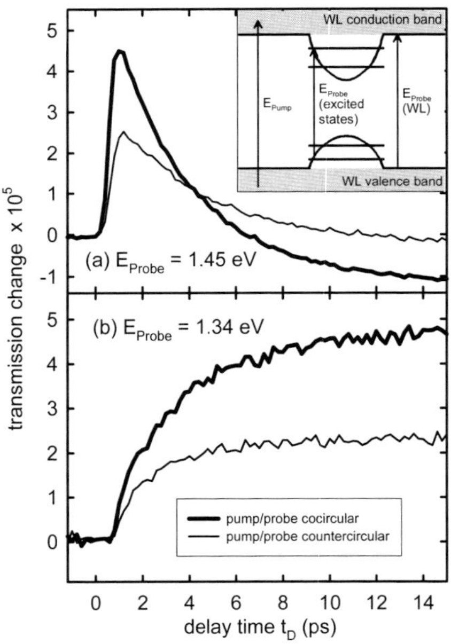

Fig. 1 (a) Transmission changes of the QD monolayer detected at the band edge of the WL after resonant carrier injection at $t_D = 0$ for co- and countercircular polarization of the excitation and probe pulses and a temperature of T = 4 K. The inset sketches the bandstructure with the vertical arrows indicating the interband transitions of the study.
(b) Corresponding transmission changes probing the excited states of the QDs.

Results for a moderate excitation density and a probe photon energy of 1.45 eV, near the band edge of the WL, are shown in Fig. 1(a). The bleaching signal decays nicely exponential with a time constant of 3 ps. These dynamics are accompanied by the buildup of a transmission change probing the QD excited levels (see Fig. 1(b)). Thus, the present measurement directly traces the population transfer from the two dimensional WL into the QDs.

More detailed insight into the carrier capture process is gained exploiting the spin selectivity of the interband transitions. Circularly polarized excitation yields a preferential spin orientation of the photogenerated carriers. While the spin orientation of holes is rapidly destroyed, the electron spin in the two-dimensional WL is expected to be preserved over the timescale of the present study. As a consequence, detecting the transmission changes of the WL with a counter-circularly polarized probe pulse yields an overall reduction of the transmission change by a factor of two (see thin line in Fig. 1(a)). More surprisingly, the nonlinear optical response of the QD interband transitions depends on the polarization configuration in the same manner as observed for the WL (see Fig. 1(b)). These observations demonstrate that the carrier capture process is a predominantly spin-preserving process.

In order to identify the nature of the capture process, the QD filling is studied for various excitation densities. To this end, we analyze the transient transmission changes of the WL band edge at a probe photon energy of 1.45 eV as a function of the excitation power. For weak excitation, we find that an exponential decay with a time constant of 3 ps provides a good fit to the experimental data. Most interestingly, the decay times extracted from exponential fits (see filled circles in Fig. 2) do not significantly depend on the excitation density indicating that carrier-carrier scattering has only a minor influence on the capture dynamics.

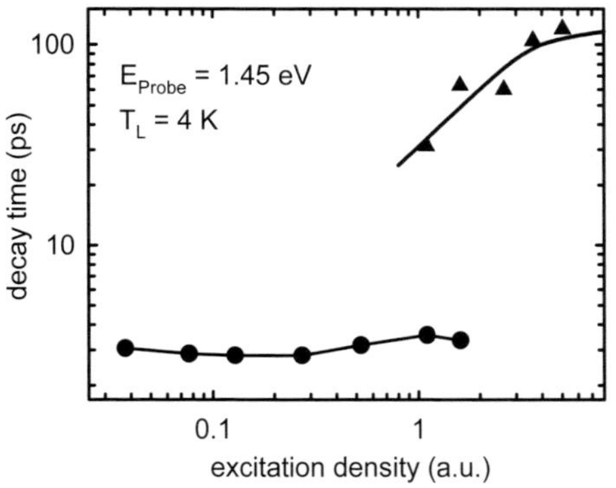

Fig.2 Decay times of the WL population for various excitation densities. The time constants are extracted from the transmission changes of the WL band edge at 1.45~eV by single and double exponential fits as discussed in the text. The abscissa may be interpreted as an estimate for the number of photogenerated electron-hole pairs per QD.

Increasing the excitation density to values comparable or larger than the QD density on the sample, the decay of the bleaching signal is effectively slowed down due to the emergence of a slower signal component. This finding is readily understood from the onset of macroscopic state filling effects in the QDs. Modelling these dynamics with a bi-exponential fit, we identify a fast capture process with a time constant of again 3 ps and a relatively long-lived component (see the filled triangles in Fig. 2). For the highest excitation densities of the study, the signal is dominated by massive QD filling and decays only on a timescale approaching typical radiative multiexciton lifetimes in QDs.

The independence of the capture time on the excitation density clearly points towards a phonon mediated process. It is interesting to relate this finding to previous experimental results claiming the importance of carrier-carrier scattering for the relaxation dynamics [1,3,4]. These time-resolved luminescence experiments are intrinsically sensitive to a combination of carrier capture and relaxation towards the QD ground state. In contrast, we study the capture into electronic states near the onset of the WL continuum. Our results are supported by an ultrafast luminescence study on large InGaAs quantum dots with a comparable level spacing [5].

In conclusion, we have analyzed the ultrafast, spin-preserving population transfer from a two dimensional WL to fully quantized QD levels in a femtosecond transmission experiment. The carrier capture time in our QDs with a level spacing comparable or smaller as the optical phonon energies amounts to 3 ps and is found to be independent of the excitation density. Taken together, these observations strongly indicate a phonon mediated process. As a result, we see no significant phonon bottleneck for the filling of the QD states. This finding may be an important ingredient for the optimization of modern QD based laser devices.

References

1. B. Ohnesorge, et al., Phys. Rev. B **54**, 11532 (1996)
2. R. Heitz, et al., Phys. Rev. B **56**, 10435 (1997)
3. M. De Giorgi, et al., Appl. Phys. Lett. **79**, 3968 (2001)
4. D. Morris, et al., Appl. Phys. Lett. **75**, 3593 (1999)
5. T. F. Boggess, et al., Appl. Phys. Lett. **78**, 276 (2001)
6. T. S. Sosnowski, et al., Phys. Rev. B **57**, R9423 (1998)
7. Müller, et al., Appl. Phys. Lett. **83**, 3572 (2003)
8. U. Bockelmann, and G. Bastard, Phys. Rev. B **42**, 8947 (1990)
9. H. Benisty, et al., Phys. Rev. B **44**, 10945 (1991)
10. T. R. Nielsen, et al., Phys. Rev. B **69**, 235314

Influence of Surfaces on the Pure Dephasing of Quantum Dots

T. Kuhn, B. Krummheuer and V. M. Axt

Institut für Festkörpertheorie, Westfälische Wilhelms-Universität Münster, Wilhelm-Klemm-Str. 10, 48149 Münster, Germany

Summary. An analysis of the influence of confined acoustic phonons on the pure dephasing of a single quantum dot located in a half space or a free-standing slab is given. We find that the proximity to a surface leaves distinct traces in the polarization and the absorption spectrum which can be traced back to the interaction of the dot with surface phonons and a reflected phonon wave packet.

1 Introduction

At low temperatures pure dephasing, i.e., a dephasing without real transitions between electronic states, is the dominant decay mechanism for the optical polarization of quantum dots (QDs) in the strong confinement limit [1]. This dephasing results from the coupling of carriers to acoustic phonons. If these QDs are excited by ultrashort pulses exact expressions for the carrier and phonon dynamics can be found [2]. The optical polarization shows a fast initial decay and a remnant polarization that does not decay any further. This behavior can be understood with the help of the phonon dynamics of the system. The optical excitation creates a phonon occupation within the QD that separates into two parts: an occupation which remains within the QD and a phonon wave packet that travels into the surroundings of the QD with sound velocity. As soon as it has left the QD, carriers and outgoing phonons can no longer interact and the polarization remains at a constant value. However, if the QD is not embedded in an infinite bulk system but in a confined structure, the emitted phonon wave packet will reach the surface where it is reflected and can thus return to the QD allowing carriers and phonons to interact again. We will show that this may leave distinct traces in the dynamics of the polarization. In this paper we will study the cases of a QD in a half-space and in a free-standing slab as prototypes for structures with one and two surfaces close to the QD.

2 Confined Phonons

Phonons in confined structures are unlike those in bulk systems. The reflection of the phonons at the surfaces features a partial conversion between longitudinal and transverse phonons. In addition one has to take into account surface phonons. These phonons can only travel along the surface and their displacement field decays exponentially with increasing distance from the surface. For a detailed description of the complete set of phonon modes in a half space and a free-standing slab we refer to Refs. [3, 4].

In this paper we concentrate on the interaction of carriers and phonons via the deformation potential as the dominant interaction mechanism in GaAs-type semiconductors. We will present the dynamics of the optical polarization after ultrafast excitation and the corresponding absorption spectra, which could be measured either in four-wave-mixing or absorption spectroscopy on single dots. The expressions for the carrier-phonon coupling matrix elements and for the optical polarization dynamics after ultrafast excitation as obtained from a generating functions approach to carrier-phonon quantum kinetics can be found in Ref. [5].

3 Quantum Dot in a Half-Space

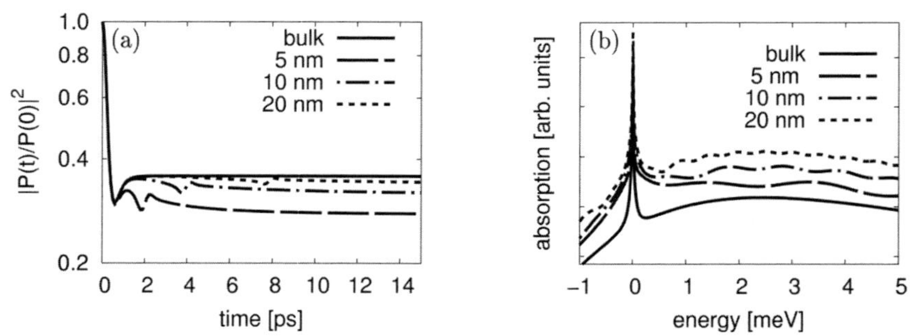

Fig. 1. (a) Optical polarization after ultrashort excitation for a QD located in a half-space at different distances from the surface. (b) Corresponding absorption spectra. For clarity the spectra have been vertically shifted. The QD has an ellipsoidal shape with sizes L_x=L_y=3 nm and L_z=1.5 nm. The temperature is T=4 K. Standard GaAs material parameters have been taken [5].

Figure 1 (a) shows the optical polarization for a single QD in a half-space at different distances from the surface. The solid line depicts the polarization for a QD in an infinite bulk system. The polarizations in the half space show two distinct differences from the situation in a bulk: a sharp dip and an additional slow decay. The sharp dip occurs when the emitted wave packet after reflection at the surface reaches the QD again. The slow decay turns out to be due to the interaction with surface phonons. Since their influence is restricted to a small distance from the surface, this decay increases with decreasing distance of the center of the QD from the surface.

The corresponding absorption spectra are given in Fig. 1 (b). The spectra look similar to those for a QD in an infinite bulk [2]: they feature a discrete zero-phonon line (ZPL) and an asymmetric acoustic background. In the half-space this background shows additional oscillations. They result from the dip in the polarization and thus correspond to the time the phonon wave packet needs to re-enter the QD. The interaction with surface phonons leads to an increased background in the vicinity of the ZPL.

4 Quantum Dot in a Slab

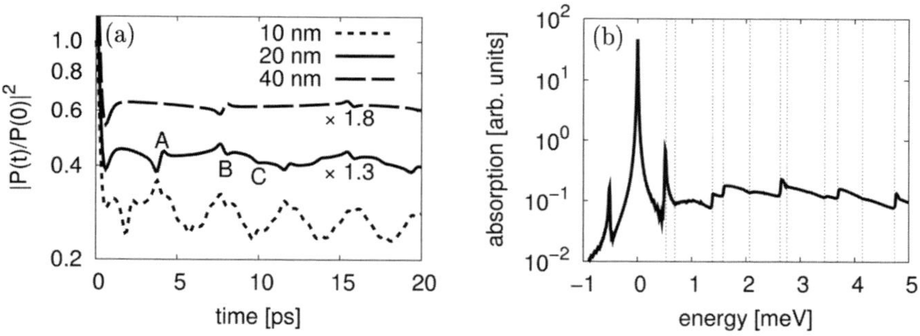

Fig. 2. (a) Optical polarization after utrashort excitation for a QDs located in the center of a slab with different widths as given in the figure. (b) Absorption spectrum for a slab width of 20 nm.

The optical polarization after ultrashort excitation for a QD in the center of a free-standing slab is shown in Fig. 2 (a). Now the emitted wave packet is repeatedly reflected from both surfaces. Its re-entrance into the QD leads to an oscillatory behavior with an elaborate fine structure. The timescale of the oscillations depends on the width of the slab. In addition each re-entrance of the wave packet into the QD leads to a sharp dip or a peak. As in the half space the first dip, marked in Fig. 2 (a) as A, occurs when the reflected wave packets re-enter the QD for the first time. The second re-entrance is marked as B and is characterized by a peak in the polarization. This different behavior is due to a phase jump which occurs each time the wave packet is reflected at the surface. The small dip marked as C is an indication of the presence of transverse phonons. These phonons do not couple to carriers via the deformation potential coupling. Thus they are neither created nor should they couple back to the QD. However, when the longitudinal waves originally emitted from the QD are reflected they are in part converted to transverse ones. In the next reflection these transverse phonons are again partially converted back into longitudinal ones, which then can interact with the QD. Since the transverse phonons travel slower than the longitudinal ones this dip occurs at a later time.

The absorption spectrum of a QD in the center of a 20 nm slab is shown in Fig. 2 (b). The acoustic background now exhibits a step-like structure. This is due to the dispersion relation of the confined phonons in a slab, which consists of several subbands (see Refs. [4] or [5]). The onset of each new subband, marked by the dotted vertical lines, leads to these step-like features. The spectrum also shows strong peaks close to the ZPL. These are van-Hove singularities caused by an infinite density of states [6].

5 Conclusion

We have shown the optical polarization after ultrashort excitation for QDs located in a half-space and a slab. In the half-space the reflected wave packet results in pronounced structures in the polarization dynamics and to an additional decay due to surface phonons. In the slab the polarization shows a complicated fine-structure superimposed on an oscillatory behaviour. The spectrum exhibits step-like features and a van-Hove singularity.
Financial support by the Deutsche Forschungsgemeinschaft is gratefully acknowledged.

References

1. Vagov, A. et al.: ‘Nonmonotonous temperature dependence of the initial decoherence in quantum dots’, *Phys. Rev. B*, **70**, 201305(R), 2004.
2. Vagov, A. et al.: ‘Electron-phonon dynamics in optically excited quantum dots: exact solution for multiple short laser pulses’, *Phys. Rev. B*, **66**, 165312, 2002.
3. Ezawa, H.: ‘Phonons in a Half Space’, *Annals of Physics*, **67**, 438, 1971.
4. Bannov, N. et al.: ‘Electron relaxation times due to the deformation potential interaction of electrons with confined acoustic phonons in a free-standing quantum well’, *Phys. Rev B*, **51**, 9930, 1995.
5. Krummheuer, B. et al.: ’Coupled polarization and acoustic phonon dynamics after optical excitation of quantum dots near surfaces’, *submitted.*
6. Debald, S. et al.: ’Control of dephasing and phonon emission in coupled quantum dots’, *Phys. Rev. B,* **66**, 041301(R), 2002.

Exploiting the Non-Markovian Nature of Carrier-Phonon Dynamics: Multi-Pulse Control of Decoherence in Quantum Dots

P. Machnikowski[1], V. M. Axt[2], T. Kuhn[2], L. Jacak[1]

[1]Institute of Physics, Wroclaw University of Technology, Wybrzeze Wyspianskiego 27, 50-370 Wroclaw, Poland
[2]Institut für Festkörpertheorie, Westfälische Wilhelms-Universität, Wilhelm-Klemm-Str. 10, 48149 Münster, Germany

Summary. We show that pure dephasing of an exciton in a quantum dot may be reduced by performing the optical excitation with a series of ultrafast pulses. Dephasing may considerably decrease already for a few pulses. In terms of the resulting coherence vs. control time, a multi-pulse approach is often more efficient than using smooth (Gaussian) pulses. These results show that a significant reduction of dephasing is achievable even with relatively limited experimental resources.

In self-assembled quantum dots (QDs) phonon-induced effects play the dominant role in the dephasing of charge states. Although the discrete nature of carrier states in QDs suppresses real phonon-assisted transitions from the ground exciton state, *pure dephasing* processes are still possible that may considerably decrease system coherence within picoseconds after an ultrafast optical excitation [1,2]. Because conservation of single-particle energies is not possible in such a process, it does not allow for a Markov limit and thus is a genuine quantum kinetic type of interaction. Studying this kind of processes allows one to test the properties of the system on the ultimate quantum level, beyond the Markovian description in terms of phonon scattering rates. Moreover, dealing with interactions that evolve over a finite time interval opens fascinating new perspectives because it becomes feasible to modify ongoing interaction processes between carriers and phonons as long as the latter are not yet completed [3].

In this contribution, we show that the non-Markovian nature of a pure dephasing process may be exploited to increase the degree of coherence of an optically created exciton in a QD. A considerable reduction of dephasing is obtained already for a few pulses and saturates quickly for a growing number of pulses. Dephasing may be further decreased by optimizing the pulse amplitudes. The pulse sequence leading to a given level of coherence is often shorter than a Gaussian pulse achieving the same goal, which is important for avoiding the destructive effect of other dephasing processes that accumulate during the control time [4].

Carrier-phonon interactions in the quantum kinetic regime have been studied extensively in extended semiconductors, with focus mostly on LO phonons. Here, the coherent control of the phonon induced polarization decay and of LO phonon quantum beats has been demonstrated [3]. The origin of the decay in these cases are real transitions to a continuum of final states. On the contrary, in the pure dephasing case, the irreversibility results from the couplings of the carriers to a continuum of phonon modes. This results in travelling phonon wave packets leaving the dot region while an acoustic polaron (coherent deformation field around confined charges) is formed after an ultrafast optical excitation [5,6]. Due to correlations built by these phonon packets between the confined carriers and the bulk of the crystal, the state of the former loses its coherence which manifests itself, e.g., in a decay of a coherent polarization signal [1].

The physical modelling of these dephasing processes is based on an independent boson model [7] with coupling constants derived from microscopic carrier-phonon interactions. This approach has successfully accounted for experimental data [2]. The system evolution is calculated within the 2nd order perturbation theory with respect to phonon couplings, including an arbitrary unperturbed evolution under a time-dependent control field exactly [8] (we restrict ourselves to resonant excitation). Comparison with exact results for infinitely short pulses [5] shows that the perturbation approach yields quantitatively reasonable results, at least for low temperatures [9]. In the calculations presented here, material parameters for a self-assembled InAs/GaAs system were used and only the deformation potential coupling to longitudinal acoustic phonons was included. Details of the calculations may be found in Ref. [9].

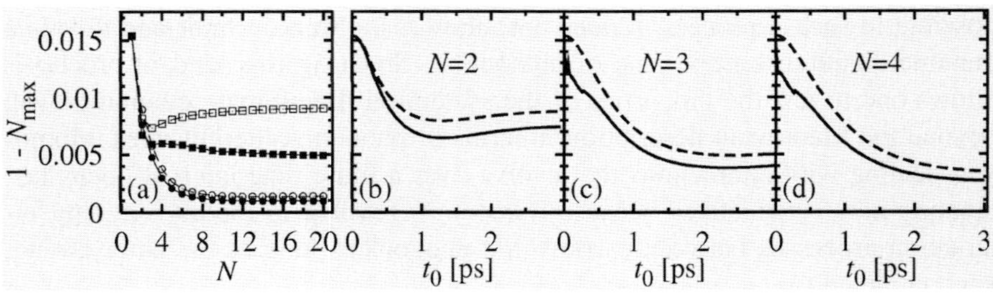

Fig. 1. (a) Decoherence as a function of the number of pulses for total durations of 1 ps (squares) and 5 ps (circles); full symbols: optimized pulse areas, empty symbols: equal pulses. (b)-(d) The same as a function of the total duration of the sequence for pulse numbers as shown (dashed: equal pulses, solid: optimized). All data at zero temperature for Gaussian pulses of width 100 fs.

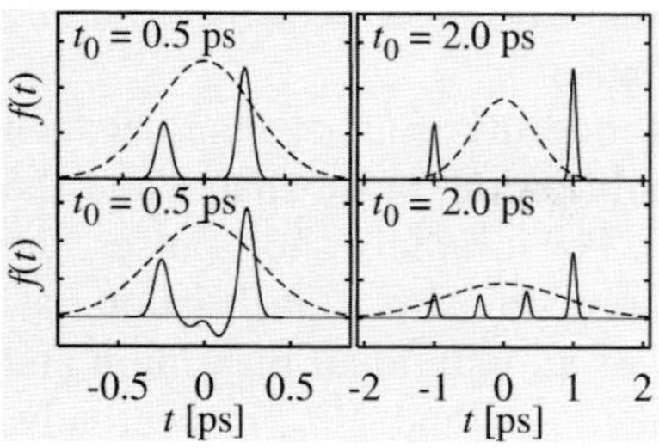

Fig. 2. Comparison between the optimized sequences of 2 and 4 pulses (solid lines) and Gaussian pulses leading to the same degree of coherence (dashed, plotted not to scale), for total durations of the sequence as shown.

We focus on the case that the total pulse area of the sequence is $\pi/2$, which brings the dot from its ground (empty) state to a superposition of the ground and the exciton state. As a figure of merit for the degree of coherence of the final state we choose the maximum exciton occupation $N_{\max}$ achievable by a Rabi rotation starting from this state performed by an optimally chosen pulse applied after a time delay longer than the reservoir memory time. This measure of coherence is based on an experimentally accessible quantity and has the advantage that it reflects only the true, irreversible loss of coherence, while the decay of coherent polarization [1,9] includes a certain contribution from the reversible polaronic effect.

In Fig. 1(a) we compare the dephasing resulting from excitations with a pulse sequence of fixed total length t_0 but with different numbers of pulses N. As can be seen, the reduction of the decoherence quickly saturates with the number of pulses, so that most of the coherence gain is achieved already for 2-4 pulses. Optimizing the pulse areas improves the result even further, especially for shorter total times t_0. Note that already for t_0=1 ps values of the decoherence around 0.5% can be reached which can be reduced to about 0.1% for t_0=5 ps. This brings the coherent control close to the tolerable error thresholds for quantum error correction schemes [10]. For a single pulse it was shown [4] that pure dephasing decreases as the excitation pulse is getting longer and vanishes in the limit of adiabatic control (infinitely long pulses). As shown in Fig. 1(b)-(d), a similar reduction is observed for increasing sequence durations t_0. However, it must be recalled that on long time scales pure dephasing competes with other decoherence mechanisms, such as radiative decay, which are usually of Markovian nature and accumulate over the excitation time. Thus, a question of importance for potential device applications of QDs is whether pure dephasing may be reduced within a given total excitation time which should be short enough to allow for many manipulations before the Markovian interactions irreversibly destroy the coherence. In Fig. 2 we compare sequences of pulses with optimized amplitudes with single Gaussians yielding the same degree of coherence. Obviously, pulse sequences can be significantly shorter than a single Gaussian pulse.

The dephasing effect may be interpreted in terms of an overlap between a nonlinear spectral function, depending only on the pulse shape, and the spectral density of the phonon reservoir [9]. For infinitely short pulses, the nonlinear pulse spectrum has the form and properties of a "diffraction pattern" in frequency domain, with the width of an individual maximum and the distance between maxima proportional to $1/t_0$ and N/t_0, respectively, explaining the dependence of the dephasing on the sequence duration and number of pulses. Broadening the pulses leads to an overall envelope that suppresses high frequency features (see Ref. [9] for a detailed discussion).

For overlapping pulses with optimized areas, the resulting pulse envelope may be very irregular. As may be expected, these rapid variations of the pulse amplitude lead to strong spectral features at high frequencies, while phonon excitations are usually modeled only within a certain frequency range (e.g., corresponding to acoustic phonons). By controlling the extent of the nonlinear spectral function one can assure that the theoretical modeling is consistent, i.e., the results are determined by the physical content of the reservoir model and not by its unphysical truncation. The freedom of optimization may be increased by extending the modeling of the phonon reservoir (e.g., by adding optical phonons) [9].

To summarize, we have shown that the quality of coherent optical control of the excitonic system can be increased by simple means, using series of phase-locked laser pulses, without the need to generate pulses of arbitrary shape. A few pulses are enough to achieve a considerable reduction of the decoherence to values typically much smaller than 1%. In many cases, multi-pulse control is favorable compared to smooth Gaussian pulses.

This work was supported by the Polish MNI (2 PO3B 085 25). P.M. is grateful to A. von Humboldt Foundation for support.

References

1. P. Borri *et al., Phys. Rev. Lett.* **87**, 1541, 2001.
2. A. Vagov, V. M. Axt, T. Kuhn, *Phys. Rev. B*, **67**, 115338, 2003.
3. M. U. Wehner, M. H. Ulm, D. S. Chemla, M. Wegener, *Phys. Rev. Lett.*, **80**, 1992, 1998; V. M. Axt, M. Herbst, T. Kuhn, *Superlattices and Microstructures,* **26**, 117, 1999.
4. R. Alicki *et al., Phys. Rev. A,* **70**, 010501, 2004.
5. A. Vagov, V. M. Axt, T. Kuhn, *Phys. Rev. B*, **66**, 165312, 2002.
6. L. Jacak, P. Machnikowski, J. Krasnyj, P. Zoller, *Eur. Phys. J. D,* **22**, 319, 2003.

7. B. Krummheuer, V. M. Axt, T. Kuhn, *Phys. Rev. B*, **65**, 195313, 2002.
8. A. Grodecka, L. Jacak, P. Machnikowski, K. Roszak, in: P. A. Ling (ed), *Quantum Dots: Research Developments*, Nova Science Publishers, Hauppauge, NY, 2005. Preprint cond-mat/0404364.
9. V. M. Axt, P. Machnikowski, T. Kuhn, *Phys. Rev. B.,* **71**, 155305, 2005.
10. E. Knill, *Nature*, **434,** 39, 2005.

Numerical Study of Weak Localization Effects in Disordered Cavities

L. Bonci, M. Macucci, G. Iannaccone, and M. G. Pala

Dipartimento di Ingegneria dell'Informazione, Università di Pisa,

Via Caruso 16 – I-56122 Pisa

Summary. We present a study of magnetoconductance through a cavity obtained by laterally shifting a section of a Si-Ge quantum wire, focusing on how randomly distributed impurities may affect its conductance behavior as a function of magnetic field.

1 Introduction

We perform a numerical simulation, of the magnetoconductance of mesoscopic cavities obtained by means of electron beam lithography and etching in a silicon-germanium heterostructure. In order to define entrance and exit constrictions smaller than the resolution limit of the e-beam system, a bended nanowire geometry was adopted [1]. The potential within the cavity is affected by the disorder resulting from the random distribution of dopants. Such fluctuations lead to a weak localization effect, determining a suppression of conductance for low magnetic fields [2].

We have included the effect of a realistic distribution of discrete dopants via a semi-analytical method [3] that takes into account screening by the 2-DEG (2-Dimensional Electron Gas).

2 Physical Model

The structure of the device is shown in Fig. 1(a): a 2-DEG is obtained by means of modulation doping, with electrons provided by a _-doping layer located 18.5 nm above the heterointerface. The width of the bended wire is about 250 nm, and the length of the shifted portion is 570 nm. The two constrictions, which separate the central cavity from the leads, have a width of 80 nm. Details about the calculation of the confining potential are provided in Ref. [2]. Donors in the doped layer supply the electrons to populate the 2-DEG, but also produce irregularities in the potential landscape. Inclusion of the effect of the donors into the Poisson solver used to compute the electrostatic potential would not be numerically feasible: an approximate semi-analytical approach is instead used, based on the theory of Ref. [4]. A detail of the thus obtained potential is shown in Fig. 1(b).

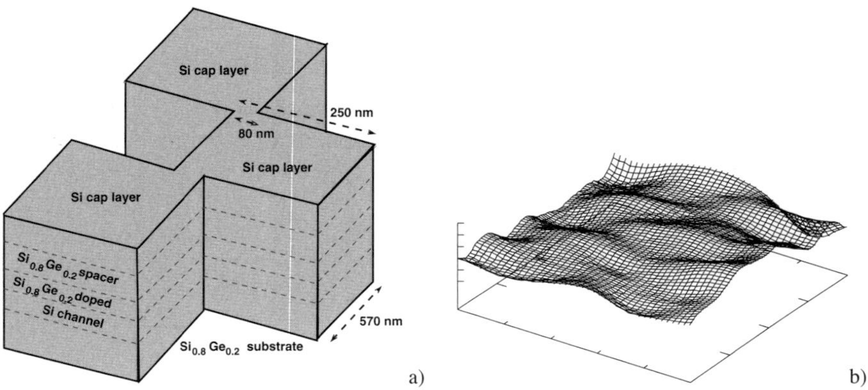

Fig. 1. Bended wire with heterostructure layer arrangement (a), and detail of the potential seen by the electrons inside the cavity (b), with the effect of the random distribution on dopants in the δ-doping layer.

3 Results

We have computed the magnetoconductance for many samples differing for the random donor distribution. Thermal averaging has been performed for a temperature of 50 mK.

Some of the results are shown in Fig. 2: for B=0 in all cases with disorder due to donors but one (thin solid curve) a minimum of the conductance is observed, corresponding to weak localization effects. The thick solid curve corresponds instead to a case with no disorder and does not exhibit a conductance minimum for B=0: therefore we attribute the weak localization effect observed in our sample only to the action of disorder and not to scattering due to the cavity walls. As shown by Akis *et al.* [5], conductance minima sometimes observed in cavities without disorder can be the result of resonance features, rather than actual weak localization effects.

To understand the anomalous case of the thin solid curve, we analyzed the corresponding distribution of impurities and we found that it is characterized by the presence of a deep local depression in the potential (dark region at the top center of Fig. 3(a)). The electron density for this case is reported in Fig. 3(b): as a result of the particular shape of the potential induced by the impurity pattern, the wave function, especially in the left part of the cavity, explores only a small region, thereby experiencing less disorder, which results in a reduction or even disappearance of weak localization effects.

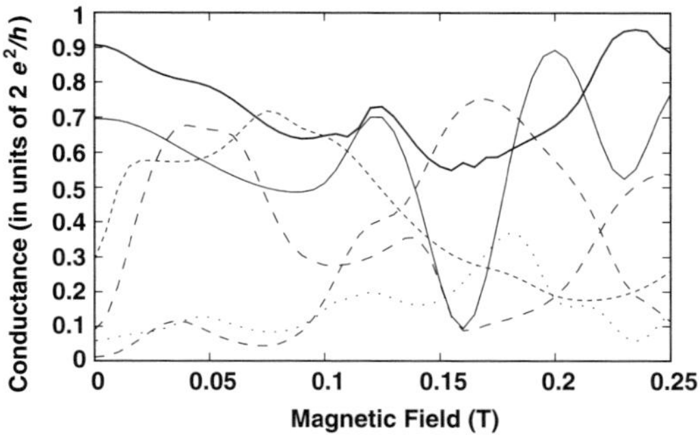

Fig. 2. Magnetoconductance vs. B for different distributions of random impurities.

To confirm this result, we have then purposely constructed impurity distributions capable of limiting the portion of the cavity explored by the wave function, in particular with the creation of a depression as a result of an artificial bias obtained multiplying the charge of each impurity by a factor $b > 1$ in a circular region with a 10 nm radius.

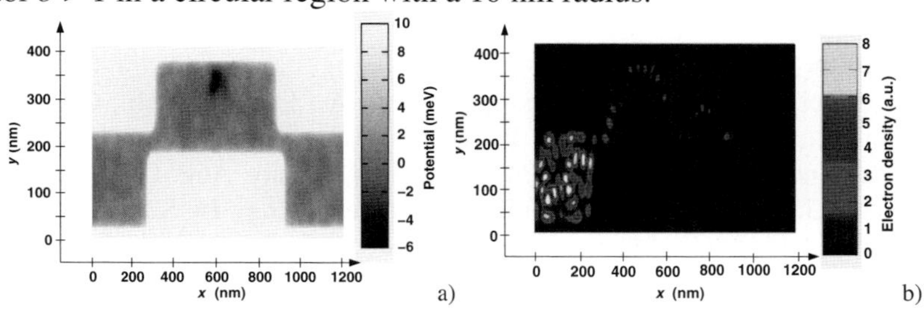

Fig. 3. Wire potential (a) and electron density for zero magnetic field (b) in the anomalous case corresponding to the thin solid curve of Fig. 2.

The location of the region with biased impurities has been selected looking at the behavior of the conductance as a function of a depression in the potential: we have chosen the position at which the largest effect on conductance was observed. Results for the magnetoconductance with such a selection are reported in Fig. 4, for $1 \leq b \leq 4$.

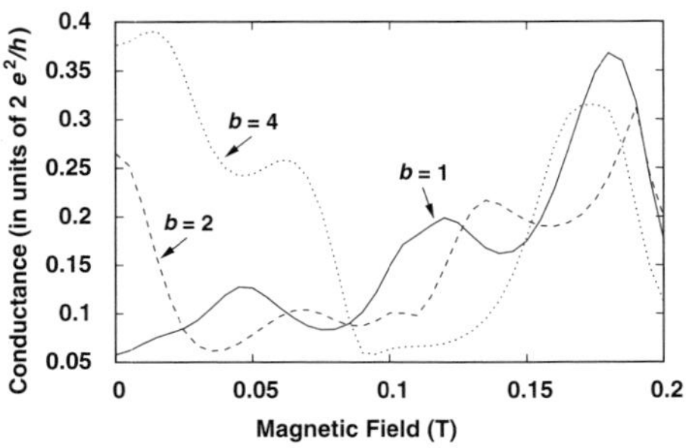

Fig. 4. Low-field magnetoconductance for the unbiased case (solid line) and for two different bias values: b=2, dashed line, and b=4, dotted line. The biased impurities are located around x = 400 nm; y = 260 nm. T = 0.05 K, E_f = 0.003 eV.

It is apparent that a suppression of weak localization is obtained, as b is increased above 1, consistently with the previous observations.

This suggests the possibility of an experiment in which such a result could be tested by introducing an artificial perturbation of the potential, whose position can be finely tuned, such as a low-temperature STM (Scanning Tunneling Microscope) tip. We acknowledge financial support from the Italian Ministry of Education, University and Research (MIUR) through the FIRB project "Nanotechnologies and Nanodevices for the Information Society" and the FISR project "Nanotechnologies for Very-High Density Memory Devices."

References

1 Scappucci G., Di Gaspare I., Notargiacomo A., Evangelisti F., Giovine E., Leoni R., Piazza V., Pingue P., Beltram F., Pala M. G., Curatola G., Iannaccone G., IEEE NANO 2004, München, Germany, Aug. 17th-19th 2004.
2 Pala M. G., Iannaccone G., Curatola G., Nanotechnology **16**, S206 (2005).
3 Bonci L., Fiori G., Macucci M., Iannaccone G., Roddaro S., Pingue P., Piazza V., Cecchini M., and Beltram F., Physica E **19**, 107 (2003).
4 Stern F. and Howard W. E., Phys. Rev. **163**, 816 (1967).
5 R. Akis, D. K. Ferry, J. P. Bird, and D. Vasileska, Phys. Rev. B **60**, 2680 (1999)

Carrier Scattering by Optical Phonons, Two-Phonon Processes in Photon Absorption, and Spontaneous Polarization in Wurtzites

Mitra Dutta, Gail J. Brown, Dinakar Ramadurai, Dwarakanath Geerpuram, Jiangyong Yang, Babak Kohanpour, Chen Chen, and Michael A. Stroscio

Electrical and Computer Engineering Department, University of Illinois at Chicago (MD, JY, BK, CC, and MAS), Physics Department, University of Illinois at Chicago (MD and MAS), Air Force Research Laboratory, Wright Patterson Air Force Base (GJB), Bioengineering Department, University of Illinois at Chicago (DR, DG, and MAS)

Summary. This paper focuses on the role of phonon scattering in wurtzites, emphasizes results indicating strong carrier-phonon scattering effects in these materials, and illustrates band-bending effects in wurtzite quantum dots due to spontaneous polarization.

1 Introduction

Wurtzite materials are of considerable interest in electronics and optoelectronics as well as in biological applications. The dielectric continuum model is applied to the wurtzite semiconductors GaN, ZnO, CdS and CdSe and it is illustrated that carrier-optical-phonon scattering rates in these materials are exceptionally large. Moreover, two-phonon effects are shown to be important in optical processes in selected würtzite crystals.

2 Optical-Phonon-Carrier Scattering Rates in Wurtzites

The dielectric continuum model [1-4] is used to illustrate that carrier-optical-phonon scattering rates for GaN, ZnO, CdS and CdSe are among the largest encountered in semiconductors of current interest in electronic and optoelectronics. Optical phonon frequencies of these semiconductors are calculated as a function of angle with respect to the c-axis. Due to the optical anisotropy of uniaxial crystals, the long-wavelength lattice vibrations may be classified in terms of orientation with respect to the c axis. In these wurtzites, the Raman and infrared (IR) active A_1 and E_1 modes split into LO and TO components with frequencies that depend on the angle, θ_k, relative to the c-axis. The Loudon model [2] is useful in describing the macroscopic equations of uniaxial polar crystals by introducing a dielectric constant associated with the direction parallel to

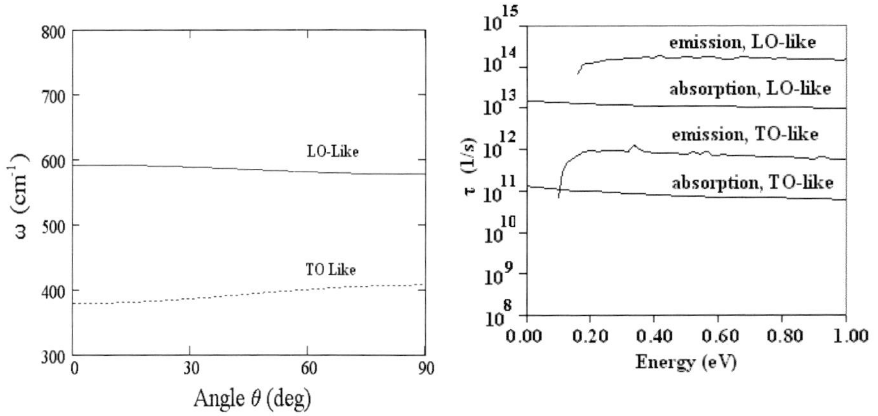

Fig. 1 The angular dependence of the LO-like and TO-like phonon in ZnO are depicted in the left portion of this figure. Scattering rates for LO-like and TO-like phonons for ZnO are depicted on the right portion.

the c-axis, ε_z, and another dielectric constant associated with the direction perpendicular to the c-axis, $\varepsilon_\perp$. Figure 1 shows the angular variation of the optical-phonon frequency for ZnO as well as the total carrier-phonon scattering rates within the Loudon model [2] for ZnO; GaN, CdS and CdSe exhibit similar forms [3]. The maximum rates for LO-like phonon emission are in excess of (a) 10^{14} sec^{-1} for ZnO and GaN and (b) 10^{13} sec^{-1} for CdS and CdSe.

3 Two-Phonon Processes in GaN

The importance of phonon effects in GaN is illustrated by the measured transmittance for a 250 μm thick GaN film as shown in Figure 2. As shown in this figure, there are actually two main absorption regions in the transmittance curve, which represent the first and second harmonics, separately. For zinc-blende semiconductors, such as GaAs the second harmonic absorption feature is normally not observed. The solid lines in the Figure correspond to the frequency region $\omega^{E1}{}_{TO} < \omega < \omega^{E1}{}_{LO}$ region and $2\omega^{E1}{}_{TO} < \omega < 2\omega^{E1}{}_{LO}$. Here we focus on the E1 mode since it is polarized transverse to the c-axis as is the incident radiation. For GaN, $\omega^{E1}{}_{TO} = 561$ cm^{-1} and $\omega^{E1}{}_{LO} = 741$ cm^{-1}. The appearance of the second harmonic absorption peaks illustrates the large two-phonon absorption processes in GaN, in agreement restralhen absorption as modeled in Ref. 5.

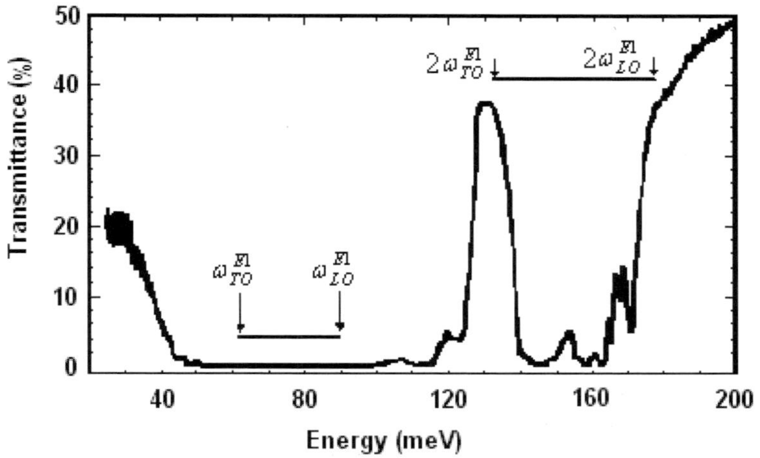

Fig. 2. Transmittance for a 250 μm thick GaN film vs. photon energy.

4 Spontaneous Polarization Effects in Wurtize Quantum Dots

The spontaneous polarization in wurtzite structures produce fields of the order a MV/cm or more. In AlN, CdS, CdSe, GaN, and ZnO these spontaneous polarizations are [6] 0.081, 0.002, 0.006, 0.029, 0.070 C/m^2, respectively. These fields are significant in wurtzite-based quantum dots [7] where they lead to band bending effects. Using the energy levels of a triangular confinement region [8], the band-bending is given by the spontaneous polarization may be expressed as,

$$E_{triangle} = E_{gap} + (h^2/8\pi^2 m_e)^{1/3}\,(9\pi F/8)^{2/3} + (h^2/8\pi^2 m_h)^{1/3}\,(9\pi F/8)^{2/3} - Fd. \quad (8)$$

The energy-level shifts predicted by Eq. 8 are in accord with experimental values reported in Ref. 7.

5 Discussion

These results indicate the importance of carrier scattering by optical phonons, two-phonon processes in photon absorption, and spontaneous polarization in wurtzites.

6 Summary

The authors are grateful to Dr. Dwight Woolard of ARO, Drs. Gerald Witt and Todd Steiner of AFOSR, and Dr. John Carrano of DARPA for their support of this research.

References

1. Stroscio, Michael A., and Dutta, Mitra, Phonons in Nanostructures, (Cambridge University Press, Cambridge, 2001).
2. Loudon, R., "The Raman Effect in Crystals," *Adv. in Phys.* **13**, 423 (1964).
3. Lee, B. C., Kim, K. W., Dutta, M. and Stroscio, M. A., "Optical Phonon Confinement and Scattering in Wurtzite Heterostructures," *Phys. Rev. B* **56**, 997 (1997).
4. Chen, Chen, Dutta, Mitra, and Stroscio, Michael A., "Electron Scattering via Interactions with Optical Phonons in Würtzite Crystals," *Phys. Rev. B* **70**, 075316-1-7, 2004.
5. Woolard, D. L., Globus, T. R., Bykhovskaia, M., Samuels, A. C., Cookmeyer, D., Hesler, J. L., Crowe, T. W., Jensen, J. O., and Loerop, W. R., "Submillimeter-wave phonon modes in DNA macromolecules," *Phys. Rev. E* **65**, 051903-1-11, 2002.
6. Stroscio, Michael A., Dutta, Mitra, Nariwani, Kavita, Shi, Peng, Ramadurai, Dinakar, and Rufo, Salvador, "Integrating and Tagging Biological Structures with Nanoscale Semiconducting Quantum-Dot Structures," pages 1-36 in Biological Nanostructures and Applications of Nanostructures in Biology: Electrical, Mechanical & Optical Properties, Stroscio, Michael A., and Dutta, Mitra, editors (Kluwer Academic Publishers, New York, 2004).
7. Ramadurai, Dinakar, Kohanpour, Babak, Alexson, Dimitri, Shi, Peng, Sethuraman, Akil, Li, Yang, Saini, Vikas, Dutta, Mitra, and Stroscio, Michael A., "Tunable Optical Properties of Colloidal Quantum Dots in Electrolytic Environments," *IEE Proceedings in Nanobiotechnology* **151**, 189-192 (2004).
8. Davydov, A. S., Quantum Mechanics (NEO Press, Ann Arbor, 1966).

Terahertz Plasma Oscillations in Nanotransistors

W. Knap[1] and J. Łusakowski[1,2]

[1]GES-UMR 5650 CNRS - Université Montpellier2, 34900 Montpellier, France
[2]Institute of Experimental Physics, Warsaw University, Hoża 69, 00-681 Warsaw, Poland

Summary Scaling of the gate length of Field Effect Transistors (FET) down to nanometer size leads to THz plasma oscillations in the transistor channel. In this work we give an overview of our recent experimental results concerning this effect. The first observation of the resonant THz detection was demonstrated on 150 nm gate length GaAs FET. The most important recent result concerning the detection is a very strong (two orders of magnitude) enhancement of the photodetection signal obtained by increasing of the drain current that drives the transistor towards the plasma instability. The first demonstration of THz plasma wave emission from a nano-transistor was observed on a 60nm InGaAs FET. The detailed mechanism of the emission is still under discussion. Some insight is given by Monte Carlo simulations, which show the importance of hot electron effects in the THz emission. The first results on room temperature THz detection in nanometer Si-MOSFETs were also demonstrated.

Introduction

THz phenomena in semiconductors involve a very large spectrum of items that are intensively studied nowadays in search for new applications and devices. One of the crucial subjects in this area is development of efficient THz sources and sensitive detectors. The aim of this paper is to show that Field Effect Transistors (FETs) are a very promising tool that could serve this purpose. The interest in using nanometer FETs as detectors and emitters of THz radiation was stimulated by the model of plasma instability developed in [1]. This theory considered the channel of a FET as a resonant cavity for plasma oscillations and predicted that a coupling of the external electromagnetic radiation with plasma instability is possible. Thus, FETs were proposed to be detectors and sources of the electromagnetic radiation in THz domain if the gate length was in nanometer range.

In the present paper we describe recent experiments on optical THz performance of nanotransistors. In Section 2 we describe results of detection experiments carried out on GaAs/GaAlAs High Electron Mobility Transistors (HEMTs) and Si MOSFETs at room and at liquid helium temperatures. Section 3 summarizes results of emission experiments on GaInAs/AlInAs HEMTs.

2. Detection

From the application point of view, the important fact is a possibility of coupling the plasma oscillations in the transistor channel to the external electromagnetic radiation. According to [1], the coupling is due to a dipole composed of an oscillating electron density in the channel and corresponding oscillations of the image charge. The coupling leads to detection and emission of the electromagnetic radiation by a FET. The detection requires application of the gate polarization only because its mechanism is based on a source – drain asymmetry accompanied with rectification of the signal originating from the incoming radiation. In such a case, a constant drain – source voltage appears as a result of the interaction of the incoming radiation with the electron plasma in the channel (the photovoltaic effect). However, as will be shown below, the sensitivity of detection grows when the drain current flows (the photoconductive effect). This is due to the fact that the effective relaxation time that defines the quality factor, $\omega\tau_{eff}$, increases with the increase of the drain current. In the following, we discuss the possibility of increasing the quality factor by technological improvements and by increasing the drain current.

Observations of plasmon resonances in 2 DEG in semiconductors started with the work of Allen, Tsui and Logan [2] who showed by Fourier transform spectroscopy resonances in transmission through a gated MOS structure. Burke *et al.* [3] performed measurements of a frequency dependent conductivity of a high mobility 2DEG in GaAs/AlGaAs quantum structures, determining the real and imaginary part of the conductivity up to 10 GHz. They found that both parts show an oscillatory character as a function of the frequency, and explained this behaviour by plasma resonances. Peralta *et al.* [4] demonstrated a voltage tuneable photoconductivity response of a double quantum well heterostructure that showed plasma resonances excited by radiation in the range between 120 GHz and 4.8 THz.

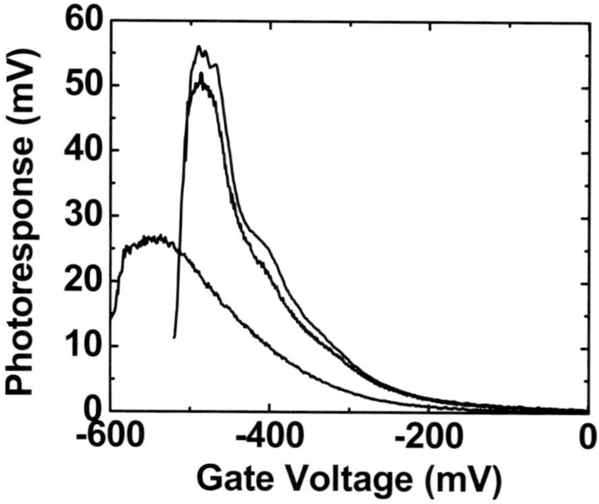

Fig. 1 Source – drain photovoltage in GaAs/GaAlAs transistor at 100 K, 20 K and 8 K (from bottom to top).

The first implementation of a sub micrometer FET as a detector of THz radiation was reported by Lü et al. [5], who exposed a commercial 0.18 μm long HEMT to a molecular laser radiation of 2.5 THz and measured the photoresponse as a function of the gate – source bias. Clear evidence of a resonant detection of THz radiation by plasma oscillation was shown by Otsuji *et al.* in InGaP/InGaAs/GaAs HEMT with the gate length of 150 nm [6]. In that experiment, the transistor was exposed to a photomixed laser beam that contained a THz difference component. The frequency of the THz component was tuned between about 1 THz and 8 THz, and two resonances in the photoresponse were observed at 1.9 THz and 5.8 THz that corresponded to the first and third harmonics of the plasma resonance described by the Dyakonov – Shur model.

THz detection in GaAs/AlGaAs and GaN/AlGaN transistors was investigated by Knap *et al.* [7, 8] in a wide range of temperature between 8 K and 300 K. In these experiments, a sub THz radiation between 100 GHz and 600 GHz was used and the experimental results were interpreted within a model based on the original Dyakonov – Shur approach. In these measurements [7, 8], a resonant plasma peak [9] appears as a small structure on a nonresonant background when the temperature is low enough

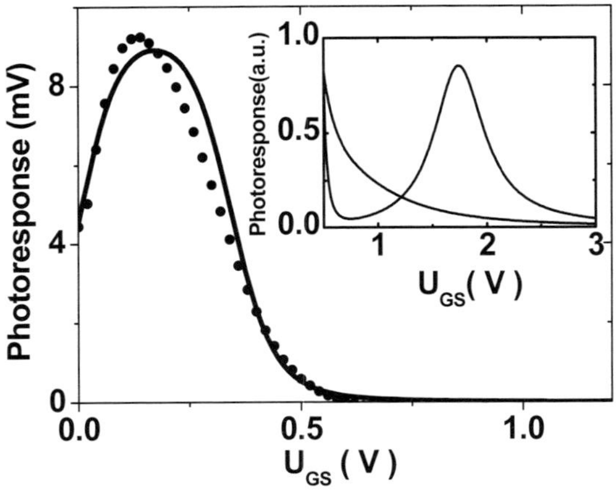

Fig. 2. Room temperature nonresonant photovoltaic detection of 119 GHz radiation by 50 nm MOSFET (symbols). The solid line shows result of theoretical calculations following Ref. [8]. Inset: development of the resonant detection with increasing the electron mobility from 100 cm^2/Vs (monotonic curve) to 500 cm^2/Vs (Lorentzian – like shape).

(Fig. 1). To verify that this structure is related to plasma excitations, the concentration of the electrons in the channel was changed by subsequent illuminations with the halogen light. This caused a shift of the threshold voltage and a corresponding shift of the position of the resonance.

A nonresonant detection signal was also observed at room temperature on nanometric Si MOSFETs with the gate length down to 30 nm exposed to 119 GHz radiation, see Fig. 2 [10]. In that experiment, the quality factor, $\omega\tau$, was smaller than 1. The shape of the detected signal was explained by the theory developed in [8]. Although the observed MOSFET response was nonresonant, the theoretical model predicts a resonant detection at room temperature for higher values of $\omega\tau$. Such predictions are shown in the inset to Fig. 2: the two curves, corresponding to different electron scattering times, were calculated for the mobilities of 100 cm^2/Vs and 500 cm^2/Vs, respectively. This gives a clear indication of a possibility of a room temperature resonant detection by Si MOSFETs of a high electron mobility.

Passing from the nonresonant to the resonant detection by increasing the quality factor can be achieved in two different ways. First, by increasing ω; this means that a transistor that is nonresonant for a low frequency may become resonant for a higher one. Also, in that respect, detection on higher harmonics of the fundamental plasma mode of the transistor channel becomes important. Second, by increasing τ; this seems to be possible only by decreasing the temperature (as in [8]) or by improving the electron mobility by technological developments. However, it was shown that a large increase of an effective scattering time can be achieved by increasing the drain current.

The first demonstration of an increase of the FET detection signal with the drain current was presented in [11]. Recently, Teppe *et al.* gave a more detailed analysis of this observation [12] showing that the current leads to an increase of the amplitude of the detection resonance and shrinkage of the resonance line.

3. Emission

The first observation of THz emission from a transistor structure was due to Tsui, Gornik and Logan [13]. The excitation of emission was due to the drain – source current. The spectral dependence of the emission was analysed with a tunable GaAs detector which performance is based upon tuning the energy of intra shallow donor transitions in the magnetic field. Deng *et al.* [14] observed an emission at 75 GHz from a GaN HEMT with the gate length of 1.5 μm and the drain – source separation of 5 μm. In this case, the spectrum of the emission was analyzed with a Fabry – Perrot interferometer.

THz radiation from InGaAs/AlInAs nanometer HEMTs was observed by Knap *et al.* [15]. The transistors used were lattice matched GaInAs/AlIAs HEMTs on InP substrate grown by molecular beam epitaxy. The transistor channel was a 20 nm $Ga_{0.47}In_{0.53}As$ quantum well surrounded by $Al_{0.48}In_{0.52}As$ barriers and the gate was 60 nm long. Other experimental and technological details can be found in [15, 16]. The measurements were carried out with a magnetic field tunable InSb detector that allowed to obtain a spectrum of the emitted signal. The results are shown in Fig. 3. In general, the spectra are broad with a sharper maximum at around 1 THz and a broad structure at about 6 THz. The low frequency peak was interpreted as resulting from the Dyakonov – Shur instability of the plasma in the transistor channel (see [15] for details). The position of the broad maximum does not change with applied gate or source bias. Monte Carlo simulations of high frequency noise spectra were calculated [16] to show a qualitative agreement with the experimental results.

According to the Dyakonov - Shur model, the plasma instability should show a threshold behavior, i.e., the intensity of the emitted signal should grow rapidly when a certain parameter increases over a threshold value. In

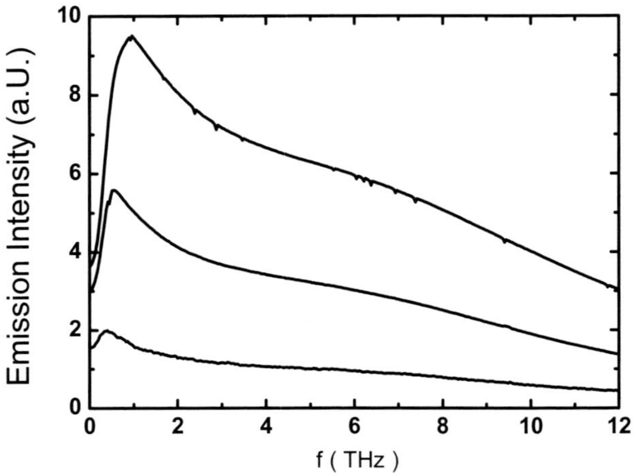

Fig. 3. Emission spectra of GaInAs/AlInAs transistor at 4. 2 K analyzed with InSb cyclotron emission detector. The curves correspond to the drain – to – source voltage equal to 0.3 V, 0.6 V and 0.8 V form bottom to top, respectively.

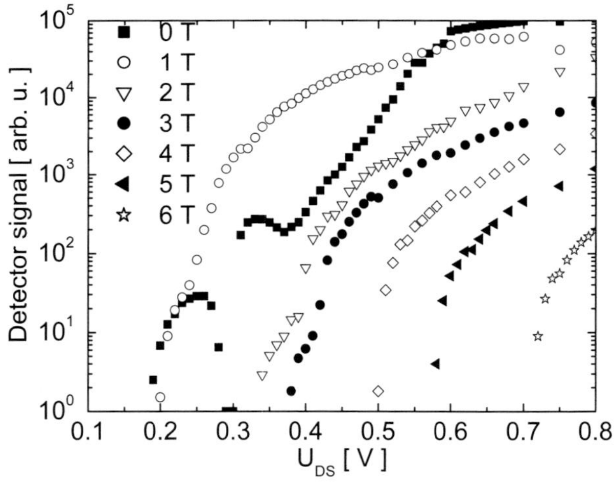

Fig. 4. Integrated emission intensity from GaInAs/AlGaAs transistor as a function of the drain – to – source voltage, U_{DS} for magnetic fields from 0 to 6 T.

the investigation of the threshold behavior, the transistor was placed in a magnetic field increasing from 0 to 6 T [17]. Then the emission signal was measured as a function of the drain – source voltage, as shown in Fig. 4. The results show an abrupt increase of the emitted signal once the drain voltage becomes higher than the threshold value. The main effect of the magnetic field is to shift the threshold voltage to higher values. As was shown in [17], this shift can be explained by the geometrical magnetoresistance of the ungated part of the channel which resistance increases in the magnetic field proportional to B^2 [18]. However, the voltage drop in the gated part of the transistor remains constant (and equal to about 50 mV) at the emission threshold for the magnetic fields up to 4 T.

The experimental results of the emission described above were interpreted within the Dyakonov – Shur instability of the plasma waves [1, 15, 16]. Other possibilities, however, should be also mentioned. First, let us notice that the threshold drain – source voltage for the emission always coincides with a current drop in the output characteristics and is close to the saturation of the drain current. Then, the transistor is in a state characterized by very strong electric fields, and hot electron phenomena make an important contribution to the device performance. It is known that such a drop of the current is connected with excitation of the Gunn effect. This effect was studied by the Monte Carlo method in the case of GaInAs/GaAlAs HEMT transistor in [19]. The mechanism leading to the Gunn oscillations in the case of a GaInAs/GaAlAs transistor structure is more complicated than in the case of a bulk material because the energy required to the transfer an electron to the L minima in the conduction band is comparable to that of a transfer to the barrier and even to band – to – band impact ionization energy in GaInAs channel. It means that different charge transfer mechanisms are coupled and might all contribute to a mechanism to the observed emission.

Second, the theory of Dyakonov and Shur considered only plasma oscillations with the wave vector in the direction of the current flow. However, the transistor channel is rather a waveguide than a resonator and oblique plasma modes can propagate in it with a component of the wave vector perpendicular to the current. The spectrum of such modes is continuous which could explain why experimentally observed emission spectrum is a broad one. The density of states of the oblique modes should also be taken into account. Additionally, the processes underlying plasma oscillations in FETs are nonlinear that leads to mixing of modes and to a broadening of the spectrum down to frequencies much lower than the THz range.

Conclusion

This paper reviews basic facts concerning experimental investigation of detection and emission of THz radiation by nanometer Field Effect Transistors and the ballistic motion in Si MOSFETs. The detection can be achieved both at room and cryogenic temperatures and its character (resonant or non resonant) depends on the quality factor of the transistor resonating cavity. The emission was observed so far at liquid helium temperatures only and is clearly related to hot electron phenomena caused by the presence of strong electric fields. Most of detection experiments can be explained by the Dyakonov – Shur model of plasma instability that was constructed for a description of plasma behavior in the presence of small electric fields. It seems, however, that a proper description of the emission process requires a generalization of this model to take into account hot electron phenomena. Mobility investigation in nanometer transistors is important in connection with the THz phenomena because the mobility – related relaxation time is crucial for determination of the quality factor of the transistor THz cavity. In that respect, mobility limitations by ballistic motion in nanotransistors become of a particular importance.

Acknowledgements

The authors are thankful to M. Dyakonov, N. Dyakonova, F. Teppe, L. Varani, S. Bollaert, T. Gonzalez and J. Mateos for helpful discussions.

References

[1] M. Dyakonov and M. S. Shur, Phys. Rev. Lett. **71**, 2465 (1993).
[2] S. J. Allen, D. C. Tsui, and R. A. Logan, Phys. Rev. Lett. **38**, 980 (1977).
[3] P. J. Burke, I. B. Spielman, J. P. Eisenstein, L. N. Pfeiffer, and K. W. West, Appl. Phys. Lett. **76**, 745 (2000).
[4] X. G. Peralta, S. J. Allen, M. C. Wanke, N. E. Haeff, J. A. Simmons, M. P. Lilly, J. L. Reno, P. J. Burke, and J. P. Eisenstein, Appl. Phys. Lett. **81**, 1627 (2002).
[5] J.-Q. Lü, M. S. Shur, J. L. Hesler, L. Sun, and R. Weikle, IEEE Electron Device Lett. **19**, 373 (1998).
[6] T. Otsuji, M. Hanabe, and O. Ogawara, Appl. Phys. Lett. **85**, 2119 (2004).
[7] W. Knap, Y. Deng, S. Rumyantsev, J.-Q. Lü, M. S. Shur, C. A. Saylor, and L. C. Brunel, Appl. Phys. Lett. **80**, 3433 (2002).
[8] W. Knap, V. Kachorovskii, Y. Deng, S. Rumyantsev, J.-Q. Lü, R. Gaska, M. S. Shur, G. Simin, X. Hu, M. Asif Khan, C. A. Saylor, and L. C. Brunel, J. Appl. Phys. **91**, 9346 (2002).
[9] W. Knap, Y. Deng, S. Rumyantsev, M. S. Shur, Appl. Phys. Lett. **81**, 4637 (2002).

[10] W. Knap, F. Teppe, Y. Meziani, N. Dyakonova, J. Łusakowski, F. Boeuf, T. Skotnicki, D. K. Maude, S. Rumyantsev, and M. S. Shur, Appl. Phys. Lett. **85**, 675 (2004).
[11] J.-Q. Lü and M. S. Shur, Appl. Phys. Lett. **78**, 2587 (2001).
[12] F. Teppe, W. Knap, D. Veksler, M. S. Shur, A. P. Dmitriev, V. Yu. Kachorovskii, and S. Rumyantsev, Appl. Phys. Lett. **87**, 052107 (2005)
[13] D. C. Tsui, E. Gornik, and R. A . Logan, Solid State Comm. **35**, 875 (1980).
[14] Y. Deng, R. Kersting, J. Xu, R. Ascazubi, X. C. Zhang, M. S. Shur, R. Gaska, G. S. Simin, M. Asif Khan, V. Ryzhii, Appl. Phys. Lett. **84**, 70 (2004).
[15] W. Knap, J. Łusakowski, T. Parenty, S. Bollaert, A. Cappy, V. V. Popov, M. S. Shur, Appl. Phys. Lett. **84**, 2331 (2004).
[16] J. Łusakowski, W. Knap, N. Dyakonova, L. Varani, J. Mateos, T. Gonzalez, T. Parenty, S. Bollaret, A. Cappy and K. Karpierz, J. Appl. Phys. **97**, 064307 (2005).
[17] N. Dyakonova, F. Teppe, J. Łusakowski, W. Knap, M. Levinshtein, A. P. Dmitriev, M. S. Shur, S. Bollaert, and A. Cappy, J. Appl. Phys. **97**, 114313 (2005).
[18] Y. M. Mezziani, J. Łusakowski, W. Knap, N. Dyakonova, F. Teppe, K. Romanjek, M. Ferrier, R. Clerc, G. Ghibaudo, F. Boeuf, and T. Skotnicki, J. Appl. Phys. **96**, 5761 (2004).
[19] G. M. Dunn, A. Philips and P. J. Topham, Semicond. Sci. Technol. **16**, 562 (2001).

High-Intensity THz Radiation From a Large Interdigitated Array Photoconductive Emitter

S. Winnerl, A. Dreyhaupt, F. Peter, D. Stehr, M. Helm, and T. Dekorsy*

Institute of Ion Beam Physics and Materials Research, Forschungszentrum Rossendorf, P.O. Box 510119, D-01314 Dresden
* University Konstanz, Physics Department, Box M700, D-78457 Konstanz, Germany

Summary. We report on the performance of photoconductive THz emitters based on an interdigitated metal-semiconductor-metal finger structure. In every second period of this structure optical excitation is inhibited. Thus carrier acceleration is unidirectional over the whole device area and the emitted THz radiation interferes constructively in the far field. Excitation with amplified laser pulses leads to THz amplitudes of 6 kV/cm. Saturation of the emission due to screening of the bias field was observed for excitation densities in the 10^{18} cm^{-3} range. However, since the emitter concept is scalable to large areas, the THz emitter offers large potential for further increase of the emitted THz field amplitude.

The generation of THz radiation pulses is of great importance for a variety of scientific and technological applications [1]. Beside methods which are based on optical rectification or difference frequency mixing [2,3], one often employed emitter concept is based on modulated photocurrents. The photoconductive emitters can be roughly divided into large-aperture emitters [4] or small-gap electrode structures coupled to antennas [5]. THz emitters taking advantage of high intensity laser pulses from amplified laser systems have to be large-aperture emitters since focusing of the amplified pulses would destroy the emitter structure. However, due to the larger electrode distances in these emitters high-voltage pulses are required for achieving high accelerating fields and the electric fields in the device are limited by the breakdown voltage in air. Here we present a scalable large-aperture emitter consisting of interdigitated metal-semiconductor-metal (MSM) structures, which combines the advantages of large-aperture emitters and small-gap electrodes [6].

The THz emitter is fabricated on semi-insulating (SI) GaAs as the photoconductive material. Interdigitated Cr-Au electrodes are patterned on top with an electrode spacing of 5 μm. Emitters with areas of 2×2 mm^2 and 10×10 mm^2 were fabricated. Biasing the interdigitated finger electrodes provides a strong electric field of alternating direction between neighbouring finger electrodes. A second metallization layer, electrically insulated from the interdigitated electrodes, covers every second gap between the finger electrodes. Hence, optical excitation is possible only in areas with the same electric field direction. In the far field the emission from unidirectionally accelerated carriers interferes constructively.

We focus on the excitation of the THz emitter with an amplified Ti:sapphire laser (wavelength 800 nm, pulse duration 25 fs, pulse energies up to 1 mJ, 1 kHz rep. rate). For experiments with pulses from a non-amplified Ti:sapphire oscillator we refer to Ref. [6]. The diameter of the pump spot on the emitter is 4.5 mm. The THz detection is based on a polarization sensitive electro-optic (eo) detection with a lock-in technique where the lock-in amplifier is locked to a bias voltage modulation applied to the THz emitter. The detection bandwidth is limited to the low-frequency THz range up to 3 THz due to the use of a thick ZnTe eo-detector crystal.

The emitted THz electric fields and their Fourier transform are shown in Fig.1 for the excitation densities 1.7×10^{17} cm^{-3} and 1.7×10^{18} cm^{-3} (emitter area 2×2 mm^2). The data are recorded at a bias voltage of 23 V corresponding to an acceleration field of 46 kV/cm. With increasing excitation density, the THz intensity increases and the spectra are blue-shifted. This effect is attributed to the faster screening of the bias field, which yields shorter current surges. It should be noted that the emitted electric field amplitude is 6.1 kV/cm at an excitation density of 1.7×10^{18} cm^{-3} (corresponding to 10 μJ pulse energy only).

In order to determine the optimum excitation densities for these large-area THz emitters we perform a variation of the excitation densities at different accelerating fields. The data are recorded on a different emitter (emitter area 10×10 mm^2) and a different THz beam path, yielding presently lower total THz amplitudes than those shown in Fig. 1. In Fig. 2 the dependence of the maximum terahertz amplitude on the excitation density is shown for different acceleration fields. The saturation of the THz emission at high excitation densities can be quantitatively described by the relationship $E_{THz}=E_0(n/n_0)/[1+(n/n_0)]$, where n is the excitation density, E_0 a parameter for scaling the electric field and n_0 a characteristic density for the field screening. The fits shown in Fig.2 were obtained with $n_0=5\times10^{17}$ cm^{-3} for all accelerating fields. A similar behaviour has been described for large aperture THz emitters [7].

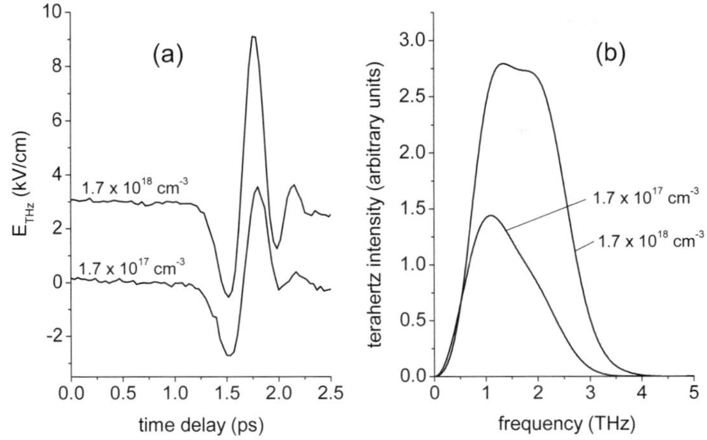

Fig. 1. (a) Time-resolved THz traces at different excitation densities as given in the figure. The data are vertically off-set for clarity. (b) Numerical Fourier transform of the data in (a).

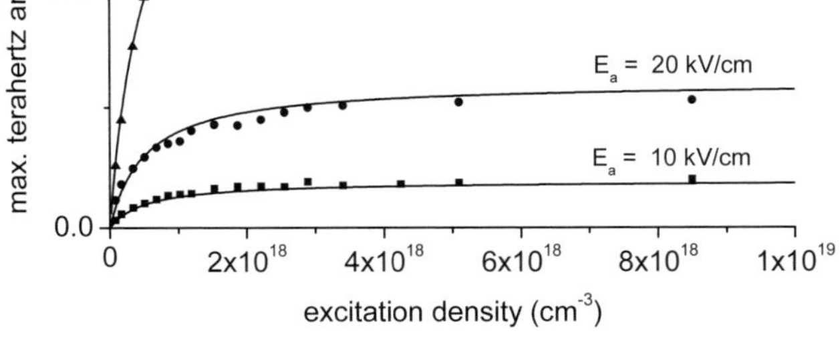

Fig. 2. Maximum THz amplitude for different acceleration fields E_a as given in the figure. Points are experimental, solid lines are theoretical fits to the data.

Photoconductive emitters are currently the most efficient THz generators for systems using optical pulse energies up to about 10 μJ [8,9]. Due to the screening in photoconductive emitters, for higher pulse energies nonlinear crystals are more efficient [9]. We suggest that keeping the carrier density moderate by optically exciting larger areas while maintaining strong acceleration fields in the scaleable emitter will extend the range where photoconductive emitters are superior for the generation of higher pulse energies.

In summary, we presented a scalable THz emitter concept which provides carrier acceleration in high electric fields in large areas. Hence it circumvents the difficulties inherent to large-aperture THz emitters and small-gap electrode emitters. Electric THz field amplitudes up to 6.1 kV/cm are achieved upon excitation of the emitter with 10 μJ/pulse from a fs Ti:sapphire amplifier. This value is presently only limited by screening of the bias field applied. By using low-temperature grown GaAs with high breakdown fields the field can be further enhanced. The device concept is as well applicable for frequency mixing of CW lasers for the generation of CW THz radiation.

References

1. J. Shan and T.F. Heinz, in *Ultrafast Dynamical Processes in Semiconductors*, ed. K.T. Tsen, Springer 2004.
2. R. A. Kaindl, F. Eickemeyer, M. Woerner, and T. Elsaesser, Appl. Phys. Lett. **75**, 1060 (1999).
3. R. Huber, A. Brodschelm, F. Tauser, und A. Leitenstorfer, Appl. Phys. Lett. **76**, 3191 (2000).
4. J.T. Darrow, X.-C. Zhang, and D.H. Auston, Appl. Phys. Lett. **58**, 25 (1991).
5. D. Grischkowsky, S. Keiding, M. v. Exter, and C. Fattinger, J. Soc. Opt. Am. B **7**, 2006 (1990).
6. A. Dreyhaupt, S. Winnerl, T. Dekorsy, and M. Helm, Appl. Phys. Lett. **86**, 121114 (2005).
7. K. Benicewicz, J.P. Roberts, and A.J. Taylor, J. Opt. Soc. Am. B **11**, 2533 (1994).
8. P.C.M. Planken, C.E.W.M. van Rijmenam, and R.N. Schouten, Semicond. Sci. and Technol. **20**, 121(2005).
9. T. Löffler, M. Kreß, M. Thomson, T. Hahn, N. Hasegawa, and H. Roskos, Semicond. Sci. and Technol. **20**, 134 (2005).

Broadband Terahertz Emission From Ion-Implanted Semiconductors

J. Lloyd-Hughes[1], E. Castro-Camus[1], M. D. Fraser[2], H. H. Tan[2], C. Jagadish[2], M. B. Johnston[1]

1. University of Oxford, Department of Physics, Clarendon Laboratory, Parks Road, Oxford, OX1 3PU, United Kingdom
2. Department of Electronic Materials Engineering, Research School of Physical Sciences and Engineering, Institute of Advanced Studies, Australian National University, Canberra ACT 0200, Australia.

Summary. The terahertz radiation emitted from Fe^{+} ion-implanted InGaAs surface emitters and InP photoconductive switches was measured. We experimentally observe an increase in the spectral width of terahertz radiation at greater ion damage, which we attribute to the ultrafast capture of photoexcited carriers. Results from a three-dimensional carrier dynamics simulation support this explanation.

1. Introduction

Single-cycle pulses of electromagnetic radiation, with spectra covering the far-infrared or terahertz (THz) range of 0.1-10 THz (3 mm-30μm), can be generated by the ultrafast separation of photoexcited carriers under an electric field.[1] The technique of terahertz time-domain spectroscopy relies upon the coherent generation and detection of such single-cycle pulses, and is proving useful in diverse areas of condensed matter physics.[2,3,4]

In order to increase the application of these emitters it is desirable to increase their spectral range. One method of decreasing the pulse duration (and thus broadening the spectrum) is to reduce the electric field decay time after excitation by using a defect-laden semiconductor. Such materials can be made either by low-temperature growth or via ion-implantation, and have sub-picosecond carrier trapping lifetimes and large carrier-defect momentum scattering rates.[1,5,6]

2. Sample Details and Experimental Setup

A tandem accelerator was used to irradiate InGaAs, GaAs and InP samples with high energy ions. By choosing the incident ion species, energy and dose defects can be created with a certain depth distribution in a target. Multi-energy ion implantations were performed in order to create a uniform damage profile (Fig. 1) extending across the absorption depth of the semiconductor.[1]

The terahertz time-domain spectroscopy setup used was based on a 10fs Ti:Sapphire laser that outputs 400mW at a central wavelength of 790nm, and was similar to that described in Ref. 1. Chopping was performed electrically at 20kHz in the photoconductive switch emitter case, and optically at 2kHz for surface emitters. All measurements were taken at room temperature, with the terahertz path length under vacuum.

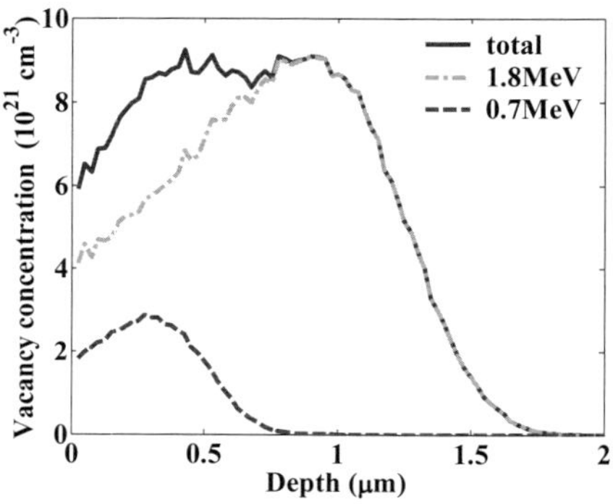

Fig. 1. The damage (vacancy) profile of $In_{0.53}Ga_{0.47}As:Fe^{+}$ calculated using the SRIM software (available at www.srim.org), extends over the absorption depth of 1.55μm photons (~1.3μm). Dual-energy (0.7 and 1.8MeV) implants of Fe^{+} ions were performed at room temperature, for different ion doses. The highest dose for the 1.8MeV implant was 1×10^{16}cm^{-2}, and the other samples had 5% and 0.1% of this dose. The 0.7MeV implants had 28% of the corresponding 1.8MeV dose. For the $InP:Fe^{+}$ samples 2MeV and 0.8MeV implants were used, producing similar damage. A post-implantation annealing step (500°C for 30 minutes) allowed the resistivity to recover.

3. Measured and Simulated Terahertz Emission

Figure 2a shows the measured THz electric field emitted from $InP:Fe^{+}$ photoconductive switches with 400μm gap, and biased by a 20kHz square wave at ±120V. The peak electric field decreases from 110Vm^{-1} to 7Vm^{-1}

between the unimplanted and highest dose samples due to a reduced electron mobility. A higher ion dose produces electric field pulses with a shorter duration (Fig. 2a), and a broader spectrum (Fig. 2b). In the highest ion dose sample the trapping lifetime of photoexcited electrons (as measured by time-resolved photoluminescence) is 130fs, and is due to deep Fe-related acceptor defects.[5] A similar trend was observed in the THz emission from $In_{0.53}Ga_{0.47}As:Fe^{+}$ surfaces (Fig. 2c and d), where the carrier trapping time of the highest dose sample is of the order of 300fs.[6]

The terahertz emission from semiconductors can be accurately modelled using a three-dimensional carrier dynamics simulation.[7,8] By including an exponential decay in the number of photoexcited carriers damaged semiconductors can also be simulated.[1] The simulated electric field from InP photoconductive switches is plotted for carrier lifetimes of 100ps and 130fs in Fig. 3. Both the time and frequency domain results are qualitatively similar to those in Fig. 2a and 2b.

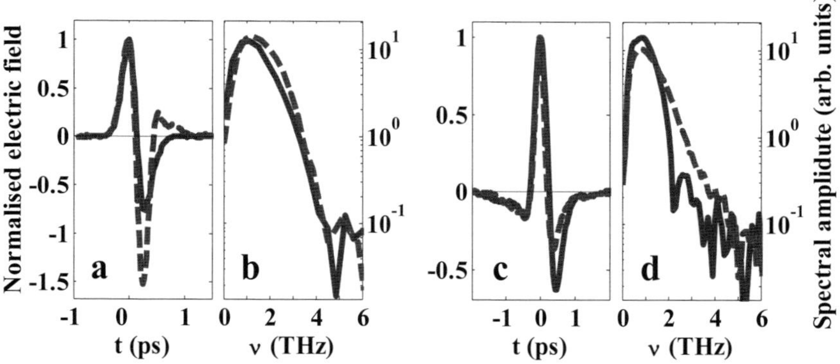

Fig. 2. (a) Normalised electric field versus time t from unimplanted InP (solid line) and $InP:Fe^{+}$ (dashed line) with an incident ion dose at 1.8MeV (0.7MeV) of $5\times10^{14}cm^{-2}$ ($1.4\times10^{14}cm^{-2}$). Measured using a 200μm <110> ZnTe crystal. **(b)** Spectra of a) as a function of frequency ν. **(c)** Normalised electric field from unimplanted $In_{0.53}Ga_{0.47}As$ (solid line) and $In_{0.53}Ga_{0.47}As:Fe^{+}$ (dashed line) with an incident ion dose at 1.8MeV (0.7MeV) of $5\times10^{14}cm^{-2}$ ($1.4\times10^{14}cm^{-2}$). The peak electric fields were $144Vm^{-1}$ and $20Vm^{-1}$ respectively. Measured using a 20μm <110> on 1mm <100> ZnTe crystal. **(d)** Spectra of c).

4. Conclusion

We observed a bandwidth increase with ion dose in $InP:Fe^{+}$ photoconductive switches, which offer benefits over GaAs based emitters in the spectral range 0-9THz owing to the higher TO phonon frequency of InP

(9.2THz, c.f. 8.1THz in GaAs). A similar bandwidth increase was seen from $In_{0.53}Ga_{0.47}As:Fe^{+}$ surfaces, which may be beneficial in terahertz systems based on 1.55μm wavelength lasers. The simulation results agree with the experimentally observed trend.

The authors would like to thank the EPSRC and the ARC.

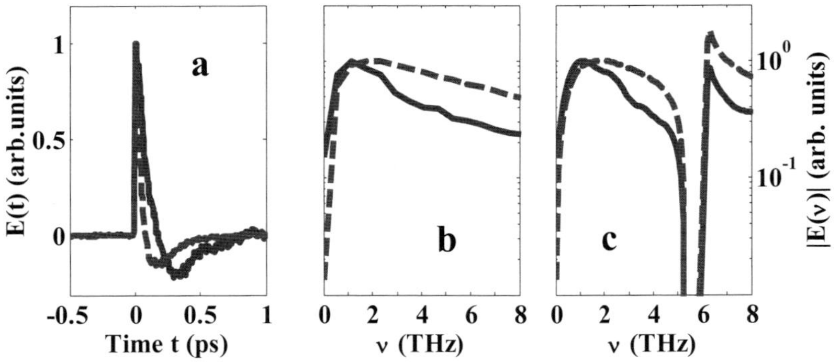

Fig. 3. (a) Normalised simulated electric field E(t) from InP photoconductive switches with a carrier trapping time of 100ps (solid line) and 130fs (dashed line). **(b)** Normalised spectra of a) as a function of frequency ν. **(c)** Normalised spectra from b) after including effect of 200μm ZnTe measurement crystal using a harmonic oscillator transmission function model with a TO (LO) phonon frequency of 5.3THz (6.2THz).

References

1. J. Lloyd-Hughes et al., *Phys. Rev. B.*, **70**, 235330, 2004.
2. R. A. Kaindl et al., *Phys. Rev. Lett.* **88**, 027003, 2002.
3. M. B. Johnston et al., *Chem. Phys. Lett.* **377**, 256, 2003.
4. R. Huber et al., *Nature* **414**, 286 2001.
5. C. Carmody et al., *J. Appl. Phys.* **94**, 1074, 2003.
6. C. Carmody et al., *App. Phys. Lett*, **82**, 3913, 2003
7. M. B. Johnston et al., *Phys. Rev. B.* **65**, 165301, 2002.
8. E. Castro-Camus et al., *Phys. Rev. B.* **71**, 195301, 2005.

THz Collective Real-Space Oscillations of Ballistic Electrons in Wide Parabolic Potential Wells: an Exotic Transport Regime

M. Betz,[1] S. Trumm,[1] M. Eckardt,[2] A. Schwanhäußer,[2] S. Malzer,[2] F. Sotier,[3] A. Leitenstorfer,[3] T. Müller,[4] K. Unterrainer,[4] and G. H. Döhler[2]

[1] Physik-Department E11, Technische Universität München
[2] Max-Planck-Forschungsgruppe für Optik, Information und Photonik, Universität Erlangen
[3] Fachbereich Physik, Universität Konstanz
[4] Institut für Festkörperelektronik, Technische Universität Wien

Summary. After femtosecond photoinjection of carriers near the boundary of a parabolic shaped 250 nm wide potential well, electrons are found to coherently oscillate across the well instead of performing the intuitively expected unidirectional relaxation towards the bottom of the well. Surprisingly, the coherence of the periodic electron motion is maintained despite multiple phonon scattering events. This novel transport regime is predicted by Monte Carlo simulations and verified by analyzing the femtosecond transmission of the heterostructure.

Coherent real-space oscillations of electronic wave packets in semiconductor heterostructures have become an important issue. Especially, they have been observed in asymmetric double quantum-well structures [1] as well as in semiconductor superlattices [2-4]. These transport phenomena are purely quantum mechanical in nature and intimately related to heterostructure dimensions much smaller than the de Broglie wavelength. In strong contrast, carrier dynamics in larger semiconductor structures is expected to be readily described in a semiclassical picture.

In this contribution, we report on the observation of a novel transport regime between purely semiclassical carrier dynamics and the evolution of a fully quantized system. Electrons perform THz real-space oscillations in parabolic shaped AlGaAs potentials wells of a width of 250 nm [5]. The coherence of the carrier motion is only weakly affected by phonon scattering within the Γ-valley of GaAs. The experimental analysis of the ultrafast carrier dynamics is achieved by a femtosecond transmission experiment.

Realistic Monte Carlo simulations provide detailed insight into this novel kind of carrier motion in semiconductor nanostructures.
Our $Al_xGa_{1-x}As$ heterostructure is sketched in Fig. 1. Part (a) depicts the Aluminum content along the growth direction of the sample. Highly p-doped $Al_{0.3}Ga_{0.7}As$ layers define a n-doped potential well. On the left hand side, we start with a narrow region of GaAs, whereas the aluminum content is gradually increased to 10% towards the middle of the well. As a consequence, resonant interband absorption allows for the generation of a well-defined electron-hole ensemble near the left boundary. Fig. 1(b) shows the conduction band profile of the potential well: The space-charge density of the depleted donors, en_D, translates into a parabolic energy dispersion with curvature $d^2V / dz^2 = 4\pi e^2 n_D / \varepsilon_0$ via Poisson's equation.
First, we study the electron dynamics within the heterostructure by extensive theoretical simulations. We employ a semiclassical Monte Carlo approach which takes into account the k-space and real-space band structure of the specimen, the local absorption spectrum as well as the properties of a femtosecond excitation pulse and the dominant carrier-phonon and intervalley scattering processes [6].

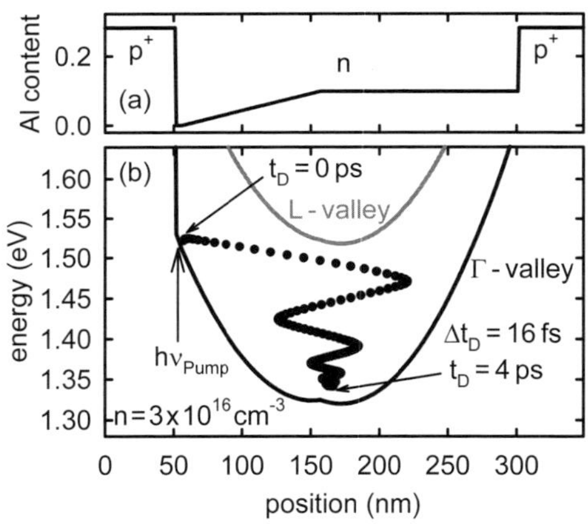

Fig. 1. (a) Aluminum content of the $Al_xGa_{1-x}As$ alloy along the growth direction. The doping level of the barriers is $n_A = 1 \times 10^{18}$ cm^{-3}. (b) Conduction band profile for a donor concentration of $n_D = 3 \times 10^{16}$ cm^{-3}. Black dots: position of the center-of-mass of the electrons and their average energy for a temperature of $T_L = 10$ K and delay times up to 4 ps with a time interval of $\Delta t = 16$ fs elapsed between two adjacent dots.

Fig. 1(b) depicts results of our Monte Carlo simulations for an electron ensemble generated at $t_D = 0$. The dots indicate the center of mass position as well as the average energy of the electron ensemble at time intervals of $\Delta t_D = 16$ fs. Importantly, the potential depth is designed to be slightly less than the energy separation $\Delta_{\Gamma L}$ to the L-sidevalley of the conduction band. Therefore, ballistic electrons never can attain the threshold energy for intervalley scattering in this well. The absence of this efficient energy- and momentum-relaxation process completely modifies the character of the trajectory of the electron ensemble as compared to the intuitive expectation of electrons simply relaxing towards the bottom of the well. In particular, the electrons are now found to perform ultrafast real-space oscillations over several periods with only moderate attenuation of the amplitude. Obviously, highly inelastic LO phonon scattering processes do not destroy the coherence of these real space oscillations of the electron ensemble, although the average energy loss corresponds to more than 2 LO phonons per period. This finding results from two important facts. First, the emission of LO phonons via the Fröhlich interaction strongly favours forward scattering in k-space. Second, the eigenfrequency of an electron exposed to a parabolic potential is independent of the kinetic energy of the charge carrier.

We now turn to the experimental verification of these novel coherent real-space oscillations. A sample with a ten times repeated heterostructure as discussed in Fig. 1(a) and (b) grown on top of each other is processed to allow for optical transmission experiments. The experimental setup utilizes a two-color Ti:sapphire laser providing two independently tunable femtosecond pulse trains. An 80 fs pump pulse of a central photon energy of 1.52 eV prepares a well-defined electron-hole ensemble at the left boundary of the potential well. A second broadband probe pulse of a duration of 20 fs is temporally delayed, spectrally dispersed after transmission through the specimen and detected with a photodiode.

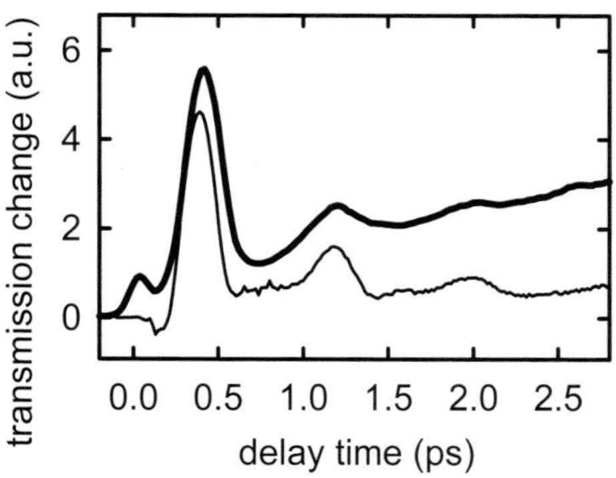

Fig.2 Transmission change of the heterostructure for a lattice temperature of $T_L = 10$ K and a probe photon energy of 1.64 eV. Thick line: experimental data, thin line: theoretical results.

The transient modification of the optical properties of the potential well predominantly arises from a modulation of the internal field after femtosecond photoexcitation of electron-hole pairs [6,7,8]. Specifically, a dipole builds up between the oscillating electron ensemble and the hole distribution that remains at the left boundary of the potential well. As a result, a space-charge field is superimposed to the internal electric field and the Franz-Keldysh absorption of the specimen is altered.

The thick line in Fig. 2 depicts the transmission change of the sample detected for a probe photon energy of 1.64 eV and a lattice temperature of T_L = 10 K. This choice of a photon energy slightly above the band gap energy of $Al_{0.1}Ga_{0.9}As$ makes the nonlinear optical response especially sensitive to the arrival of the nonthermal electron distribution in the right part of the potential well. The transmission change shows a maximum at t_D = 400 fs. Then, the signal decreases again as may be expected from electrons moving backwards at this time (see Fig. 1 (b)). At t_D = 1.2 ps, a second maximum is clearly resolved corresponding to a second entry of the electron ensemble on the right hand side of the potential well. Even a weak third maximum at t_D = 2 ps is visible.
We also directly simulate the transient absorption properties of the heterostructure from the results for the time-dependent electron distribution and the corresponding screening fields within the potential well. Subsequently, the corresponding modifications of the well known Franz-Keldysh absorption spectra are summed up over the heterostructure [9]. The temporal shape of the simulation result (compare thin line in Fig. 2) agrees well with the experiment and confirms the observation of a periodic electron motion of a frequency of 1.2 THz.
In conclusion, we have predicted and observed a formerly unexpected transport regime in wide parabolic shaped AlGaAs potential wells [5]. Excluding side-valley scattering by a sophisticated heterostructure design, electrons quasi-ballistically oscillate over a distance as large as 180 nm. The frequency may be custom tailored by the choice of the well width and its doping level. Moreover, the heterostructure concept may serve as a source for picosecond pulses in the far-infrared.

References

1. K. Leo, et al., Phys. Rev. Lett. **66**, 201 (1991).
2. J. Feldmann, et al., Phys. Rev. B **46**, 7252 (1992).
3. K. Leo, P. Haring Bolivar, F. Brüggemann, R. Schwedler, Solid State Comm. **84**, 943 (1992).
4. C. Waschke, et al., Phys. Rev. Lett **70**, 3319 (1993).
5. M. Eckardt, et al., Europhys. Lett. **70**, 534 (2005).
6. A. Schwanhäußer, et al., Phys. Rev. B **70**, 085211 (2004).
7. C. V. Shank, et al., Appl. Phys. Lett. **38**, 204 (1981).
8. M. Wraback, et al., Appl. Phys. Lett. **79**, 1303 (2001).
9. L. Robledo, et al., Physica E **13**, 708 (2001).

Effect of Injector Doping on Non-Equilibrium Electron Dynamics in Mid-Infrared GaAs/AlGaAs Quantum Cascade Lasers

V. D. Jovanović[1], D. Indjin[1], N. Vukmirović[1], Z. Ikonić[1], P. Harrison[1], E. H. Linfield[1], H. Page[2], X. Marcadet[2], C. Sirtori[2], C. Worrall[3], H. Beere[3], D. A. Ritchie[3]

[1]School of Electronic and Electrical Engineering, University of Leeds, Leeds LS2 9JT, United Kingdom

[2]Thales Resarch and Technology, Domaine de Corbeville, 91404 Orsay, France

[3]Cavendish Laboratory, University of Cambridge, Madingley Road, Cambridge CB3 0HE, United Kingdom

Summary. The aim of this work is to report a detailed theoretical investigation, supported by the experimental measurements, of the influence of the injector doping level on the output characteristics and the working range limits in mid-infrared GaAs/AlGaAs quantum cascade lasers. A fully self-consistent Schrödinger - Poisson analysis based on the scattering rate equation approach was employed to simulate the above-threshold electron transport in the device. An onset of v-shaped local band edge bowing has been observed, preventing the resonant sub-band level alignment in the high pump current regime. The observed saturation of the maximal current, together with an increase of threshold current, limits the dynamic working range for higher doping levels.

1 Introduction

Since the first realization [1], the GaAs-based quantum cascade laser (QCL) has demonstrated an impressive increase of the operating wavelength range in the infrared, extending up to ~160μm. To improve the performance of GaAs QCLs and understand better the physical limitations of particular designs, further investigation of the influence of relevant physical and technological parameters are required. The doping level in the active region is an important parameter with particular influence on the dynamic working range of QCLs. Until now, a very few experimental investigations have been presented to include the influence of the injector doping on GaAs QCL threshold currents [2]. Moreover, to our knowledge, there are virtually no studies of the injector doping on the saturation current or maximal gain behaviour under high injection in mid-infrared GaAs-based QCLs. The aim of this work is to report a detailed theoretical investigation, supported by experimental measurements, of the influence of the injector doping level on the output characteristics and the working range limits. Injector sheet doping densities in the range between 2-14 x 10^{11}cm^{-2} have been analysed in well established QCL designs with 33% [1] and 45% [3] Al mole fractions in barrier layers.

2 Model, results and discussion

The theoretical approach for the electron transport in QCLs was based on a fully self-consistent quantum scattering rate equations model. In order to obtain steady state non-equilibrium electron distribution over quasi-discrete states of the injector/collector miniband and over states of the active region, a system of rate equations needs to be solved. As the scattering rates are electron density dependent, the solution of the system has to be found in a self-consistent manner, until the carrier distribution converges. As the non-equilibrium carrier distribution is not predefined, the Schrödinger and Poisson equations, as well as the system of scattering rate equations, are intrinsically coupled. As a consequence, only convergence of both processes (self-self consistency) will give the accurate solution for the carrier distribution. All relevant scattering mechanisms have been taken into account, including electron-phonon, electron-electron and ionised impurity scattering.

In contrast to the conventional doping of 4 x 10^{11}cm^{-2} shown in Fig. 1 (left panel (a)), when a higher doping is used the band profile becomes quite different, see Fig. 1 (left panel (b)). This is especially important in the working regime around the resonant alignment. In the case of a lower doping, for the applied field of 60 kV/cm (i.e. voltage per QCL period), the coupling between the lowest injector state (dashed in Fig. 1) and upper laser level (bold in Fig. 1) is quite strong and the system is close to resonance. Quite the opposite happens at the higher doping, and for the same value of the applied field these levels are widely separated and the system is far from reaching the resonant condition. This is a direct consequence of the electron ionized donor separation within the each QCL period, forming the v-shaped local field domains. The domain formation is especially pronounced at high doping levels. Furthermore, we have extracted the field-current density characteristics for all other doping densities and the results are shown in Fig. 1, right panel. A good overall agreement (within 15% discrepancy) with the experimental measurement was found (see inset in Fig. 1), where the I-V curves for the 6 and 8 x 10^{11}cm^{-2} are presented. The data suggest that the increase in current density with the increase of doping is quasi-linear in the range of lower doping values and lower electric fields, while the saturation was observed for higher fields and doping levels.

The calculated waveguide losses were found to increase from ~15cm^{-1} for the doping of 2 x 10^{11} to ~35cm^{-1} for 12 x 10^{11}cm^{-2}. This can be used for estimating the threshold current density J_{th} found according to $G_M(J_{th})=\alpha_W+\alpha_M$, where the mirror losses are assumed to be $\alpha_M \sim$ 5cm^{-1}. The predicted dependence of the threshold current on the sheet carrier density follows an exponential functional form i.e. $J_{th}(N_s) \sim e^{N_s/N_0}$, with the critical doping density $N_0 \sim 6.5$ x 10^{11}cm^{-2}. This value marks the doping density at which a stronger increase of the threshold current is predicted.

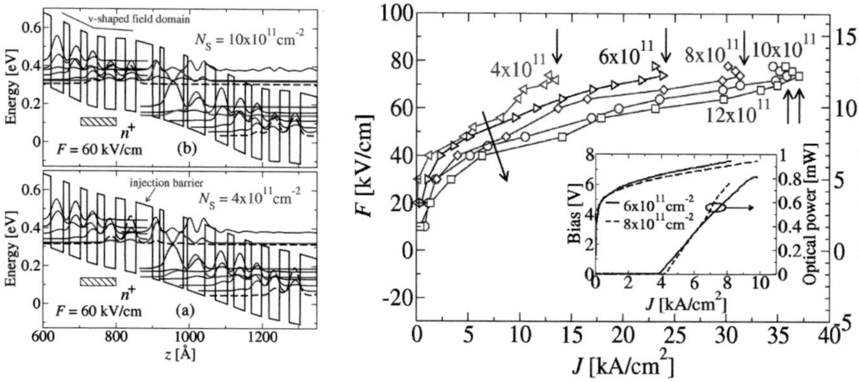

Fig. 1. ***Left panel***: A schematic diagram of calculated self-self-consistent conduction band profile, quasi-bound energy levels and wavefunctions squared for an injector active region-collector segment of $GaAs/Al_{0.45}Ga_{0.55}As$ QCL [3] for sheet carrier densities of (a) 4 x 10^{11} and (b) 10 x $10^{11}cm^{-2}$, and an applied external electric field (i.e. voltage drop per QCL period) of 60 kV/cm. The laser levels are presented by bold and the lowest injector state by dashed lines. The doped region of the injector is also indicated. ***Right panel***: Simulated field-current density characteristics for the range of doping densities between 4 x 10^{11} to 12 x $10^{11}cm^{-2}$ at 77 K. *Inset*: Measured bias- and optical power-current density characteristics for doping densities of 6 x 10^{11} and 8 x $10^{11}cm^{-2}$.

In Fig. 2, left panel, the saturation current density and the maximal gain corresponding to this current are presented as a function of the injector doping density level. The saturation current exhibits a linear dependence for the doping up to 8 x $10^{11}cm^{-2}$ and clear saturation for higher doping levels. The influence of the saturation of the maximal current reflects on the maximal gain as well. The calculations of the modal gain, shown also in Fig. 2, suggest that the performance of the laser (i.e. its output power) should deteriorate for doping densities above 10 x $10^{11}cm^{-2}$. Having in mind the predicted increase of the threshold current, one could choose an optimal value for the doping level. In the particular case, this is suggested to be between 6 and 8 x $10^{11}cm^{-2}$ in order to achieve a significant gain and at the same time avoid a considerable increase in the threshold current. The electron temperature in the single temperature approximation is calculated as a function of current density at 80K, for different doping densities. The dependences are well fitted by a quadratic function. However, for the range of working current densities, the quadratic bowing is rather small, thus a linear functional form ($T_e = T_{latt} + \alpha_{e\text{-}l} J$) can be adopted and characterised by a electron temperature-current coupling constant $\alpha_{e\text{-}l}$ [4,5]. For a fixed value of the current density, a decrease of the electron temperature with doping has been observed. This can be quantified with a decrease of $\alpha_{e\text{-}l}$ coupling constant, shown in Fig. 2, right panel. A more macroscopic explanation can be presented in terms of an effective decrease of input electrical power P_E i.e. the same value of the current density at higher doping corresponds to the lower applied bias than in case of a lower doping (see Fig. 1). Also, the power per electron decreases as the number of electrons increases. Hence, for the same current density, the electrons in the QCL, in the higher doping regime, need to heat up less than for lower doping, in order to facilitate a LO-phonon emission and efficient heat dissipation.

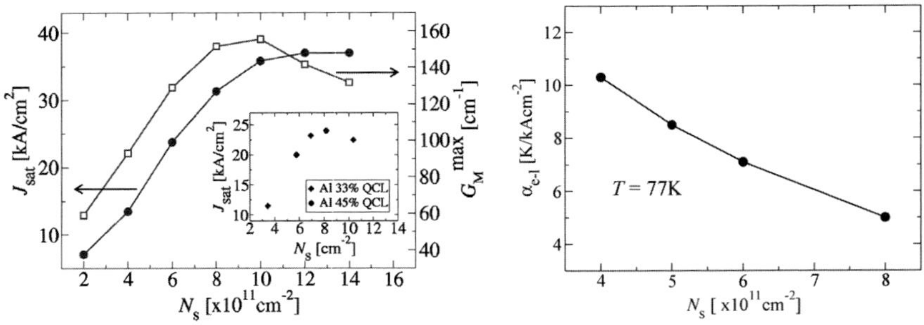

Fig. 2. ***Left panel***: Simulated saturation current density (circles) and maximal modal gain (squares) as functions of the doping density. Inset: Experimental measurements of the saturation current for 45% Al [3] (circles) and 33% Al [1] (diamonds) QCL devices for different doping densities. ***Right panel***: Calculated electron temperature–current density coupling constant as a function of the injector doping density.

3 Conclusions

We have presented a detailed study of the influence of injector doping densities on the performance of GaAs/AlGaAs QCL. The saturation of the maximal current is observed both in the calculation and the corresponding measurement. The gain reaches a maximal value for the sheet electron density between 8 and 10 x 10^{11}cm^{-2}, showing a decrease for higher doping levels, associated with the local band bowing.

References

[1] C. Sirtori, P. Kruck, S. Barbieri, P. Collot, J. Nagle, M. Beck, J. Faist, and U. Oesterle, Appl. Phys. Lett., **73** , 3486 (1998).
[2] M. Giehler, R. Hey, H. Kostial, S. Cronenberg, T. Ohtsuka, L.Schrottke, H. T. Grahn, Appl. Phys. Lett., **82**, 671 (2003).
[3] H. Page, C. Backer, A. Robertson, G. Glastre, V. Ortiz, and C. Sirtori, Appl. Phys. Lett., **78**, 3529 (2001).
[4] P. Harrison, D. Indjin and R. W. Kelsall, J. Appl. Phys, **92**, 6921 (2002).
[5] V. Spagnolo, G. Scamarcio, H. Page, and C. Sirtori, Appl. Phys. Lett., **84**, 3690 (2004).

Experimental Investigation of Hot Carriers in THz and Mid-IR Quantum Cascade Lasers

G. Scamarcio, V. Spagnolo, M. S. Vitiello, C. Di Franco

CNR-INFM Regional Laboratory LIT[3] and Dipartimento Interateneo di Fisica "M. Merlin", Università and Politecnico di Bari, Italy

Summary. We compare the electronic temperatures and the electron lattice coupling in mid-infrared and terahertz GaAs-based quantum cascade lasers (QCLs). Thermalized hot-electron distributions are found in both classes of QCLs. The results illustrate the influence of the quantum design on the electron-lattice energy relaxation rates, that ultimately determine the device thermal performance.

1 Introduction

The nature of the electronic distribution among the different subbands in quantum cascade lasers (QCLs) is a subject of great relevance for the development of these compact sources operating either in the mid-infrared or in the THz range. A hot-electron distribution may arise from the detailed balance between the injection and the energy relaxation rates, i.e. inter- and intra-subband electron-electron (e-e), electron–LO phonon, electron-impurity, and interface roughness scattering [1]. Also, thermally induced leakage of electrons into delocalized continuum-like states competes with the injection into the upper laser level and may contribute to the establishment of a non-equilibrium electronic distribution. The above phenomena may increase the thermal backfilling and weaken the population inversion. In this paper, we compare experimental results on the electronic temperatures and the electron-lattice coupling of mid-infrared and THz QCLs. This information has been obtained using micro-probe band-to-band photoluminescence [2, 3] that allows the investigation of the laser front facet down to a spatial resolution of ~ 1 μm. The local lattice temperature (T_L) is obtained by comparing the PL peak energy shift induced by heating with a calibration curve obtained by probing the device at zero-current while varying the heat sink temperature. The electronic temperature (T_E) is extracted by the lineshape analysis of the high energy slope of the PL bands.

2 Experimental Technique and Results

Three GaAs-based QCL devices operating in the mid-IR have been considered. Sample A includes a three-quantum-well $GaAs/Al_{0.45}Ga_{0.55}As$ active region designed for emission at $\lambda = 9.0$ µm. Sample B is based on a $GaAs/Al_{0.45}Ga_{0.55}As$ chirped superlattice and operates at 12.6 µm. Sample C employs a GaAs/AlAs chirped superlattice designed for emission at 11.8 µm. Figure 1 shows a set of PL spectra as a function of the electrical power up to 7 W for sample A.

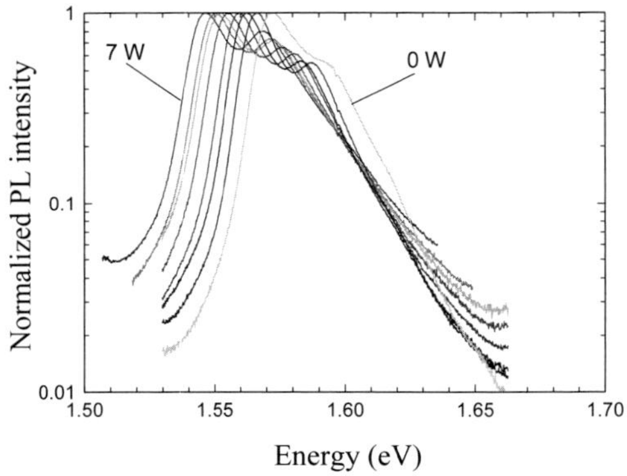

Fig. 1. Photoluminescence spectra from the active region of device A as a function of the electrical power (P), at 1 W steps. Rightmost spectrum: device off; leftmost spectrum: 7 W. The heat sink temperature is $T_H = 80$ K.

Self-consistent band structure calculations show that the main PL peaks are due to transitions involving the lower confined level of the conduction band in the active layer and two different valence subbands. Similar results are found for samples B and C.

Figure 2a shows that the temperatures T_L and T_E extracted from the analysis of the PL spectra increase linearly with the electrical power (P), with the slopes R_L (thermal resistance) and R_E, respectively. The difference ΔT between the electronic and lattice temperatures in the active region as a function of the injected current density J is reported in Fig. 2b. A linear correlation between ΔT and J is found for all three mid-IR samples A-C. This trend confirms the establishment of a thermalized hot-electron energy distribution [1].

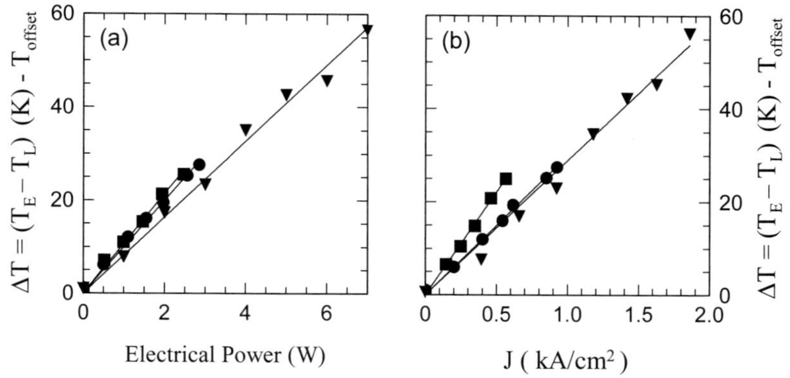

Fig. 2. a: Difference between the electronic and lattice temperature (T_E-T_L) for sample A (▼), sample B (●) and sample C (■) as a function of P, at T_H =80 K. The lines are linear fits to the data. b: T_E-T_L as a function of the current density J measured at T_H = 80 K for sample A (▼), sample B (●) and sample C (■). The slope of the linear fit gives the electron–lattice coupling constant. Offset values of 22 K for sample A and 17 K for samples B-C are subtracted to account for the Kr^+ laser induced heating of the electronic ensemble.

The proportionality constant $1/\alpha_{E\text{-}L} = J/\Delta T$ measures the strength of the electron–lattice coupling. Comparison among devices suggests that the three-quantum-well active region is the most effective design for the electron-lattice coupling. The coupling is also improved by using larger band offsets, due to the reduction of thermally activated electron leakage in the continuum [4]. The expression $(R_E - R_L)^{-1}$ gives the strength of the electron-lattice energy relaxation rate $(\tau_E)^{-1}$. Simple rate equation arguments lead to τ_E^{-1}=P/($N_e N k_B(T_E$-T_L)) where N_e is the number of electrons per stage and N is the number of stages. If the vast majority of electrons share the same temperature, one can rewrite $\tau_E^{-1}=(N_e N k_B(R_E-R_L))^{-1}$. From the measured R_E and R_L values we obtained τ_E^{-1}= 1- 1.6 ps^{-1} for devices A-C.

Figure 3a shows a set of PL spectra for a THz QCL device (sample D) based on the resonant-phonon scheme for the depopulation of the lower radiative state, and designed for emission at 2.8 THz. Note that the P values are sufficiently low to exceed the lasing threshold (P_{th} ~ 1.5 W, J_{th} ~ 400 A/cm^2). Each spectrum shows a main peak that corresponds to the transition 1→2 between the injector ground state (see Fig. 3b) and a valence subband [3]. The structure on the high energy tail of the peak 1→2 is due to the allowed transitions j→k between conduction and valence subbands by comparison with the calculated energies E_{jk}.

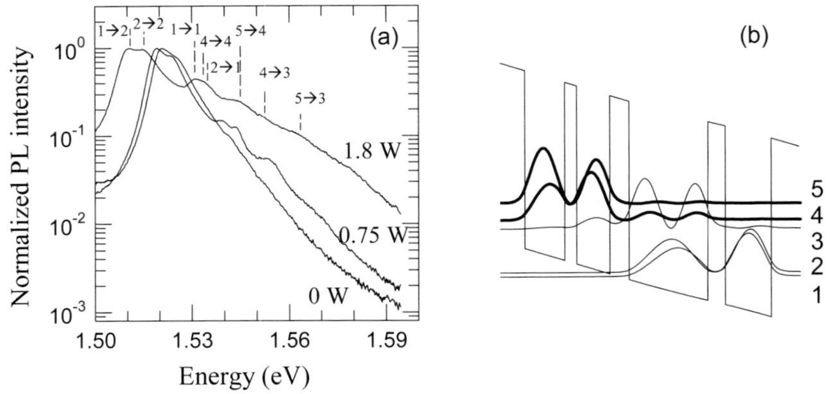

Fig. 3. a: Representative PL spectra of sample D as a function of P at $T_H = 50$ K. The dashed vertical lines labelled j→k mark the energies of the transitions between levels in the conduction (j) and valence (k) bands. b: Conduction band structures of sample D calculated with a voltage drop of 65 mV per stage. The energy levels are labelled using increasing integers starting from the ground state.

The analysis of the PL lineshape is based on the following expression:

$$I_{PL}(E) \propto \sum_{j=1}^{5} \sum_{k=1}^{4} A_{jk} E_{jk}^{4} \left| \langle \psi_{\mathrm{j}} | \psi_{\mathrm{k}} \rangle \right|^{2} \cdot L(E)$$

where $A_{jk} = n_j \cdot p_k$, n_j and p_k are the populations of the conduction and valence subbands. The term $\langle \Psi_j | \Psi_k \rangle$ is the overlap integral of the envelope functions. The lineshape function L(*E*) is obtained joining a Lorentzian with a phenomenological broadening $\Gamma/2 = 3.2$ meV on the low energy side and an exponential decay $\propto \exp(-E/k_B T_E^j)$ on the high energy side. T_E^j is the electronic temperature of the conduction j^{th} subband. The difference between the fitting parameters T_E^j and T_L are plotted in Fig. 4 as a function of *P*. We found that the electronic temperatures of the subbands j = 1-4 remain very close to the lattice temperature. Instead, the temperature of the upper laser level T_E^5 exceeds T_L by ~ 100 K. In our case, we tentatively ascribe the large difference between T_E^5 and $T_E^{1,2}$ to the reduced efficiency of inter-subband e-e scattering channels coupling electrons in the j=5 and j=1,2 levels, with respect to intra-subband e-e processes, as calculated for prototype THz QCLs structures. One important implication of our findings is that the high T_E^5 values lead to relatively fast non-radiative relaxation times $\tau_{5\rightarrow 4,3} \approx 1.3$ ps, and thus it is a key limiting factor for the operation at high temperatures of the investigated THz QCLs. Similar results are found for two THz QCLs emitting at 3.2 THz (sample E) and 3.8 THz (sample F) [8, 9]. From the measured R_e and R_L values we obtained τ_E^{-1}(d)=4.9 ps^{-1}, τ_E^{-1}(e)=1.67 ps^{-1}, and τ_E^{-1}(f)=1.05 ps^{-1} for devices (d), (e) and (f), respectively.

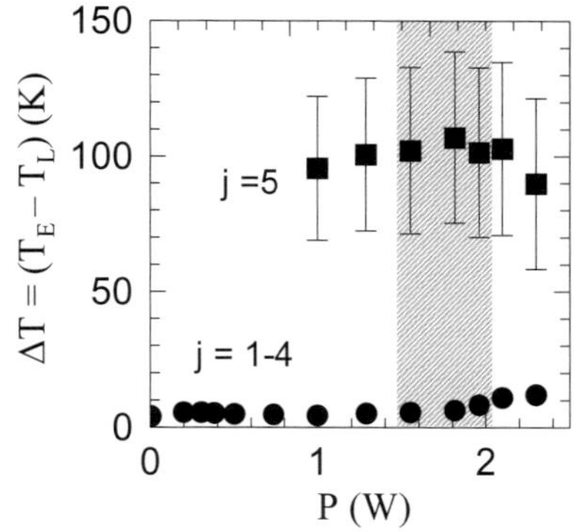

Fig. 4. $T_E^j - T_L$ as a function of P, measured for sample D at T_H =50 K. The shaded area marks the lasing region.

The more efficient carrier thermalization in sample D reduces thermal backfilling of the subband j = 4, helps in keeping the optical gain closer to the designed value, and thus improves optical performance. In fact, the larger τ_E values in samples E and F are reflected in the measured laser thresholds $J_{th}(D) = 400$ A/cm^2; $J_{th}(E) = 450$A/cm^2; $J_{th}(F) = 630$ A/cm^2.

This work was partly supported by MIUR, project FIRB-RBAU01E8SS and the ANSWER Project STRP 505642-1. We acknowledge H. Page, C. Sirtori, B. S. Williams, Q. Hu, G. Strasser, for useful discussions and the selection of suitable QCL devices.

References

1. Lurji, S.: *Hot Electron in semiconductor: Physics and devices*, Clarendon Press, 1998.

2. Spagnolo, V., Scamarcio, G., Page, H., Sirtori, C.: 'Simultaneous measurement of the electronic and lattice temperatures in GaAs/$Al_{0.45}Ga_{0.55}As$ quantum-cascade lasers: Influence on the optical performance', *App. Phys. Lett.*, **84**, 3690-3692, 2004.

3. Vitiello, M.S., Scamarcio, G., Spagnolo, V., Williams, B.S., Kumar, S., Hu, Q., Reno, J.L.: 'Measurement of subband electronic temperatures and population inversion in THz quantum-cascade lasers', *App. Phys. Lett.*, **86**, 111115-111117, 2005.

4. Spagnolo, V., Scamarcio, G., Schrenk, W., Strasser, G.: 'Influence of the band-offset on the electronic temperature of GaAs/Al(Ga)As superlattice quantum cascade lasers', *Semicond. Sci. Technol.*, **19**, S110-S112, 2004.

Time- and Spectrally Resolved THz Photoconductivity in Quantum Hall Devices

C. Stellmach[1], Y.B. Vasilyev[2], R. Bonk[1], A. Hirsch[1], N.G. Kalugin[3], G. Hein[4], C. R. Becker[5] and G. Nachtwei[1]

[1] Institute of Applied Physics, TU Braunschweig, Germany, iap@tu-bs.de
[2] Ioffe Physicotechnical Institute, St. Petersburg, Russia
[3] Department of Physics, Texas A&M University, College Station, USA
[4] Physikalisch-Technische Bundesanstalt, Germany
[5] Physical Institute, University Würzburg, Germany

Summary. The Terahertz photoconductivity is measured on GaAs/AlGaAs quantum Hall systems. Time resolved investigations show relaxation times form 10 to over 200ns and a dependence on the sample mobility and source-drain voltage. The latter is explained with a heating effect. In addition, spectrally resolved measurements are discussed. The spectral resolution is a function on the sample mobility. Finally, similar measurements on a HgTe/HgCdTe quantum well are presented. All results are important in respect to a possible THz detector application.

1 Introduction

Quantum Hall (QH) systems [1] interact effectively with THz radiation, because the Landau gap energy is of the same order as the photon energy (≈10meV). Measurements of the THz-photoconductivity are interesting both with respect to the basic understanding of excitation mechanisms and with respect to the application of these systems as detectors. It was shown that QH systems are promising to serve as sensitive, tunable and fast detectors (see e.g. [2, 3]), which is hard to realize in the THz range with other techniques [4].

In this study we present time and spectrally resolved measurements of the THz photoconductivity of GaAs/AlGaAs QH samples. We find that the relaxation time is a function of the sample mobility and the applied source-drain voltage. A simple picture is suggested to explain the results. The spectral response depends on the sample mobility, too.

In addition, first photoconductivity measurements on HgTe/HgCdTe quantum wells under QH conditions are shown. We think, that this system is a candidate to utilize the good detector properties of QH systems at moderate magnetic fields, because of the small effective mass m^* and the possibility of low electron concentrations.

2 Experimental Details

The properties of the used samples, are summarized in the following table. All GaAs/AlGaAs heterostructures are patterned in circular Corbino geometry (inner and outer radius: 500μm and 1500μm). The HgTe/HgCdTe quantum well [5] is patterned in Hall-bar structure (300·200 μm^2).

A pulsed *p*-Ge laser [6] is used as source, which can be tuned from 1.7THz to 2.5THz (corresponding to wavelengths of 180μm to 120μm). The laser and the QH sample are mounted in the same He bath cryostat (T=4K) and connected by a waveguide (see [3] for setup details).

sample number	type	electron concentration [m^{-2}]	mobility [T^{-1}]
#8447	GaAs/AlGaAs	$2.7 \cdot 10^{15}$	10
#8788	GaAs/AlGaAs	$2.0 \cdot 10^{15}$	50
#8815	GaAs/AlGaAs	$1.9 \cdot 10^{15}$	150
Q1960	HgTe/HgCdTe	$2.5 \cdot 10^{15}$	6.3

3 Time and Spectrally Resolved Measurements

The time-resolved measurements are performed on the GaAs/AlGaAs samples at the filling factor ν=2. A fast pulse generator for the laser and a suitable detector circuit is used. This allows us to resolve intrinsic time-scales.

In Corbino samples the current is proportional to longitudinal conductivity: $I_{SD} \sim j_{SD} = E_{radial}\, \sigma_{xx}$. The photoresponse (PR) corresponds to the difference ΔI_{SD} measured with and without illumination. Fig. 1(left) shows a typical time resolved PR measurement with an increase, when the laser is switched on, a constant PR during the illumination (system in dissipative state) and a relaxation back in the QH state, when the laser is turned off. The relaxation process is fitted by a function of the type $\exp(-t/\tau)$.

The results for different source drain voltages V_{SD} are shown in Fig. 1(right). Relaxation times τ from 10ns to over 200ns are observed (much shorter than earlier results on meander structures [7]). Samples with smaller mobility are typically faster. The samples #8447 and #8788 have a maximum slightly under the breakdown voltage V_c. We explain this with the following picture based on the heating effect of I_{SD} in the dissipative state: At low V_{SD} the current and the heating are low and the relaxation is fast. With increasing V_{SD} the heating effect increases and τ becomes longer. For $V_{SD}>V_c$ there is no QH state and therefore τ is short.

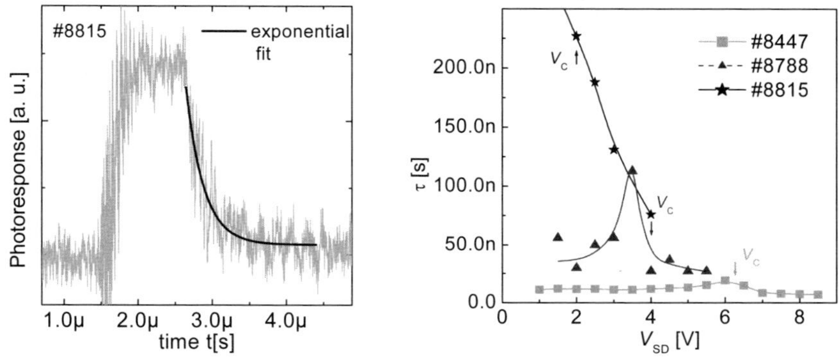

Fig. 1. Left: Time-resolved photoresponse with exponential fit of the relaxation. Right: Relaxation time versus source-drain voltage.

Further, detailed spectrally resolved measurements are performed on the GaAs/AlGaAs samples near the filling factor ν=2. The PR is measured for different laser energies E_{photon} while sweeping the magnetic field B (see Fig. 2, left). We find clearly a bolometric signal [7] with maxima at the flanks of the QH plateau region. E_{photon} influences the maxima. If the cyclotron resonance (CR) condition $E_{photon}=\hbar\,\omega_c=\hbar\, eB/m^*$ is fulfilled, the corresponding maximum is enhanced. For the high mobility sample #8815 well distinguishable CR and bolometric peaks are observed (not shown). To determine the spectral resolution, the PR at constant ν at the flank of the plateau is plotted versus E_{photon} and fitted by a Lorentzian function in the form $\Gamma/(4(E-E_{photon})^2+\Gamma^2)$ with the full width at half maximum Γ. Γ corresponds to the spectral resolution. We find Γ = 2meV for the sample #8447 and smaller values for the higher mobility samples.

5 Measurements on HgTe/HgCdTe Quantum Wells

Further, we made PR measurements on the HgTe/HgCdTe sample under QH conditions. In addition to the laser excitation, the PR was measured with a Glowbar source and a ΔT curve (difference of two ρ_{xx} measurements at different temperatures) is shown in Fig. 2(right). All curves are similar and show bolometric shape with maxima at the flanks of the QH minima. A clear CR signal has not been found yet. It is planned to investigate the PR mechanisms in more detail and possibly to develop a QH based THz detector, which operates at moderate magnetic fields.

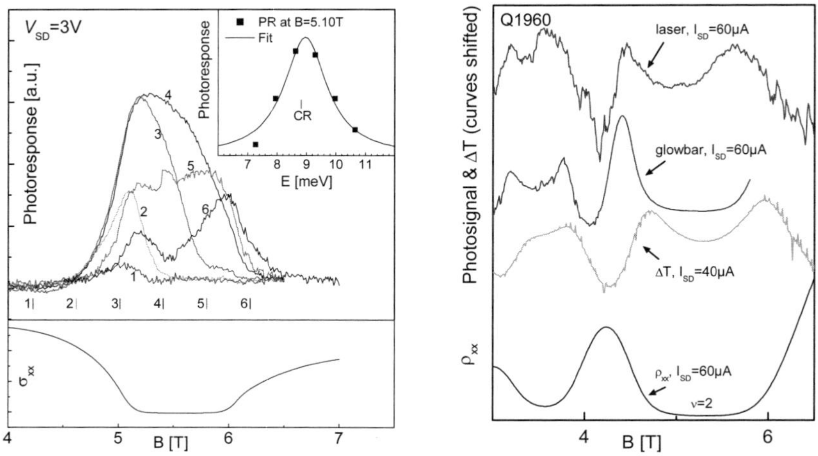

Fig. 2. Transport curves in the lower part and PR in the upper part versus magnetic field. Left: For sample #8447 for different laser energies (positions of the laser energies are marked by short vertical lines, the digits link this lines with the PR curves). The inlay shows the spectral resolution. Right: PR for the HgTe/HgCdTe sample. For details see text.

Acknowledgments

This work was supported by the Deutsche Forschungsgemeinschaft. We thank the MBE group of the Max-Plank Institute Stuttgart for the wafers.

References

1. K. von Klitzing, Rev. of Mod. Phys. **58**, 494 (1986)
2. Y. Kawaguchi *et al.*, Appl. Phys. Lett. **80**, 136 (2002)
3. C. Stellmach *et al.*, Semicond. Sci. Technol. **19**, 454 (2004)
4. P.H. Siegel, IEEE Trans. Microwave Theory Tech. **50**, 910 (2002)
5. F. Goschenhofer, thesis, Universität Würzburg, 1998
6. E. Gornik, A.A. Andronov (Ed.), Opt. Quantum Electron. **23**, 111 (1991)
7. N.G. Kalugin *et al.*, Phys. Rev. B **66**, 085308 (2002)

Transport Properties and Terahertz Emission in Narrow Minigap GaAs-GaAlAs Superlattices

A. A. Andronov[1], E.P.Dodin[1], A.Yu.Klimov, V. V.Rogov, Yu. N. Nozdrin[1], D. I. Zinchenko[1], A. A. Marmalyuk[2], A. A. Padalitsa[2],

[1] Institute for Physics of Microstructures, RAS, Nizhny Novgorod GSP-105 603600, Russia; [2] SIGM plus, Moscow, Russia

Summary. Low temperature transport and THz emission in GaAs-GaAlAs narrow minigap superlattices are studied. The THz emission due to interminiband transition, the Bloch oscillation, bremsstrahlung and reststrahlen (in heated by current pulse lattice) are observed by Ge:Ga, Ge:Au, n-GaAs and n-InSb detectors.

The main drive behind these studies is a desire to fabricate a THz SL Bloch oscillator basing on the dynamic negative differential conductivity (NDC) above the BO frequency [1] in the narrow minigap superlattices (SL) with the lowest miniband below an optical phonon energy. And in this work we measure the integral spontaneous THz emission by the detectors placed in liquid helium nearby the SL mesa. To the best of our knowledge none has measured the spontaneous THz BO emission: only the coherent BO emission excited by optical pulses was observed (see e.g. [2, 3])

MOCVD – grown SLs on n+ substrates there studied. The SLs parameters are : **1** - SL 698 (500 periods): 155 Å GaAs, 20 Å $Al_{0.1}Ga_{0.9}As$; **2** - SL 502 (150 periods): 163 Å GaAs, 10 Å $Al_{0.1}Ga_{0.9}As$; **3** - SL 426 (100 periods): 186 Å GaAs, 10 Å $Al_{0.1}Ga_{0.9}As$.

Below data on SL 698 (the thickest one) are presented. In the simulation the model used earlier [1] was employed with the Bragg scattering and the interminiband Zener tunnelling treated as scattering processes. Isotropic constant elastic scattering was introduced to model suppressed mobility in our SLs. In Fig 4 E_0 is the optical phonon scattering field ($E_0 \approx 6.5$ kV/cm that corresponds to about 5.7 volts applied to SL 698).

Both in the observation and in the simulation static NDC exist in the low voltage region. In all of the SLs studied current oscillation in 10 – 200 MHz range were observed in the NDC regions (shown in Fig 2).

The emission observed by the Ge:Ga detector is a sum of the inter-miniband and the BO emissions. At the low voltage saturation of the Ge:Ga response occurs (marked in Fig 2). Here the emission is presumable interminiband one because the BO emission is out of the Ge:Ga response

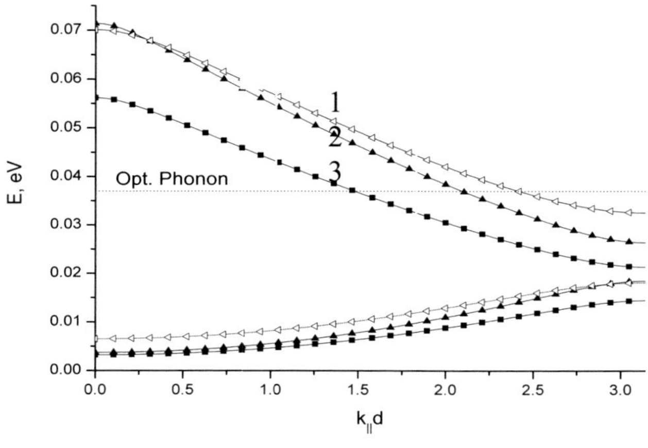

Fig 1 The two first miniband of the superlattices studied

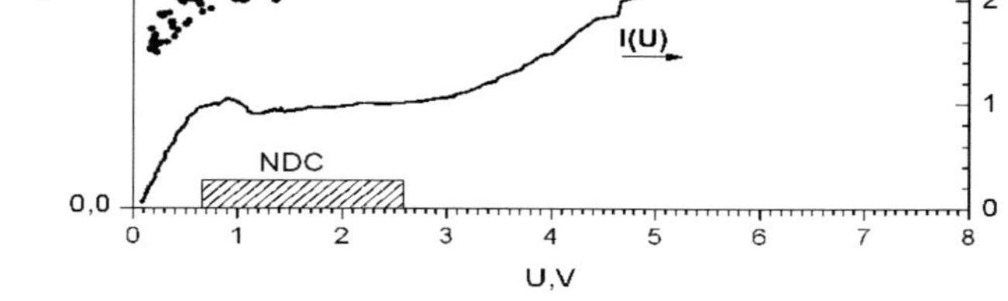

Fig 2 I – V curve and Ge:Ga detector response for 300 micrometers mesa at 4K

band while fraction of electrons in the second miniband (as simulation shows) is almost constant. Beyond the saturation region the BO emission enters the Ge:Ga response band and is stronger than the interminiband one.

Indeed, emission powers of an oscillating dipole inside a mesa with metallic contacts at distance **L** is: $\mathbf{P = \pi e^2 x^2 \omega^3 / 2 L c^2}$, where **e** is an electron charge, $\boldsymbol{\omega}$ is the emission frequency, **x** is the dipole spatial oscillating amplitude and **c** is the light velocity. We take $x = \Delta / e E$ (Δ is the miniband width) for the BO and $\mathbf{x^2} = (\hbar^2/ 2m\ \hbar\omega)\, f$, **m** is electron effective mass and **f** is the oscillator strength for interminiband emission. For narrow minigap SL for quanta near minigap one can get $f = 4\,\Delta_0 / \hbar\omega$ (cf. [4]), Δ_0 is the energy in the middle of minigap. Now ratio of the BO $\mathbf{P_B}$ and the interminiband $\mathbf{P_I}$ emission powers is: $\mathbf{P_B / P_I} = (\pi^2/4)\,(\Delta / \Delta_0)^2\ \mathbf{(1/\delta)}$, δ is the fraction of electrons in the second miniband. Simulation shows the fraction $\delta \approx \mathbf{0.1}$ at 1 – 6 kV/cm. That implies that the interminiband emission should also be constant. So the rise of the detector signal at about 3 – 5 kV/cm should be due to BO emission. At even higher fields due to rise in electron energy (Fig 4) and δ produced by rise in tunnelling rate interminiband emission rise again. The interminiband contribution to the Ge:Ga detector response is supported also by observation of the "afterglow" behind the voltage pulse (see Fig 3). Presumably it is the equilibrium interminiband emission in the heated by voltage pulse mesa. For the Ge:Au detector the afterglow is due also to equilibrium emission in lattice reststrahlen band in the heated mesa. The latter is supported by appearance in the detector response a shoulder at the high voltage pulse (Fig 3). The dependence of the Ge:Au detector response on voltage is similar to that of the Ge:Ga with relatively suppressed the low voltage value because here the

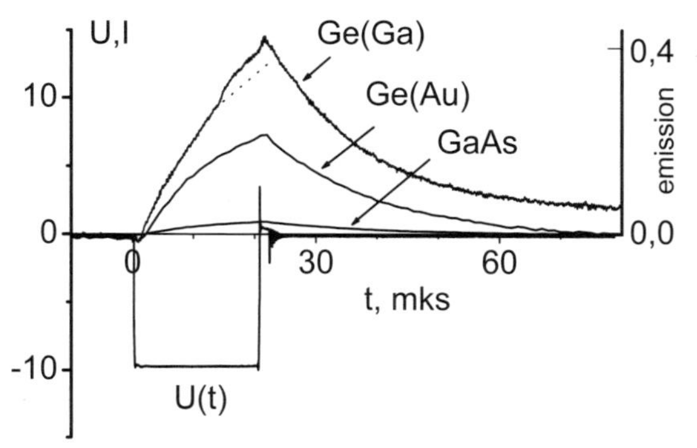

Fig 3. Voltage and the detector response pulses at 4K

BO and the interminiband emissions are out of the Ge:Au response band. The dependence of the n-GaAs detector response on voltage is eventually a straight line without any saturation. We think that the response is due to the electron bremsstrahlung.

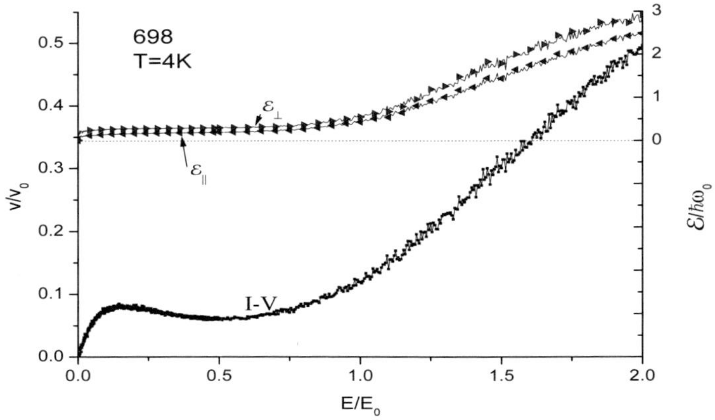

Fig 4. Simulated ratio V/V_0 of electron drift velocity to the electron velocity V_0 at an optical phonon energy $\hbar\omega_0$, and ratio of average electron energies $\varepsilon_{\parallel}$ along and perpendicular $\varepsilon_{\perp}$ to the superlattice axis to an optical phonon energy $\hbar\omega_0$.

The n-InSb detector response has sharp threshold and sustained long after voltage pulse: we attribute the response to the microwave equilibrium emission of the heated mesa.

To summarize: we have observed different types of THz spontaneous emission from weak barrier superlattices subject to pulsed voltage. The Bloch oscillation emission has been identified.

This work is supported by RAS Programs "Low dimensional structure" and "Semiconductor lasers" and by RFBR Grant N 05-02-16468

References

1. Andronov A., Nefedov I. and Sosnin A. "Transport in weak barrier superlattices and the problem of the Tterahertz Bloch generator", *Semiconductors*, **37**, pp.378 -384, 2003.

2. Waschke C., Roskos H.G., Schwedler R., Leo K. and Kurz H. "Coherent Submillimeter-Wave Emission from Bloch Oscillations in a Semiconducting Superlattice", *Phys.Rev.Lett.*, **70**, pp 3319-3322, 1993.
3. Y.Shimada, N.Sekine and K.Hirakawa "Terahertz emission due to interminiband Zener resonant tunneling in wide miniband $GaAs/Al_{0.3}Ga_{0.7}As$ superlattices", *Appl.Phys.Lett.*, **84**, pp 4926-4928 , 2004
4. Helm M. : `Infrared spectroscopy and transport electrons in semiconductor superlattices`, *Semicon. Sci.Techn.,* **10**, pp 557-575, 1995

Investigation of Antenna-Coupled MOM Diodes for Infrared Sensor Applications

Balázs Rakos, Heng Yang, Jeffrey A. Bean, Gary H. Bernstein, Patrick Fay, Árpád I. Csurgay, and Wolfgang Porod

Center for Nano Science and Technology, Department of Electrical Engineering, University of Notre Dame, Notre Dame, IN 46556 USA

Summary. We report the fabrication and experimental study of uncooled, unbiased infrared sensors suited for the detection of optical and near IR-radiation. The detector consists of a metal-oxide-metal junction fabricated on an oxidized silicon wafer by a single electron beam lithography step combined with double-angle evaporation. We fabricated symmetrical and asymmetrical diodes with junction areas around 50 nm x 50 nm, which is smaller than those used in previous work. In this paper we present the fabrication procedure of the sensors, dc I-V measurements of the diodes, and the optical and near IR measurements.

1 Introduction

In this work, we propose quantum tunneling nanoantennas as uncooled sensors for near infrared focal plane arrays (FPA). The sensors are frequency selective, CMOS compatible and facilitate high speed processing exceeding 10,000 frames per second.

The infrared detection of metal-oxide-metal (MOM) diodes is based on the rectification of the high-frequency currents induced by the incident radiation in the surrounding metal structure, which serves as an antenna. Instead of semiconductor-based infrared detectors, these devices can operate at room temperature even in the mid-infrared and far infrared range. Although detectors based on microbolometers can also operate at room temperature, those devices are much slower than the antenna-diode structures.

Point-contact metal-oxide-metal (MOM) diodes combined with wire antennas were first used for detection and mixing at submillimeter wavelengths in 1966 [1]. Point-contact devices are not suitable for commercial applications, because of their mechanical instability and irreproducibility. With the application of photolithography, stable and reproducible thin-film Ni-NiO-Ni diode-antenna structures were integrated on a substrate [2].

The use of electron beam lithography (EBL) has made it possible to fabricate diodes with very small contact areas (around 110 nm x 110 nm) combined with antennas suited for the 10.6 um radiation [3-5]. Since a diode with a smaller contact area has a higher cutoff frequency, these devices have a better performance than the structures made by photolithography.

We have fabricated antenna-MOM diode detectors on a silicon wafer with contact areas around 50 nm x 50 nm, which is smaller than in previous reports for these devices [2-5].

In this study we present the fabrication and dc measurements of symmetrical and asymmetrical tunnel junctions made by double-angle evaporation. The curvature coefficients $\left(\gamma = \frac{\partial^2 I}{\partial V^2} \Big/ \frac{\partial I}{\partial V}\right)$ of the asymmetrical diodes are not zero even at zero bias voltage. These structures do not require biasing, and since their noise level is lower than that of the symmetrical diodes (due to the absence of dark current as well as elimination of 1/f noise in unbiased detectors), better detector performance is expected.

2 Fabrication

The nanoantenna-MOM diode sensors were fabricated on an oxidized (oxide thickness: 1.5 um), around 625 um thick, 11-16 Ω-cm, p-type silicon wafer with a single electron beam lithography step combined with double-angle evaporation and oxidation. Tunnel junctions for the realization of single electron transistors have already been fabricated with this method [6]. During double-angle evaporation, the EBL-defined pattern is shifted, and overlap between the two metal layers forms the diode (Fig. 1).

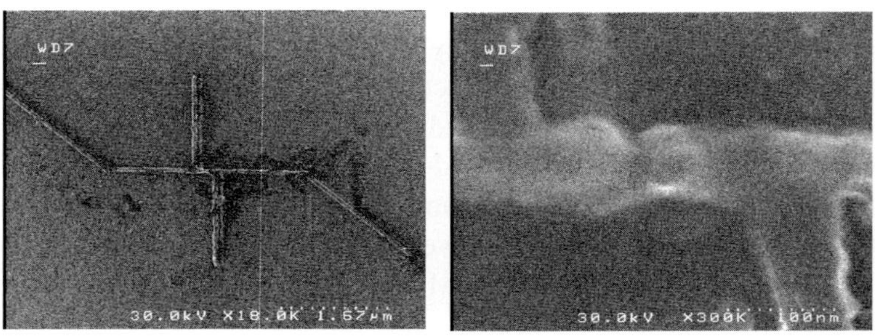

Fig. 1. Electron micrograph of an Al-Al_2O_3-Ti antenna-diode sensor (left). The overlap between the Al and Ti layers forms the MOM junction.

Table 1. Curvature coefficients of MOM diodes.

Diode	Zero bias curvature (1/V)	Peak curvature (1/V)
Al-Al_2O_3-Al	0	1.5
Al-Al_2O_3-Ti	0.3	2
Al-Al_2O_3-Pt	1	1.3
Al-Al_2O_3-Ni	0.5	1
Ni-NiO-Pt	3	13

3 Experimental Results

The current-voltage characteristics of the diodes were measured at a probe station. The first and second derivatives of the I-V curves were obtained from a third-order polynomial fit.

We fabricated Al-Al_2O_3-Al diodes, that show symmetrical I-V characteristics (Fig. 2). Furthermore, we fabricated asymmetrical Al-Al_2O_3-Ti, Al-Al_2O_3-Pt, Al-Al_2O_3-Ni, and Ni-NiO-Pt diodes. The I-V curve of a Ni-NiO-Pt diode is also displayed in Fig. 2. The oxidation was performed in air in all cases. The curvature coefficients of the diodes are shown in Table 1.

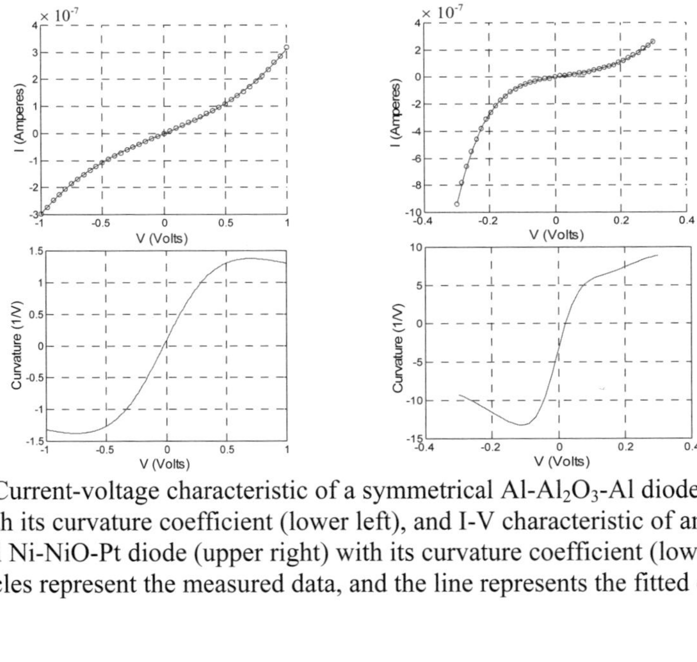

Fig. 2. Current-voltage characteristic of a symmetrical Al-Al_2O_3-Al diode (upper left) with its curvature coefficient (lower left), and I-V characteristic of an asymmetrical Ni-NiO-Pt diode (upper right) with its curvature coefficient (lower right). The circles represent the measured data, and the line represents the fitted curve.

The visible and IR radiation were supplied by separate visible and infrared sources, respectively. The wavelength of the radiation was adjusted using a monochromator (bandwidth of radiation: 10 nm). The responses of the detectors were measured using a current preamplifier connected to a lock-in amplifier. The frequency-dependent response can be altered by changing the length of the vertical lines of the structure (Fig. 3).

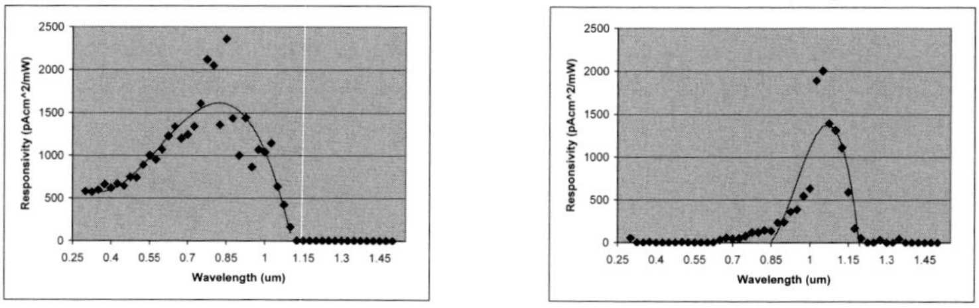

Fig. 3. Frequency-dependent responses of sensors with 0 um long (left) and 3 um long vertical lines (right).

4 Conclusions

In this paper we presented the fabrication method of ultrasmall, thin-film tunnel-junction infrared sensors. We found that a single electron beam lithography step combined with double-angle evaporation is a simple and stable process for realizing such devices.

We measured the dc I-V responses of the diodes, and strongly nonlinear current-voltage characteristics have been achieved in nanoscale MOM diodes. These devices are promising for nanoantenna-based optical detection.

References

1. J. W. Dees: Microwave J. 9 (1966) 48-55.
2. M. Heiblum, S. Wang, J. R. Whinnery, T. K. Gustafson: IEEE J. Quantum Electron. QE-14 159 (1978).
3. I. Wilke, W. Herrmann, F. K. Kneubühl: Appl. Phys. B 58 (1994) 87-95.
4. I. Wilke, Y. Oppliger, W. Herrmann, F. K. Kneubühl: Appl. Phys. A 58 (1994) 329-341.
5. C. Fumeaux, W. Herrmann, F. K. Kneubühl, H. Rothuizen: Infrared Physics & Technology 39 (1998) 123-183.
6. G. L. Snider, A. O. Orlov, I. Amlani, X. Zuo, G. H. Bernstein, C. S. Lent, J. R. Merz, and W. Porod *J. Vac. Sci. Technol.* A 17(4), 1999.

Transport and Noise in Ultrafast Unipolar Nanodiodes and Nanotransistors

[1] T. González, [2] A. M. Song, [1] B. G. Vasallo, [1] D. Pardo, and [1] J. Mateos

[1] Departamento de Física Aplicada, Universidad de Salamanca,
Plaza de la Merced s/n, 37008 Salamanca, Spain

[2] Department of Electrical Engineering and Electronics,
University of Manchester, Manchester, M60 1QD, United Kingdom

Summary. By means of a 2D Monte Carlo simulator, we analyze the static, dynamic and noise behavior of self-switching diodes and transistors, and provide considerations for the optimization of their downscaling process.

1 Introduction

Recently, a nanoscale unipolar rectifying diode based on electrostatic effects, so called self-switching diode (SSD), has been proposed [1]. This device provides a rectifying behavior without the use of any doping junction or barrier structure and can be fabricated with a simple single-step lithographic process. By adding a lateral gate contact, a nanometer self-switching transistor (SST) can also be easily fabricated [2]. The downscaling of SSDs and SSTs to the deca-nanometer range is so simple that, together with the intrinsically high electron velocity of InGaAs channels, the fabrication of devices working in the THz range can be envisaged, as predicted by Monte Carlo (MC) simulations for the intrinsic cutoff frequency of SSDs [3]. Even if parasitic resistances and capacitances degrade the intrinsic behavior, microwave detection with good enough sensitivity up to 110 GHz has been demonstrated experimentally at room temperature in arrays of SSDs [4]. When reaching so high frequencies, noise may become a limiting factor for applications and requires special attention. In this work we will make use of a 2D MC simulator (employed previously to successfully model other nanoscaled ballistic devices [5, 6]) to analyze the high-frequency and noise performance of SSDs and SSTs.

2 Device and Monte Carlo Model

Fig. 1(a) shows an AFM image of a fabricated SSD, consisting essentially of a narrow channel, defined by etched trenches on an InGaAs/InAlAs wa-

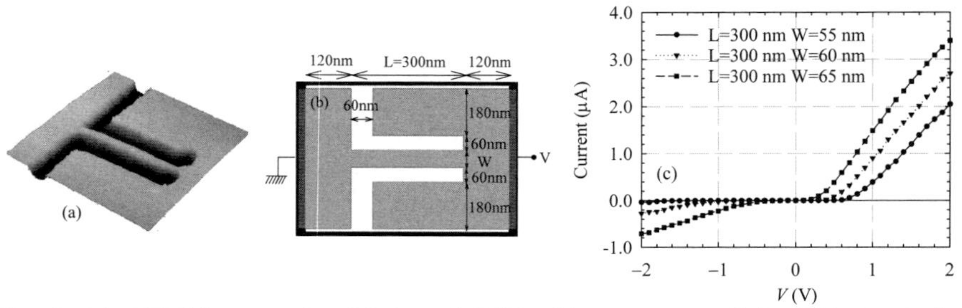

Fig. 1. (a) AFM image of a fabricated SSD, (b) geometry of the TV simulation domain, and (c) *I-V* curves for several channel widths.

fer, with symmetry broken by the continuation of the trenches to the device boundary. Fig. 1(c) shows the *I-V* characteristic of the SSD sketched in Fig. 1(b). At equilibrium the channel is closed due to the depletion (potential barrier) induced by the surface charges located at the lateral walls. When $V>0$, the positive potential reaches the lateral regions of the channel, so that the barrier is lowered (or even removed) and the electron flow is allowed (open channel). On the contrary, when $V<0$ the potential profile in the right part of the device is almost unchanged with respect to the equilibrium situation and the channel remains closed (inverse leakage current appears for too wide channels) [3].

For the analysis of these devices we make use of an ensemble MC simulator self-consistently coupled with a 2D Poisson solver. The model takes into account degeneracy, electron heating, surface charges appearing at the boundaries of the semiconductors in contact with dielectrics and accurate contact injection with appropriate statistics. 2D top-view (TV) simulations are used [5]. Here, the value of two important parameters must be carefully chosen: the background doping N_{Db} (10^{17} cm^{-3}) and the lateral surface charge density σ (0.3×10^{12} cm^{-2}). The non-simulated dimension Z is estimated as $Z=n_s/N_{Db}=10^{-5}$ cm, with $n_s=10^{12}$ cm^{-2}, typical value of sheet electron density in InGaAs channels [1]. Other details about the model can be found in Ref. 5.

3 Dynamic and Noise Behavior

The *I-V* curves of SSDs with W=50 nm and different channel lengths [Fig. 2(a)] are shown in Fig. 2(b). To improve the current control, the width of the trenches is reduced to 5 nm (near the limits of up-to-date technology), thus achieving higher forward currents. Short-channel effects (like those of traditional FETs) appear when the aspect ratio of the channel (L/W) decreases, leading to the presence of an inverse leakage current, as observed in Fig. 2(b) for L=100 nm [also in Fig. 1(c)]. The excellent intrinsic dy-

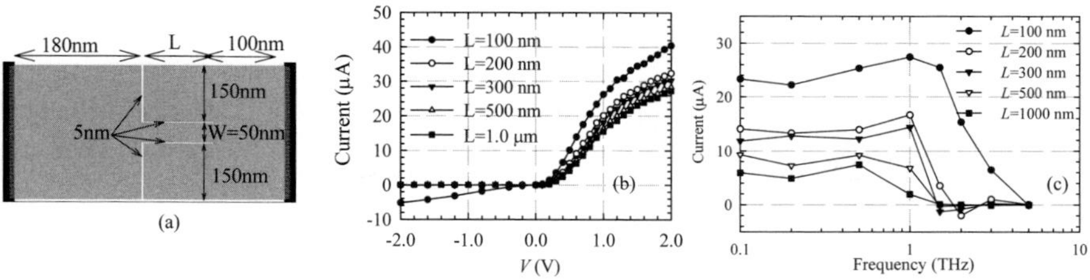

Fig. 2. (a) Geometry of the simulated SSDs, (b) *I-V* curves for different channel lengths, and (c) dc current response to a sinusoidal voltage with amplitude of 0.5 V as a function of frequency.

namic behavior of these devices is illustrated in Fig. 2(c), which shows the dc current response as a function of the frequency of a sinusoidal input voltage with 0.5 V of amplitude. The intrinsic cut-off frequency of the rectifying behavior of SSDs enters the THz range and, as expected, depends on the channel length, being lower for longer channels. Even if the parasitics degrade this intrinsic behavior, good performance in real devices operating as detectors up to 110 GHz at 300 K has already been reported [4].

As an example of the noise behavior of SSDs, Fig. 3(a) shows the low-frequency current spectral density $S_I(0)$, compared to $2qI$, in an SSD with L=250 nm, W=50 nm and 5 nm wide trenches. Fig. 3(b) shows the Fano Factor, $F=S_I(0)/2qI$, together with the potential barriers for electrons flowing from cathode to anode and from anode to cathode (right and left directions, respectively). The barrier in the left direction is always large, while that in the right direction decreases until vanishing when the applied direct potential increases, thus providing the explained rectification. As long as the barrier has a significant value (higher than about 0.1 eV), the SSD displays full shot noise (both under direct and inverse bias) with F=1. When the direct bias is increased, the barrier from cathode to anode is lowered, the channel resistance decreases, and the diffusive accesses to the channel become more and more important in the total noise of the device,

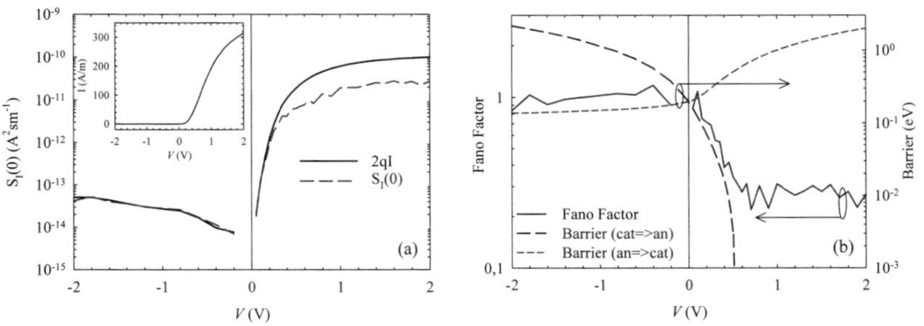

Fig. 3. (a) $S_I(0)$ and $2qI$, and (b) Fano factor and potential barrier for electrons flowing from cathode to anode and from anode to cathode as a function of the potential applied to the anode in an SSD with L=250 nm, W=50 nm and 5 nm wide trenches, whose *I-V* curve is shown in the inset of (a).

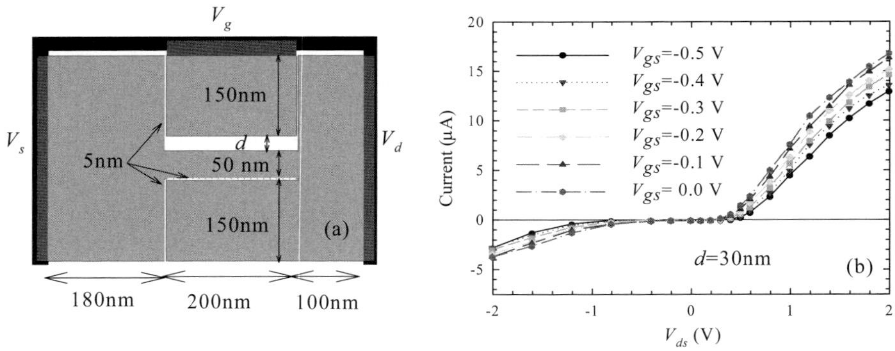

Fig. 4. (a) Geometry of the simulated SST and (b) output characteristics.

$S_I(0)$ thus showing an increasingly suppressed value with respect to full shot noise (F<1). When the barrier disappears, F takes a constant value of about 0.3, just related to the total resistance of the device. This specific value depends on the geometry of the SSD. In contrast, the appearance of full shot noise when the barrier is significant is independent of the device design as long as transport is ballistic or quasiballistic.

With a small modification of the geometry of SSDs, a control terminal (lateral gate) can be implemented in the device, as shown in Fig. 4(a), thus becoming an SST. This third terminal, which acts as the gate of a typical FET transistor, controls the current flow through the channel, and leads to the output characteristics of Fig. 4(b). In the case shown, a wider trench between gate and channel is considered to avoid an excessive control by the gate that would cancel the rectifying behavior of the device.

References

1. Song A. M., Missous M., Omling P., Peaker A. R., Samuelson L., and Seifert W.: 'Unidirectional electron flow in a nanometer-scale semiconductor channel: A self-switching device', *Appl. Phys. Lett.* **83**, 1881-1883, 2003.
2. Song A. M., Maximov I., Missous M., and Seifert W.: 'Diode-like characteristics of nanometer-scale semiconductor channels with a broken symmetry', *Physica E* **21**, 1116-1120, 2004.
3. Mateos J., Vasallo B. G., Pardo D., and González T.: 'Operation and high-frequency performance of nanoscale unipolar rectifying diodes', *Appl. Phys. Lett.* **86**, 212103, 2005.
4. Balocco C., Song A. M., Aberg M., Forchel A., González T., Mateos J., Maximov I., Missous M., Rezazadeh A. A., Saijets J., Samuelson L., Wallin D., Willliams K., Worschech L., and Xu H. Q.: 'Microwave detection at 110 GHz by nanowires with a broken symmetry', *Nano Lett.* **5**, 1423-1427, 2005.

5. Mateos J., Vasallo B. G., Pardo D., González T., Galloo J. S., Bollaert S., Roelens Y., and Cappy A.: 'Microscopic modeling of nonlinear transport in ballistic nanodevices', *IEEE Trans. Electron Devices* **50**, 1897-1905, 2003.
6. Mateos J., Vasallo B. G., Pardo D., González T., Pichonat E., Galloo J. S., Bollaert S., Roelens Y., and Cappy A., 'Nonlinear effects in T-branch junctions', *IEEE Electron Device Lett.* **25**, 235-237, 2004.

Monte Carlo Study of Coupled SO Phonon-Plasmon Scattering in Si MOSFETs with High κ- Dielectric Gate Stacks: Hot Electron and Disorder Effects

J. R. Barker, J. R. Watling, A. Brown, S. Roy, P. Zeitzoff*, G. Bersuker*, A. Asenov

Device Modelling Group, Department of Electronics and Electrical Engineering, University of Glasgow, Glasgow G12 8LT, Scotland, UK.

*SEMATECH, 2706 Montopolis Drive, Austin, Texas 78741, USA.

Summary. A Monte Carlo scheme is described for simulating electron-phonon-plasmon scattering in realistic high-κ gate stack Si MOSFETs that accounts for hot electron effects, modulation of the electron-phonon-plasmon scattering rates by the interface boundary roughness and inhomogeneity of the dielectric layers.

1 Introduction

There is strong industrial interest in implementing high-κ gate stacks with metal gates into decanano silicon MOSFET technology to allow physically thicker dielectrics in order to reduce the gate leakage (tunnelling) current prevalent in ultra-thin silicon dioxide in present Si MOSFETs. However, it has been pointed out [i, ii] that severe mobility degradation ensues because of the strong coupling of carriers in the channel to soft, surface optical (SO) phonons [iii, [iv] in the vicinity of the high-κ dielectric interface. Recently, we have shown [ii] that for ideal interfaces and homogeneous dielectrics it might be possible to offset the mobility degradation by a mobility enhancement that derives from using a strained silicon channel. Unfortunately, the electron-phonon-plasmon scattering in such structures is a complex process that involves the entire gate stack via the coupling of plasmons in the gate and substrate to the phonon modes. Because the renormalised phonon/plasmon energies and screening are sensitive to layer thicknesses the electron-phonon-plasmon scattering strength is modulated by the roughness of the interfaces.

Both the gate-dielectric interface and the dielectric/interfacial layer-semiconductor interfaces will contribute to remote surface roughness effects. In addition there is evidence that many of the high-κ gate stacks display significant phase separation that will distort our assessment of the spatial dependence of the scattering rates (Fig.1). An understanding of the coupling problem also requires an evaluation of the effects of the variation of the carrier velocity dispersion (via electron temperature) between source and drain that acts to modulate the Landau damping of the coupled system.

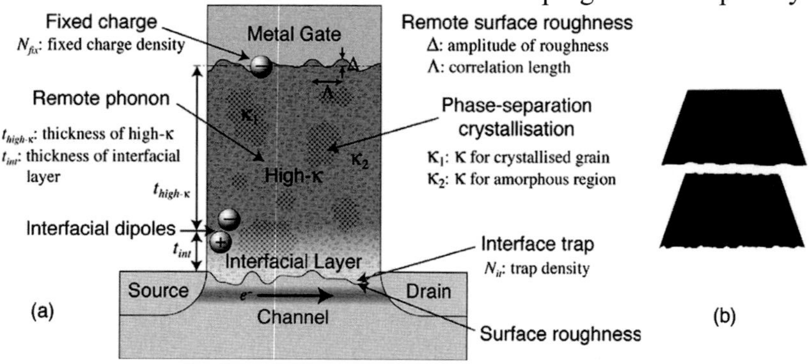

Fig. 1. (a) Features of realistic high-κ gate stack MOSFETs – adapted from [v]. (b) Layer inhomogeneities at different cross sections in the stack.

2 Monte Carlo Scheme

We extend the formalism developed by Fischetti *et al* [i] to include an inhomogeneous dielectric, interfacial layer and surface roughness at the different interfaces. The solution domain is partitioned into vertical rectangular columns located by discrete coordinates (X, Y) in the stack plane (see Fig. 2). Each column comprises blocks in which the material parameters are approximated as constant. Because of interface roughness and separated phases the blocks are disjoint. Each vertical column labelled by (X, Y) is then treated as a slice through a stack of uniform layers from gate down to the channel. The screened SO phonon scattering rates for each column $R(\mathbf{k},\mathbf{k}';X,Y)$ are then determined by the formalism in [1] assuming a bare channel plasmon energy $\hbar\omega_p(X,Y)$ and local electron temperature $T_e(X, Y)$ derived respectively from the averages of the electron density $n(x, y)$ and temperature $T_e(x, y)$ over the channel block at (X, Y). Landau damping is computed from a hot electron Lindhard model.

The full Monte Carlo scheme proposed here is compute intensive. However, simplifications may be made by using Padé approximants for the plasma dispersion function following [vi]. Coarse-graining the column widths also significantly reduces the complexity. This procedure is a good approximation for the mobility-degrading SO phonon scattering but fails to capture the elastic boundary scattering off the rough silicon channel to insulator interface. Boundary scattering is therefore treated by standard methods [vii]. The lowest order simulation uses a single column with averaged interfacial layer dimensions and parameters to form the stack blocks.

3 Polycrystallinity and Intrinsic Parameter Fluctuations

Preliminary results show that strong parameter fluctuations in nano-scale MOSFETs are introduced by non-uniformity of the dielectric properties of the high-κ material due to random grain orientation (polycrystalline HfO_2 with different orientations of grains resulting in varying dielectric properties) and phase separation (HfO_2 crystallised in 'islands' within an amorphous SiO_2 matrix). Such variations in the film structure are shown in Fig. 2a, which shows a plan-view TEM image of a HfSiO film, with phase separation. A 2D Fourier-Fractal model is shown in Fig. 2b.

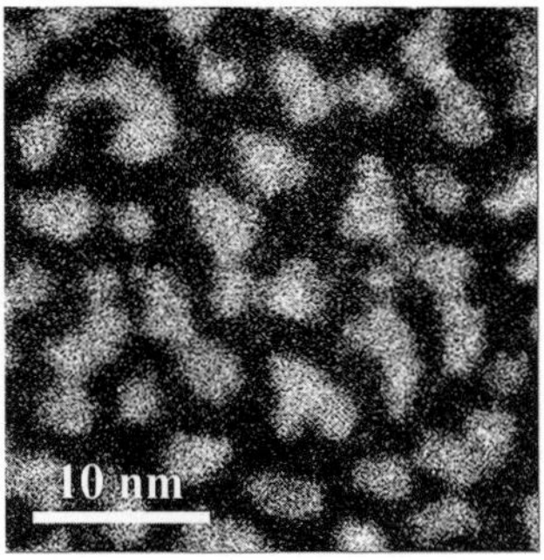

Fig. 2a. Plan-view TEM image of a HfSiO film illustrating phase separation of the Hf and Si oxides.

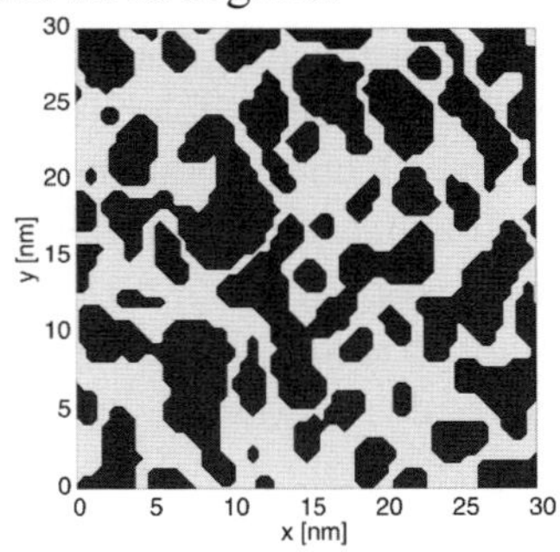

Fig. 2b. Stochastic dielectric pattern from 2-D Fourier-Fractal synthesis

The structural fluctuations in the dielectric of the oxide lead to corresponding fluctuations in the surface potential as illustrated in Fig. 3a. The particular random dielectric pattern in this particular case is shown in the plane above the device. The regions of high dielectric producing higher surface potentials, leading to fluctuations in the threshold voltage of nano-scale devices as can be seen from Fig. 3b, where we show the I_D-V_G characteristics for ten 50×50 nm MOSFETs clearly demonstrating the spread in threshold voltages, modelled using Drift Diffusion. Also shown are the curves for a pure SiO_2 gate oxide and pure high-κ (above an interfacial layer) the limiting cases for the fluctuations. In all cases the physical gate oxide is 4 nm thick and degradation in subthreshold slope due to the reduced gate oxide capacitance when SiO_2 is employed is evident.

Fig. 3a. Electrostatic potential in a 50×50 nm MOSFET showing the fluctuations in surface potential due to the spatial variations in gate dielectric, shown in the plane above.

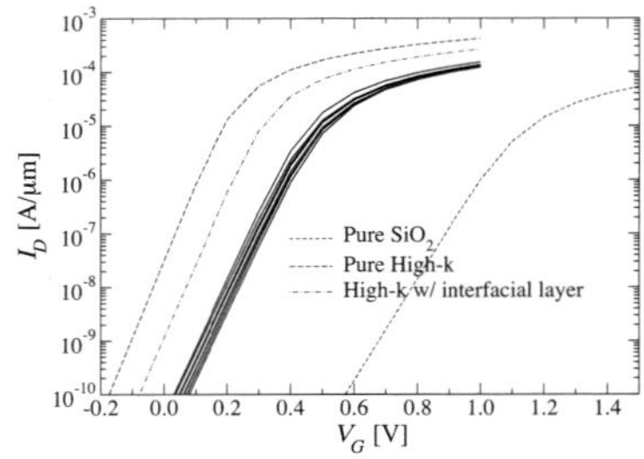

Fig. 3b. I_D-V_G characteristics for 10 devices with random dielectric pattern, and the limiting cases.

4 Interfacial Oxide Layer

We find that the rough interface boundaries gives rise to a random variation of the effective dielectric thickness that carries through into a modulation of the electron-phonon-plasmon coupling and local phonon energies. This is illustrated by a simple single column Monte Carlo simulation for a gate stack consisting of a HfO_2 high-κ and a SiO_2 interfacial layer, similar to that illustrated in Fig 1. Fig. 4. shows the variation in computed mobility versus oxide interfacial layer thickness with a constant inversion layer field of 1MV/cm.

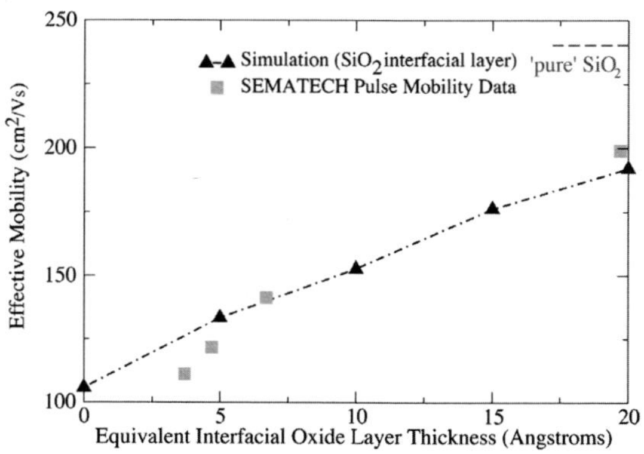

Fig. 4. Effective mobility within the channel as a function of the interfacial (SiO_2) layer thickness.

5 Electron temperature

Electron heating at various locations in the device severely alters the influence of Landau damping on the plasmon modes. The variation of electron temperature along the channel thus provides an additional spatial modulation of the scattering rates and quanta. The typical variation of T_e is shown in Fig. 5 showing the Monte Carlo computed electron temperature for a degenerate channel in a 67nm effective channel length Si *n*-MOSFET [8].

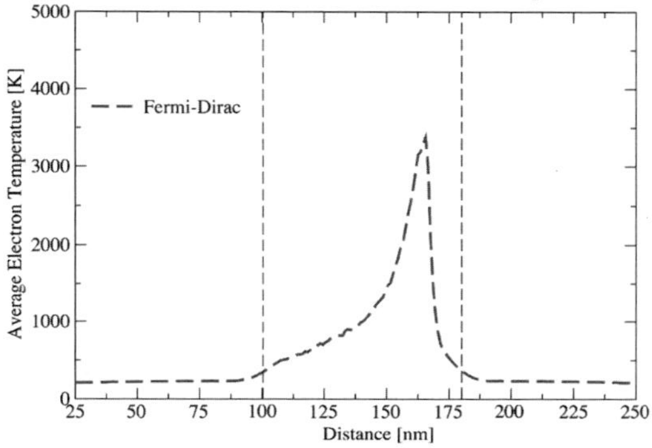

Fig. 5. Variation in electron temperature along channel for degenerate electrons in a Si n-MOSFET. $V_G=V_D=1V$. Vertical dashed lines indicate the start and end of the metallurgical gate.

6 Acknowledgements

This work is supported by SEMATECH under their Advanced Gate Stack Program and EPSRC grant GR/S80097/01 "Meeting the material challenges of nano-CMOS electronics". One of us(JRW) also acknowledges the support of EPSRC through an Advanced Research Fellowship.

7 References

[1] M.V. Fischetti, D. A. Neumayer, E. A. Cartier, J. App. Phys. 90, 4587, 2001.

[2] L. Yang, J. R. Watling, F. Adamu-Lema, A. Asenov and J. R. Barker, IEDM 2004 Tech. Dig., pp. 597-600, 2004.

[3] K. Hess and P. Vogl, Solid State Commun. 30, 807,1979.

[4] B.T. Moore and D.K. Ferry, J.Appl.Phys. 51, 2603, 1980.

[5] S. Saito *et al*, IEDM 2003 Tech. Dig, pp. 796-799, 2003.

[6] D. Lowe and J.R. Barker, J. Phys. C: Solid State Phys. **18**, pp. 2507, 1985.

[7] J. R. Watling, L. Yang, M. Boriçi, R. C. W. Wilkins, A. Asenov, J. R. Barker and S. Roy, Solid State Electronics, **48**, pp. 1337-1346, 2004.

[8] K. Rim et al, Symposium on VLSI Technology Digest of Technical Papers, pp. 98-99, 2002.

Implementation of Separable Scattering Mechanisms in Three-Dimensional Quantum Mechanical Simulations of Devices

M.J. Gilbert, R. Akis and D.K. Ferry

Department of Electrical Engineering and Center for Solid State Electronics Research, Arizona State University, Tempe, AZ 85287-5706

Summary. We present results of a numerical scheme through which different phonon scattering processes can be incorporated into the quantum mechanical simulation of semiconductor devices. This method allows the effects of different phonon processes to be seen individually and without resorting to the use of fitting parameters to approximate the correct physics. We find that phonon processes allow the carriers to reach higher energies than in the ballistic case modifying the output current. Further, we use the scattering model to approximate the ballistic-to-diffusive crossover and find the distance to be much shorter than anticipated.

In order to understand the operation of any type of semiconductor device, one must consider the effects of scattering. For many years, Monte Carlo has been the workhorse that has yielded great insight into the operation of a wide variety of semiconductor devices. The Monte Carlo technique handles scattering in a very natural way where quantum mechanical rates are derived from Fermi's Golden Rule and then applied to a distribution of electrons. However, Monte Carlo cannot account for the quantum effects that are seen in future devices. Therefore, other device simulation techniques, such as Green's functions and scattering matrices [1,2], have been proposed which can account for the quantum mechanical effects in these devices in three-dimensions.

In a manner similar to Green's function approaches, but more amenable to the site representation used in these methods, scattering can be computed on a mode basis and then transformed to the site basis. Here, we present results of the first implementation of separable phonon scattering rates in a three-dimensional, fully quantum mechanical, self-consistent device simulation.

To simulate the devices in this paper, we begin by discretely doping the devices using a method that previously presented in [2]. Once the devices are discretely doped, the dopant atoms are mapped back onto the solution mesh and the Poisson equation is solved for the initial guess at the potential. To include scattering in this approach, as in most semiconductor theory, we treat the scattering as weak. Hence, we can use the Fermi golden rule expression, equivalent to a first-order, non-self-consistent Born approximation, for each scattering process and generate a real space self-energy from it [3]. The scattering rates are then entered into the Hamiltonian for the three-dimensional recursive scattering matrix algorithm [2]. The quantum kernel is started with these scattering rates included and then the density obtained is then fed back into the Poisson equation for updating. This process is continued until a desired level of self-consistency is obtained.

We now examine a system—a quantum wire transistor—where the quantum mechanical nature of the scattering matrices is needed to correctly characterize the system. In a system such as this, the phonon scattering should have a more profound effect on the transport, as the reduced area of the wire should give rise to an enhanced overlap integral in the scattering calculations. In the system under consideration, the thickness of the silicon layer is 6.51 nm and the width of the channel is 8.15 nm. The length of the quantum wire that comprises the channel is 9.77 nm in length. Oxide barriers are placed on either side of the channel to simulate the appearance of a hard wall boundary that would be present in an actual experimental system. The source and drain of the device are 18.47 nm wide and 10.32 nm in length. The latter is discretely doped *n*-type with a doping concentration of $1\times10^{20}cm^{-3}$, while the channel is undoped. The quantum wire that forms the channel of the device has metal gates on three sides to form a tri-gate type transistor. The gate oxide thickness (SiO_2) on this device is 1 nm. In this simulation, the temperature is taken to be 300 K and the bias across the device is $V_{sd} = 0.6$ V.

In Fig. 1, we plot the I_d-V_g curves for the ballistic and quasi-ballistic (only impurity scattering is present) case for an identical doping distribution. The main effect of the phonons on the system is to allow the electrons propagating in the system to scatter to higher subbands, and then back again. This scattering allows the electrons to cross the source-channel barrier at slightly lower gate voltage than in the ballistic case. As the gate voltage increases, the strong effect of the contacts on the charged quantum wire is that quantum interference begins to form in the ballistic case resulting from oscillations in the conductance. With phonon scattering present, electrons still show these oscillations, but they are damped by scattering. There are more conductance fluctuations of reduced magnitude as the phonon assisted curve has a slight dip near $V_g = 0.45$ V before it reaches a peak and begins to decline again. However, the ballistic curve increases in typical exponential fashion before the carriers reach levels where they tunnel through the longitudinal states set up in the finite length of the quantum wire.

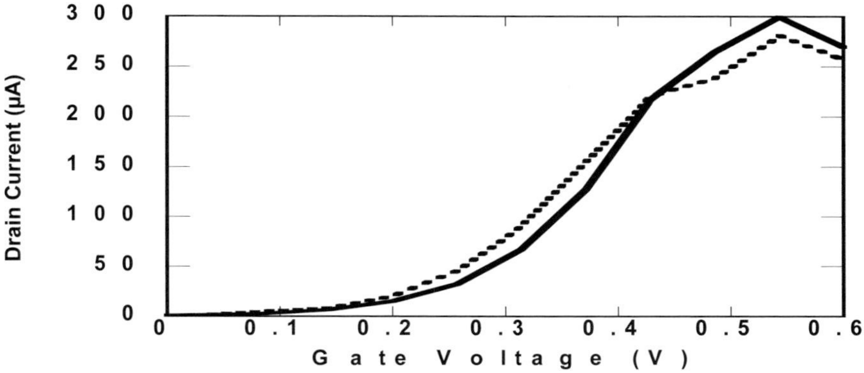

Fig. 1. Plot of the I_d-V_g curves for the ballistic (solid) and quasi-ballistic (dashed) for an identical doping distribution with the drain voltage set to be $V_d = 0.6$ V. Here we see that the inclusion of scattering allows electrons to obtain higher energies that in the ballistic case leading to increased current flow at lower gate biases.

In Fig. 2, we plot the phonon assisted ballistic to diffusive crossover length for two different drain voltages. In each device, the gate voltage of the channel has been fixed to $V_g = 0.5$ V to ensure that an inversion layer is present in the channel of the device. We find that at a drain voltage of $V_d = 10$ mV, the ballistic to diffusive crossover occurs when the channel reaches a length of approximately 1.42 nm (about 3 silicon atoms). This length is much smaller than previously thought which adds additional emphasis to the importance of scattering in these ultra small devices. Further, we find that when the drain voltage is raised to $V_d = 100$ mV, the ballistic to diffusive crossover occurs at approximately 0.87 nm, or just over 1 silicon atom. This decline in the ballistic crossover length is due, in part, to the increased scattering of higher energy carriers as the drain voltage is increased. However, the reduced ballistic length is also due to scattering in the source of the device. The key part of the device is the source-channel barrier. This barrier is not solely located in the channel of the device. Therefore, the carriers are undergoing some diffusive transport before they reach the channel. This undoubtedly effects the mean free path in the bulk material where the entire distribution function participates in the transport.

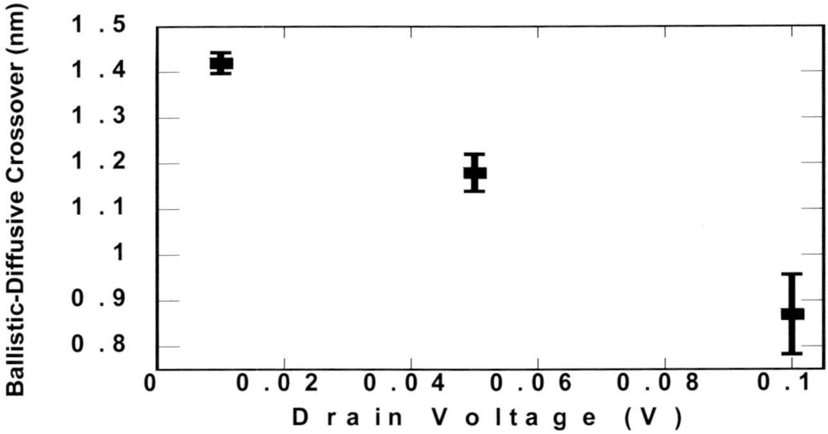

Fig.2. Ballistic to diffusive crossover length for three different drain voltages.

In this paper, we presented results of a simple physical approach to including separable scattering mechanisms as a self energy in real space. We find that the scattering due to phonons causes the electrons to scatter to/from higher subbands than previously available during simple ballistic transport. This modifies the output current by allowing electrons to reach and tunnel through longitudinal states set up in the finite length of the system which causes oscillations in the conductance of the system, and the output current. We also find that the phonon assisted ballistic to diffusive crossover is approximately 1.42 nm at low drain voltages and decreases further at higher drain voltages to 0.87 nm due to enhanced scattering of the higher energy particles.

References

1. M.J. Gilbert and D.K. Ferry, J. Appl. Phys **95**, 7954 (2004).
2. J. Wang, E. Polizzi, and M.S. Lundstrom, J. Appl. Phys. **96**, 2192 (2004).
3. M.J. Gilbert, R. Akis, and D.K. Ferry, J. Appl. Phys., Submitted for publication.

A 2D-NEGF Quantum Transport Study of Unintentional Charges in a Double Gate Nanotransistor

A. Martinez[1], J. R. Barker[1], A. Svizhenko[2], M. Bescond[1], M. P. Anantram[2] A. R. Brown[1] and A. Asenov[1]

[1]Dept of Electronics & Electrical Engineering, University of Glasgow, Glasgow G12 8LT, UK
[2]NASA Ames Research Center, Moffet Field, California, USA

Summary. The aim of this paper is to study the effect on a Double Gate (DG) MOSFET of unintentional dopants located in the channel region using a ballistic 2D self-consistent Non-equilibrium Green's function (NEGF) approach. The effect of mass anisotropy and the particular configuration of the unintentional dopant atoms on the I_D-V_G characteristic have been studied.

1 Introduction

Fabrication of CMOS transistors down to the nanometre scale is now possible [1]. At these dimensions quantum effects play an important role, introducing confinement effects and tunnelling. Unfortunately, intrinsic fluctuations in device characteristics, for example due to discrete dopants, do not average out on these scales. Therefore a self-consistent 2D Non-Equilibrium Green's Function (NEGF) real space approach [2] has been used here to study some of the effects associated with randomly situated charges in the channel of a double gate MOSFET. The subject has been studied in the literature using other methods, however rigorous quantum mechanical techniques were not applied [3].

2 Results and Conclusions

The simulated device is a Double Gate (DG) MOSFET with 1nm oxide thickness. The channel length of the transistor is 12 nm which justifies the use of a ballistic approach. The channel body thickness is 5nm, the doping in the source and drain regions, N_D, is 10^{20} cm^{-3}, and the background channel doping, N_A, is 10^{14}cm^{-3}. Four devices which differ in the arrangement of three point charges in the transistor channel have been considered in the calculation. We have used both multi-valley anisotropic (**A**) ([1 0 0] orientation) and isotropic (**I**) effective mass models and compared the results. Fig. 1 shows the potential for four different configurations of three dopants in the channel of the transistor at V_G=1.5 V. The source and drain are located at the left and right of every panel. The two gates are located at the top and the bottom of the panels.

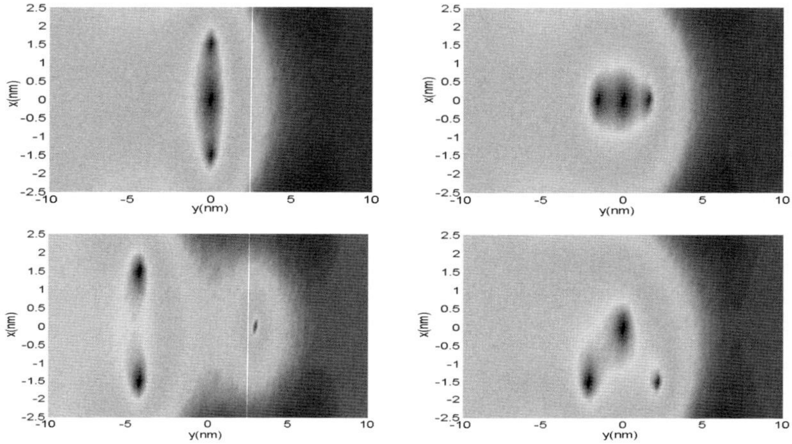

Fig. 1 The potential landscape (at V_G=1.5V) for the four arrangements of the three charges in the channel: upper left panel (vertical configuration *VC*), upper right panel (horizontal configuration *HC*), lower left panel (symmetric triangle configuration *ST*), and lower right panel (triangle configuration *TC*)

Fig. 2 shows the I_D-V_G characteristics of the four devices using isotropic and anisotropic mass models at V_D = 0.6 V. The threshold voltage for these devices with unintentional doping is over 1V higher than the base device with a dopant-free channel (hence the lateral scale in Fig.2), however the magnitude of this shift is overestimated due to the 2D nature of the simulations. The device with *VC* arrangement produces the lowest current at high bias because the channel is completely blocked by the potential barrier created by the impurities, however there is a substantial tunnelling current in the subthreshold. In the *HC* arrangement of dopants the blocking of the current is very effective at low gate voltages but at high bias screening of the impurities reduces the effect and the current flow is concentrated close to the gates. The blocking by *TC* is very effective at low gate voltages but as the gate voltage increases the potential barrier decreases substantially because the charges are mainly located in the middle of the channel. The *ST* has two point charges very close to the source that block the current effectively and are less sensitive to the gate voltage. Therefore the I_D-V_G curve for the symmetric triangle is crossed by the I_D-V_G curve for the case of the asymmetric triangle as can be seen in fig. 2.

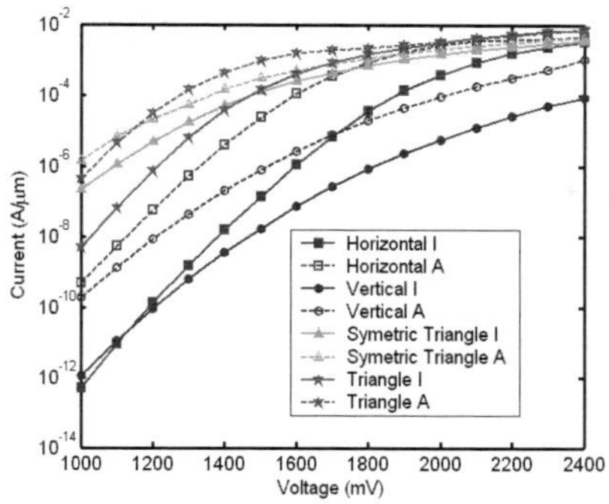

Fig. 2 Shows the I_D-V_G characteristics of the four devices (see Fig. 1) using the isotropic (I) and the anisotropic (A) mass models at V_D=0.6 V.

The current for the **A** mass models are larger (up to 2 orders of magnitude in the subthreshold) than for the **I** mass models due to a smaller threshold voltage shift associated with the large effective mass components in the confining direction and the increase of tunnelling due to small effective mass components in the transport direction. The *HC* displays the largest difference in currents between the two effective mass models because of the high degree of confinement. At high gate voltage the differences in the I_D-V_G curves between **I** and **A** models becomes smaller as the impurities are screened more and the confinement effects become less severe. For *ST* configuration Fig. 3 shows a comparison of the current density vector fields between the **I** and the **A** effective mass models. In the **A** mass model the current flow is more horizontal compared to the **I** mass model The current density $J_y(x,y)$ at low gate voltage (V_G=1.5 V) is illustrated in the right panel of Fig. 3 for the symmetric triangle configuration. It shows that the majority of the current flow is between the two foremost dopants followed by bifurcation of the flow around the farthest dopant. At high gate voltage (V_G=2.4 V) the current flow concentrates close to the gates in two parallel streams that avoid the dopants. The left panel of Fig. 4 shows the electron density in the channel. A charge density of approximately 10^{19} cm^{-3} is present in the dopant locations indicating electron tunnelling.

In conclusion, we have correlated and compared (qualitatively) in detail the I_D-V_G characteristic of DG MOSFET with different dopant configurations using NEGF technique. The threshold voltage and the sub-threshold I_D-V_G slope are strongly dependent on the particular arrangement of unintentional dopants. The differences between the **I** and **A** models (up to 2 orders of magnitude) are strongly dependent on the specific charge configurations. There is significant tunnelling through the potential barrier created by the unintentional doping charge.

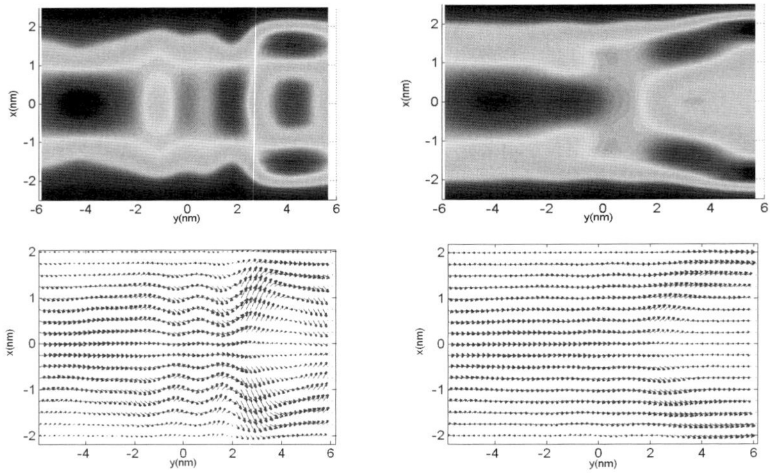

Fig. 3. The upper panels represent the current landscape (*ST* configuration at V_G=1.5 V) for **I** and **A** mass models respectively; the two lower panels show the current density vector field **J**.

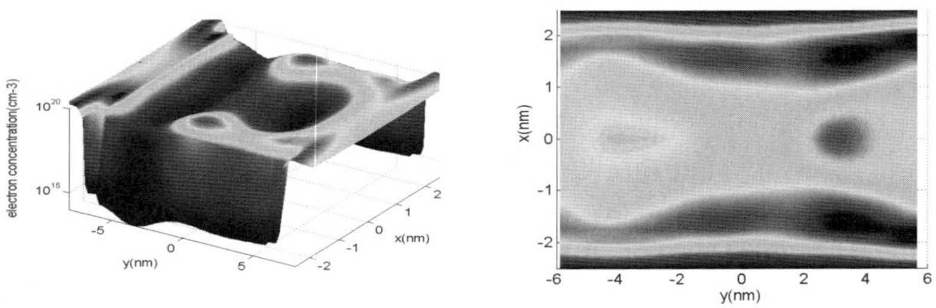

Fig. 4 The electron density and the current density profile J_y respectively for the *ST* device using the **A** model at V_G=2.4 V.

References

1. *IEDM Technical Digest*, 2003, 2004
2. A. Svizhenko *et al*, *J. App. Phys.* **91**, 2343, 2002
3. W. J. Gross *et al*, *J. App. Phys.* **91**, 3738, 2002, A. R. Brown *et al*, *IEEE Trans. Nanotechnology* **1**, 195, 2002

Wigner Function RTD Simulations with DMS Barriers

H. L. Grubin

Department of Physics, University of Connecticut

Summary. We present spin dependent Wigner function simulations of barrier devices containing dilute magnetic semiconductor (DMS) regions. Two situations are considered: the DMS layer transforms a single barrier structure into an effective two-barrier structure with NDC, and second the emitter and collector barriers both contain DMS layers. Transient behavior is also addressed.

1 Introduction

Interest in diluted magnetic semiconductors, whose lattice includes substitutional magnetic ions, e.g. Mn ions in II-VI or III-V materials, is due to the huge Zeeman splitting of conduction and valence band electrons, and the consequent transport properties of high 'g' factor devices. These include the observation of spin and magnetic field dependent IV characteristics of double barrier devices [1], suggesting the magnetic field can function as a 'pseudo' third terminal to two-terminal wired devices; the proposed spin superlattice [2]; a magnetic bipolar transistor [3]; a metal-ferromagnetic semiconductor Schottky diode [4].

Understanding these devices requires measurement, theoretical analysis and simulation. The transport equations receiving most attention are the spin drift and diffusion equations [5] (SDDE). But for structures containing a few DMS monolayers 'spin-heterostructures' form and the SDDE without quantum corrections [6] is inadequate. Further, for barrier devices, and when transients are of interest, quantum statistical transport, e.g., the transient spin dependent Wigner or spin density matrix equations are required. It is within this context that is paper is written. We present new self-consistent Wigner simulation results for two n-type barrier devices: (i) a double barrier RTD with emitter and collector DMS barriers, and (ii) a single barrier device with an embedded DMS layer that transforms spin up transport into double barrier transport. Transients are also discussed.

2 The Spin Density Matrix and the Spin Wigner Function

The spin Wigner functions can be obtained from [6], the density matrix [7], or other procedures. For the situation where the Zeeman [8] splitting lowers the spin up carrier energy, Schrödinger's equation, which governs the time dependence of the pure state Wigner function [9], is:

$$i\hbar\frac{\partial}{\partial t}\begin{pmatrix}\psi(\mathbf{x},+)\\ \psi(\mathbf{x},-)\end{pmatrix}=\left(-\frac{\hbar^2}{2m}\nabla^2+E_C(\mathbf{x},t)-\phi(\mathbf{x},t)\begin{pmatrix}1 & 0\\ 0 & -1\end{pmatrix}\right)\begin{pmatrix}\psi(\mathbf{x},+)\\ \psi(\mathbf{x},-)\end{pmatrix}$$

The potential energy includes the barrier and Poisson potential energies, $\phi(\mathbf{x})\equiv\mu_e g(\mathbf{x})B/2$, the vector potential contribution is ignored, as is the Landau energy level quantization, and the results are predicated on high 'g' factors coupled to small magnetic fields. The + (-) symbol designate spin up (down) states. Generalizing the pure state Wigner function to include mixed states; treating electron transport within a simple relaxation time approximation with each spin distribution relaxing to an equilibrium spin distribution; ignoring spin-flip coupling for this study but including spin coupling through Poisson's equation; the Wigner function equations, which are solved using procedures in [10] are:

$$\frac{\partial f_\pm(\mathbf{k},z,t)}{\partial t}+\frac{f_\pm(\mathbf{k},z,t)-f_{0,\pm}(\mathbf{k},z)}{\tau}+\frac{\hbar k_z}{m}\frac{\partial f_\pm(\mathbf{k},z,t)}{\partial z}=$$

$$\frac{1}{i\hbar\pi}\left(\int d\zeta dk_z'\exp[2i(k_z-k_z')\zeta]f_\pm\left(k_x,k_y,k_z',z,t\right)\begin{bmatrix}E_C(z-\zeta)-E_C(z+\zeta)\\ \mp(\phi(z-\zeta)-\phi(z+\zeta))\end{bmatrix}\right),$$

3 The Results

The structures studied were (i) 70nm long with 5nm 300 meV barriers separated by 5nm, and (ii) 100nm long with a single 15 nm 300 meV barrier, containing a 5nm DMS region. Both structures incorporated central $10^{22}/m^3$ regions surrounded by $10^{24}/m^3$ regions. GaAs parameters were used. For the double barrier DMS with modest, but identical Zeeman splitting in the barrier heights (the emitter and collector barriers were 20 meV smaller (larger) for the spin up (down) carriers), we observed unequal concentrations of spin up and spin down carriers in the quantum well, and modest quantitative alterations in the IV, figure 1 (left panel).

For the 100 nm single barrier device, figure 1 (second panel) with a centrally placed DMS layer, calculations were performed for a combination of 'g'-factor and magnetic fields such that the DMS region's barrier was reduced (spin up), increased (spin down) by 100meV, 200meV, 300meV, 310meV, and 400meV. The calculations with 300meV and 310meV showed double barrier spin up resonances; the others didn't.

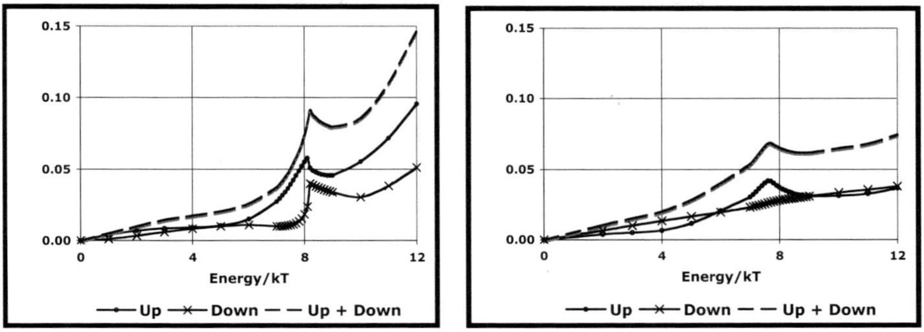

Fig. 1: IV for spin up carriers (circles) spin down carriers (crosses) and sum (dash line) for: (left panel) DBRTD with 20 meV modified barriers and (right panel) for a single barrier structure with a 5 nm embedded DMS layer with a 300 meV Zeeman split. Current is normalized to 10^{10}A/m^2.

The results of figure 1 (second panel) display resonance for the spin up carriers but none for the spin down carriers. The charge distribution prior to the current drop back for the spin up carriers displays the standard characteristics of double barrier resonance: as the bias increases there is a build up of charge in the quantum well as well as an accumulation of charge on the emitter side of the first barrier. The quantum well charge buildup continues as the bias increases. Resonance and the consequent drop in current are accompanied by a loss of charge in the quantum well and a corresponding increase in the accumulated charge on the emitter side of the first barrier (the 'notch' potential region). For the spin down carriers, there is no accumulation of charge under the barrier; there is charge build-up on the emitter side of the barrier.

NDC as displayed here is within steady state. For the single barrier structure we explored the NDC time dependence by attaching an oscillator circuit to the collector contact. The configuration studied was not conducive to relaxation oscillations. For these calculations several cycles of the calculation were recorded and there was a continual decrease in the amplitude of the oscillation. For a finite (zero) magnetic field and an oscillation period of 5 ps, the particle current displays (does not display) NDC. Weaker negative conductance was displayed at an oscillation period of 3 ps. We have not determined the shortest period at which negative conductance is sustained.

4 Conclusions

We have presented calculations indicating how a single barrier diode can be transformed into a resonant double barrier diode. Transient calculations indicate that NDC should persist for periods as low as 3 ps. The transition from single to double barrier transport is a indication that barrier diodes with DMS layers are capable of enhanced functionality.

5 Acknowledgements

This study was supported by ARO (D. L. Woolard) and ONR.

References

1. Gruber, T., et al., *Electron Spin Manipulation Using Semimagnetic Resonant Tunneling Diodes.* Appl. Phys. Letts., 2001. **78**: p. 1101.
2. Awschalom, D.D. and N. Samarth, *Spin dynamics and quantum transport in magnetic semiconductor quantum structures.* Journal of Magnetism and Magnetic Materials, 1999. **200**: p. 130-147.
3. Fabian, J., I. Zutic, and S.D. Sarma, *Magnetic bipolar transistor.* Applied Physics Letters, 2004. **84**(1): p. 85-87.
4. Lebedeva, N. and P. Kuivalainen, *Modeling of ferromagnetic semiconductor devices for spintronics.* Journal of Applied Physics, 2003. **93**(12): p. 9845-9864.
5. Zutic, I., J. Fabian, and S.D. Sarma, *Theory of spin-polarized transport in inhomogeneous magnetic semiconductors.* Phys. Rev. Letts, 2002. **88**: p. 066603.
6. O'Connell, R.F. and E.P. Wigner, *Manifestations of Bose and Fermi statistics on the quantum distribution function for systems of spin-0 and spin-1/2 particles.* Physical Review A, 1984. **30**(5): p. 2613-2618.
7. Grubin, H.L., *Density Matrix Simulations of Semiconductor Devices*, in *Quantum Transport in Ultrasmall Devices*, D.K. Ferry, et al., Editors. 1995, Plenum Press: New York. p. 241-280.
8. Erickson, R.P., K.B. Hathaway, and J.R. Cullen, *Mechanism for non-Heisenberg-exchange coupling between ferromagnetic layers.* Phys. Rev. B, 1993. **47**(5): p. 2626-2635.
9. Slonczewski, J.C., *Conductance and exchange coupling of two ferromagnets separated by a tunneling barrier.* Phys. Rev. B, 1989. **39**(10): p. 6995-7002.
10. Grubin, H.L. and R.C. Buggeln, *Wigner Function Simulations of Quantum Device-Circuit Interactions*, in *Terahertz Sensing Technology-Volume 2: Emerging Scientific Applications and Novel Device Concepts*, D.L. Woolard, W.R. Loerop, and M.S. Shur, Editors. 2003, World Scientific Publishing Co. Pte. Ltd.: Singapore. p. 353-383.

High Field Transport in GaN and AlGaN/GaN Heterojunction Field Effect Transistors

S. Yamakawa, J. Branlard,[1] M. Saraniti,[1] and S. M. Goodnick

Department of Electrical Engineering, Arizona State University, Tempe, AZ, 85287-5706, USA; [1]Department of Electrical and Computer Engineering, Illinois Institute of Technology, Chicago, IL, 60616-3793,USA

Summary. Here we report on high field transport in GaN and GaN field effect devices, based on the rigid-ion model of the electron-phonon interaction within the Cellular Monte Carlo (CMC) approach, including quantum-mechanical effects. The calculated velocity is compared with experimental data from pulsed I-V measurements, where good agreement with experiment is found. We have applied the CMC transport kernel above to the simulation of the DC and high frequency characteristics of GaN MESFETs and AlGaN/GaN HFET devices. Various effects are considered, such as thermal heating and nonequilibrium phonons, in comparing with the dc and high frequency behavior of these devices.

1 Introduction

The group III-nitrides have received considerable attention in recent years for short wavelength optoelectronic and high power, high frequency electronic applications. These materials are generally wide-band gap materials, and can crystallize in both wurtzite and zinc-blende polytypes, although the former is the dominant equilibrium structure [1]. The high field transport properties of these materials are still relatively less well understood compared to cubic semiconductors such as Si and GaAs. Recent experimental studies of high field transport have indicated peak velocities in excess of $2\text{-}3\times10^7$cm/s in bulk GaN [2] and AlGaN/GaN heterostructures [3].

There are a number of theoretical investigations into the transport properties of wurtzite phase GaN, in particular based on full-band Monte Carlo simulation [4]. While these studies were the first full-band simulations of GaN, they used electron-phonon scattering rates based on deformation potentials taken for other III-V zincblende materials in the absence of such information for the nitrides.

GaN/AlGaN heterostructure (or high mobility) field-effect transistors (HFETs or HEMTs) have been demonstrated [5], which have shown promising performance as microwave power devices. In this heterostructure system, there is a difference in spontaneous polarization between AlGaN and GaN, in addition to the strain due to the difference of lattice constants, giving rise to piezoelectric polarization. As a result, the discontinuity of the polarization at the interface results in a high density carrier without any intentional doping of the barrier layer. Typical sheet densities are on the order of $1\text{-}2\times10^{13}$ cm^{-2}, which is an order of magnitude higher than conventional AlGaAs/GaAs HFETs. An understanding of the limitations on the performance of such polarization induced channel devices due to transport and other non-equilibrium phenomena is critical to the improvement of this technology for high power high frequency applications.

In the present paper, we use a full-band Cellular Monte Carlo method to investigate high field transport in wurzite GaN using a microscopic approach to the electron-phonon interaction based on the rigid ion model, which is discussed in Section 2. We then combine this model with a multigrid field solver and an effective potential approach for quantum confinement to simulate the performance of AlGaN/GaN HFETs in Section 3. Finally, we look at the frequency response of such devices, and compare to experimental data.

2 Cellular Monte Carlo Model

The electron-phonon interaction is obtained via the rigid pseudo-ion model [6]. The rigid pseudo-ion model calculates deformation potentials from the electronic band, the atomic pseudopotential and the phonon dispersion relationship. This method gives full anisotropic deformation potentials with no arbitrary parameters. Here, the band structure of wurtzite phase GaN is calculated by the empirical pseudopotential method, with parameters taken from [3]. The full phonon dispersion is calculated using both a Keating potential and valence-shell model.

An anisotropic polar optical phonon scattering rate for wurtzite GaN is included in the simulation [7]. In wurtzite phase materials, both longitudinal optical (LO)-like and transverse optical (TO)-like phonons interact with the electrons. Moreover, the high frequency dielectric constants are different along the a and c axes, resulting in anisotropic scattering rates. Piezoelectric scattering are also known to be strong in nitride materials due to the lack of the inversion symmetry, which is included as well. Crystal dislocations are another important effect for GaN semiconductor devices.

Elastic scattering rate due to threading dislocations, as well as ionized impurity scattering, are included in the transport model.

A Cellular Monte Carlo (CMC) method is used to calculate the transport properties, in which the scattering rates for every initial and final state are tabulated over the Brillouin zone [8]. Figure 1a) shows the calculated velocity-field characteristics for bulk GaN at 300K, with various mechanisms included, compared with the experimental data of Barker *et al.* [2]. While the peak velocity is mainly determined by phonon scattering (deformation, piezo- and polar mode scattering), the low field region is primarily dominated by elastic mechanisms including ionized impurity and dislocations. Recently, Ardaravičius *et al.* [9] have shown the potential importance of nonequilibrium hot phonons in GaN high field transport. We have included hot phonons within the full band CMC using the same procedure, with the calculated results shown in Fig. 1b. As can be seen, at high carrier densities, a strong effect is predicted due to hot phonons on the velocity field characteristics at high fields.

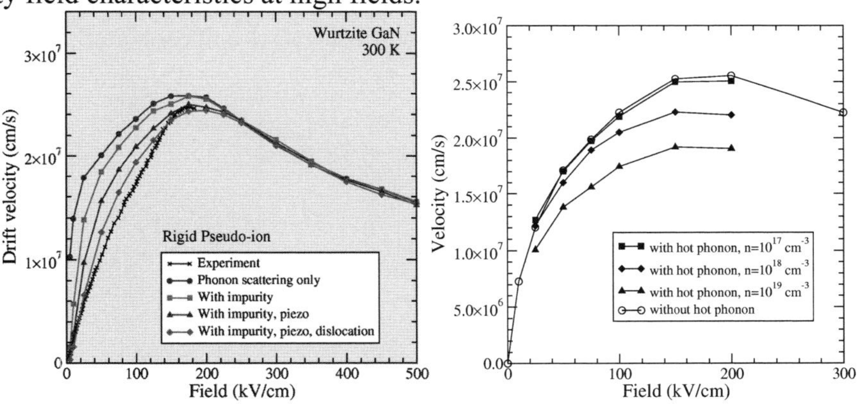

a) b)

Fig. 1. a) Calculated velocity-field characteristics including various scattering mechanisms based on the fully anisotropic rigid pseudo-ion model and b) effect of including nonequilibrium hot phonons for several different bulk densities.

3 AlGaN/GaN HFET Simulation

Figure 2a shows a schematic diagram of an AlGaN/GaN HFET device, simulated using the CMC model described in Section 2. This device corresponds to the experimental microwave power device reported by Lee *et al.* [10] grown on a SiC substrate. Due to the strong polarization fields at the heterointerface in this system, strong carrier confinement and quasi-two-dimensional effects are important.

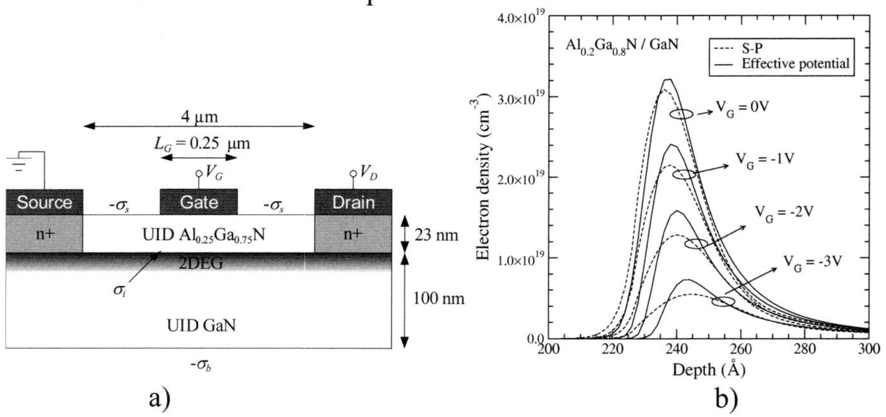

Fig. 2. a) Schematic diagram of the simulated HFET structure, b) Comparison of the interface charge density for various gate biases calculated self-consistently, and that calculated using the effective potential using a_0 as a fit parameter.

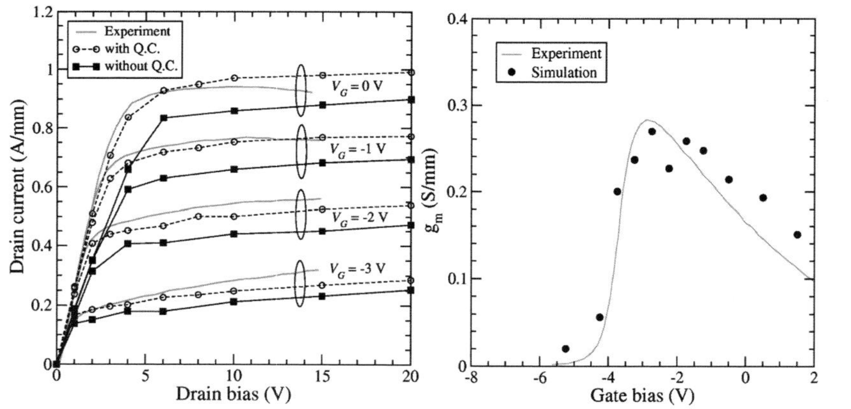

Fig. 3. a) Calculated I-V with and without the effective potential compared with experiment [10] b) Transconductance versus gate bias for V_d=10 V.

In the present approach, we account for quantum effects semi-classically using the so-called 'effective potential' approach [11], in which the effective potential, V_{eff}, is calculated as a convolution of a Hartree potential, V, obtained from Poisson's equation, and a Gaussian function. In principle, the half-width of the Gaussian, a_0, may be calculated from first principles. Here it is treated as a parameter to obtain the best least square fit between the charge density coming from the solution of the self-consistent 1D Schrödinger-Poisson equation, and the density using the effective potential. Figure 2b shows the comparison for several different gate voltages. Replacing the solution to the 2D Poisson equation with the effective potential at each time step, the current-voltage characteristics were simulated and compared with the experimental DC characteristics as shown in Fig. 3. The experimental curves show some evidence of self-heating at high source-drain bias due to power dissipation; however, thermal simulation shows that the temperature rise in the channel is relatively weak for this particular structure. As can be seen, good agreement is obtained when the effective potential model is included, both in the I-V and transconductance versus gate voltage.

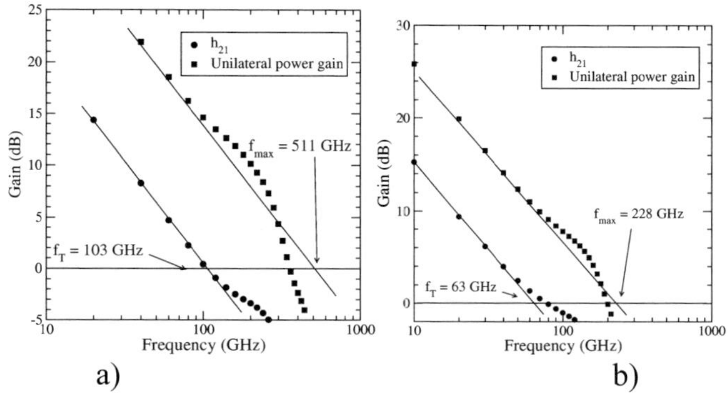

Fig. 4. Simulated short-circuit and unilateral power gain versus frequency a) without thermal or hot phonon effects b) with hot phonons in channel only.

We have simulated the frequency response of the device simulated in Fig. 3a using both Gaussian pulse and sinusoidal excitation in the CMC. The experimentally measured cutoff frequency, f_T = 44GHz, and maximum frequency of oscillation, f_{max}=84 GHz, are considerably lower than that simulated in Fig. 3a. f_{max} is dependent on several parasitic effects not included in the CMC simulation explicitly, particularly non-zero gate resistance at high frequency. However, the degradation of f_T is not well understood. We have attempted to account for the effective reduction of the channel velocity through inclusion of hot phonon effects at the interface. This effect is difficult to incorporate as it is not only strongly k-space dependent, but strongly spatially varying as well. Here we overestimate this effect by assuming one average phonon distribution over the entire channel, with the resulting frequency response shown in Fig. 3b, which is closer. However, inclusion of hot phonons also degrades the DC I-V characteristics at the same time, hence cannot provide a universal explanation for the discrepancy between predicted and measured cutoff frequency.

Acknowledgements

The authors acknowledge the support of the National Science Foundation Grant ECS-0115548 and the DoD HPCMP Program Environment Training program.

References

1. Pearton, S.J., Ren, F., Zhang, A.P. and Lee, K.P.: 'Fabrication and performance of GaN electronic devices', *Materials Science and Engineering* **R30**, 55-, 2000.
2. Barker, J.M., Akis, R., Ferry, D.K., Goodnick, S.M., Thornton, T.J., Koleske, D.D., Wickenden, A.E., and Henry, R.L.: 'High-field transport studies of GaN', *Physica B*. **314**, 39-41, 2002.
3. Barker, J.M., Ferry, D.K., Goodnick, S.M., Koleske, D.D., Wickenden, A.E., and Henry, R.L.: 'Measurements of the velocity-field characteristic in AlGaN/GaN heterostructures', *Microelectronic-Engineerin*, **63**, 193-197, 2002.
4. Oguzman, I.H., Kolník, J., Brennan, K.F., Wang, R., Fang, T., and Ruden, P.P.: 'Hole transport properties of bulk zinc-blende and wurtzite phase of GaN based on an ensemble Monte Carlo calculation including a full zone band structure', *J. Appl. Phys.*, **80**, 4429-4436, 1996.
5. Ibbetson, J.P., Fini, P.T., Ness, K.D., DenBaars, S.P., Speck, J.S., and Mishra, U.K.,: 'Polarization effects, surface states, and the source of electrons in AlGaN/GaN heterostructure field effect transistors', *Appl. Phys. Lett.* **77**, 250-252, 2000.
6. Zollner, S., Gopalan, S., and Cardona, M.: 'Microscopic theory of intervalley scattering in GaAs: k dependence of deformation potentials and scattering rates', *J. Appl. Phys*. **68**, 1682-1693, 1990.
7. Bulutay, C., Ridley, B.K., and Zakhleniuk, N.A.: 'Full-band polar optical phonon scattering analysis and negative differential conductivity in wurtzite GaN', *Phys. Rev. B*, **62**, 15754-15763, 2000.
8. Saraniti, M., and Goodnick, S.M.: 'Hybrid Fullband Cellular Automaton/Monte Carlo Approach for Fast Simulation of Charge Transport in Semiconductors, ' *IEEE Trans. Elec. Dev*,. **47**, 1909-1906, 2000.
9. Ardaravičius, L., Matulionis, A., Liberis, J., Kiprijanovic, O., Ramonas, M., Eastman, L.F., Shealy, J.R. and Vertiatchikh, A.: 'Electron drift velocity in AlGaN/GaN channel at high electric fields', *Appl. Phys. Lett.*, **83**, 4038-4040, 2003.
10. Lee, C. Saunier, P., Yang,, J., and Khan, M.A.: 'AlGaN-GaN HEMTs on SiC with CW power performance of > 4W/mm and 23% PAE at 35 GHz', *IEEE Electron Device Letters*, **24**, 616-618, 2003.
11. Ferry, D.K.: 'The onset of quantization in ultra-submicron semiconductor devices', *Superlatt. Microstruct.*, **27**, 61-66, 2000.

Impact Ionization and High-Field Electron Transport in GaN

A. Kuligk, N. Fitzer and R. Redmer

Institute of Physics, University of Rostock

Summary. *Ab initio* band structure calculations were performed for GaAs, GaN, and ZnS within density functional theory to determine the impact ionization rate and the high-field electron transport characteristics. The drift velocity, mean kinetic energy, valley populations, and the ionization coefficient are gained from full-band ensemble Monte Carlo simulations. A pronounced influence of the band structure is found for all materials. Results are shown here for GaN.

1 *Ab initio* Impact Ionization Rate

We have performed electron structure calculations for GaAs, GaN, and ZnS within density functional theory (DFT) by using an exact exchange formalism with a local density approximation for correlations (EXX-LDA) as outlined in Refs. 1-3. All calculations use norm-conserving Troullier-Martins-type EXX-LDA pseudo-potentials. The $3d$ electrons are part of the frozen core. Results for the band structure and density of states (DOS) for GaN are compared with the empirical pseudopotential method (EPM) in Fig. 1. The EXX-LDA schema reproduces the experimental band gaps and also yields good agreement with recent GW results based on EXX-LDA calculations in the optimized-effective potential approach (OEPx) [4] for the first conduction band. The OEPx method includes the d electrons as valence electrons. Pronounced differences between the EXX-LDA, EPM, and OEPx schemas are obtained for the location of the valleys and the higher conduction bands which will affect the high-field electron transport characteristics. The EXX-LDA and EPM band structures are used in calculations of the microscopic impact ionization rate and in full-band ensemble Monte Carlo simulations.
In the process of impact ionization, an energetic conduction band electron impact ionizes a valence band electron. We apply Fermi`s golden rule to

evaluate the impact ionization rate (IIR) and consider direct, exchange, and umklapp processes up to 6th order. The respective integrals are calculated on a grid in the Brillouin zone taking into account a total of 11901 points. The resulting IIR is strongly asymmetric in k-space; for details, see Ref. 3. We show here the energy-averaged IIR for GaN in Fig. 2.

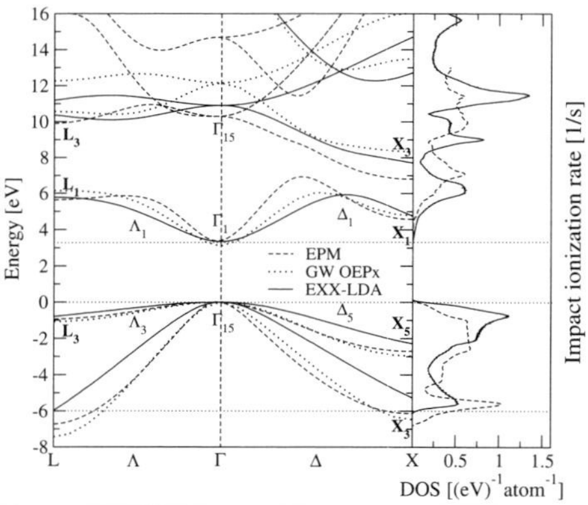

Fig. 1. Band structure in GaN: EXX-LDA and EPM [3] are compared with recent GW OEPx results [4].

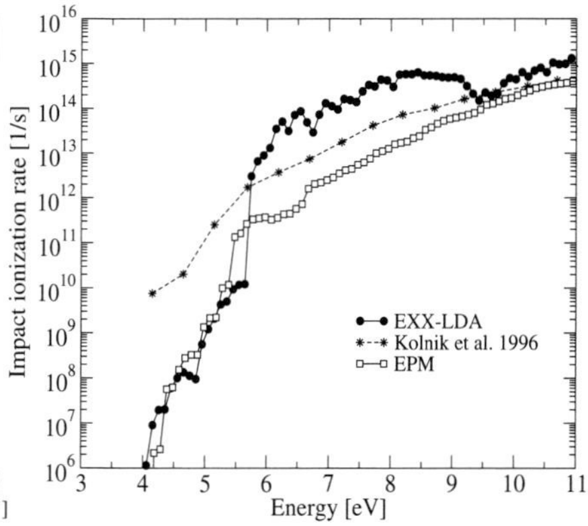

Fig. 2. IIR in GaN: EXX-LDA and EPM results [3] are compared with EPM calculations of Kolnik et al. [5].

For low energies near the threshold, the EPM and EXX-LDA results coincide. The steep increase of the EXX-LDA rate at about 5.75 eV stems from contributions of higher bands and the second maximum of the DOS in this region. The EXX-LDA rate is still higher than the EPM rate for higher energies which leads to an efficient `cooling` of hot electrons.

2 High-Field Transport and Ionization Coefficient

The high-field electron transport properties are studied by full-band ensemble Monte Carlo (EMC) simulations; for details, see Refs. 6-7. Again, discretization is performed in the Brillouin zone such that a total of 33861 points is considered in the simulations. The ensemble is represented by 64000 electrons in an external electric field. Scattering processes are treated by a Monte Carlo method at every time step; the average is performed over the steady final state at the end of the simulation. We take into account electron-phonon interactions (acoustic and optical nonpolar, polar optical, nonpolar intervalley) as well as inelastic electron-electron scattering (impact ionization), see above. The simulations provide electron distribution functions, drift velocities, average energies, and valley populations as function of the applied electric field strength; for details, see Ref. 8.

As an example, we show the drift velocity in GaN in Fig. 3. The EXX-LDA yields an effective mass which is about three times higher than that derived from the EPM. Furthermore, the neighbouring valleys at the Γ and X point are not distinctive as in the EPM. As a consequence, the drift velocities are smaller than the other data [9-11] shown in Fig. 3 and the population of the neighbouring valleys sets in only at high field strengths. This leads to a less pronounced maximum of the drift velocity which is also shifted to higher field strengths.

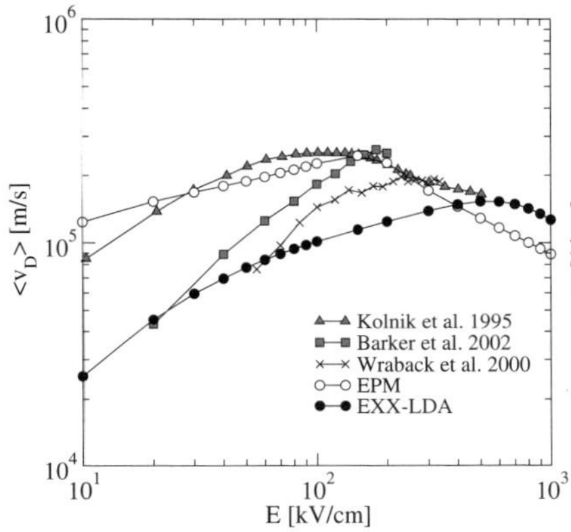

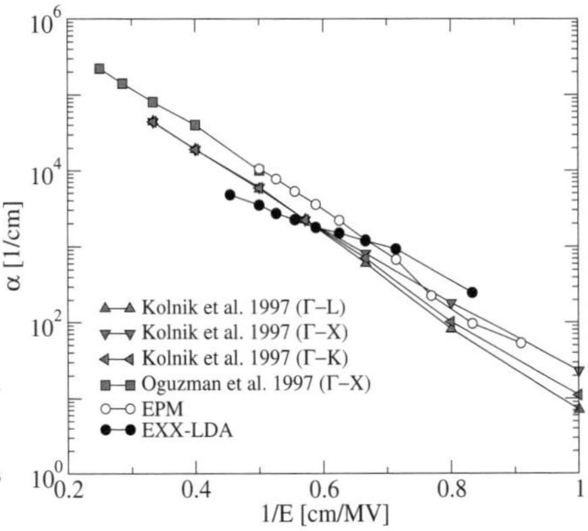

Fig. 3. Drift velocity in GaN [8]: EXX-LDA and EPM results are compared with EPM calculations of Kolnik et al. [9] and experimental data of Barker et al. [10] and Wraback et al. [11].

Fig. 4. Ionization coefficient in GaN [8]: EXX-LDA and EPM results are compared with EPM calculations of Kolnik et al. [12] and Oguzman et al. [13] along highly symmetric lines.

The calculated IIR and the electron distribution function derived from the full-band EMC simulations are used to determine the ionization coefficient α dependent on the applied field strength. Results are shown for GaN in Fig. 4. The EXX-LDA results show a slightly smaller increase with the field than the EPM-based calculations which were performed along highly symmetric lines [12-13]. Although the microscopic IIR in EXX-LDA is higher than the EPM rate, the high-energy tail of the electron distribution function (which is relevant in calculating α) is depopulated accordingly, so that the respective ionization coefficient is smaller than the EPM result for high field strengths.

We conclude that details of the band structure, especially the effective masses and the location of the neighbouring valleys and of the higher bands, are crucial for the IIR, the drift velocity, and the ionization coefficient in wide band gap semiconductors. Results for GaAs and ZnS are given in [8]. The influence of the *d* electrons which are part of the frozen core in the present calculations has to be studied in more detail. GW calculations within the LDA [14] show a lowering of the gap energies when including the *d* electrons as valence electrons. The same behaviour is also found in GW calculations within the OEPx schema [4], see also Fig. 1.

References

1. Städele, M., Majewski, J.A., Vogl, P., Görling, A.: `Exact Kohn-Sham Potential in Semiconductors`, *Phys. Rev. Lett.* **79**, 2089-2092, 1997.
2. Städele, M., et al.: `Exact exchange Kohn-Sham formalism applied to semiconductors`, *Phys. Rev. B* **59**, 10031-10043, 1999.
3. Kuligk, A., Fitzer, N., Redmer, R.: `Ab initio impact ionization rate in GaAs, GaN, and ZnS`, *Phys. Rev. B* **71**, 085201, 2005.
4. Rinke, P., et al.: `Combining GW calculations with exact-exchange density-functional theory`, *New J. Phys.* **7**, 126, 2005.
5. Kolnik, J., et al.: `Calculation of the wave-vector dependent interband impact-ionization rate in wurtzite and zinc-blende phases of bulk GaN`, *J. Appl. Phys.* **79**, 8838-8840, 1996.
6. Reigrotzki, M., et al.: `Impact ionization rate and high-field transport in ZnS with nonlocal band structure`, *J. Appl. Phys.* **80**, 5054-5060, 1996.
7. Dür, M., et al.: `High-field transport and electroluminescence in ZnS phosphor layers`, *J. Appl. Phys.* **83**, 3176-3185, 1998.
8. Kuligk, A.: `Stoßionisation und Hochfeldtransport in GaAs, GaN und ZnS`, PhD Thesis, University of Rostock, 2005.
9. Kolnik J., et al.: `Electronic transport studies of bulk zincblende and wurtzite phases of GaN`, *J. Appl. Phys.* **78**, 1033-1038, 1995.
10. Barker, J.M., et al.: `High field transport studies of GaN`, *phys. stat. sol. (a)* **190**, 263-270, 2002.
11. Wraback, M., et al.: `Time-resolved electroabsorption measurement of the electron velocity-field characteristic in GaN`, *Appl. Phys. Lett.* **76**, 1155-1157, 2000.
12. Kolnik J., et al.: `Monte Carlo calculation of electron initiated impact ionization in bulk zincblende and wurtzite GaN`, *J. Appl. Phys.* **81**, 726-733, 1997.
13. Oguzman, I.H., et al.: `Theory of hole initiated impact ionization in bulk zincblende and wurtzite GaN`, *J. Appl. Phys.* **81**, 7827-7834, 1997.
14. Rohlfing, M., Krüger, P., Pollmann, J.: `Role of semicore *d* electrons in quasiparticle band-structure calculations`, *Phys. Rev. B* **57**, 6485-6492, 1998.

Studies of High Field Transport in a High-Quality InN Film by Ultrafast Raman Spectroscopy

K. T. Tsen[1], D. K. Ferry[2], H. Lu[3] and W. J. Schaff[3]

[1]Department of Physics and Astronomy, Arizona State University, Tempe, AZ 85287
[2]Department of Electrical Engineering, Arizona State University, Tempe, AZ 85287
[3]Department of Electrical and Computer Engineering, Cornell University, Ithaca, NY 14853

Gallium nitride (GaN), aluminum nitride, indium nitride (InN), and their alloys have long been considered as promising materials for device applications. Recently, growth of high quality InN as well as $In_xGa_{1-x}N$ have been demonstrated. In particular, progress in the manufacturing of very high quality, single-crystal InN thin films has opened up a new challenging research avenue in the III-nitride semiconductors. In contrast to earlier beliefs, it has recently been found that InN has a relatively narrow bandgap, only ~0.8 eV. Consequently, it is expected that InN has the smallest effective mass of the III-N semiconductors. As a result, very high electron mobility and a very large saturation velocity are expected. Recent single-particle Raman scattering, supported by ensemble Monte Carlo simulations suggest that steady velocities of the order of 5×10^7 cm/s can be found in high quality, single crystal wurtzite films of InN [1]. Here, we report on these calculations for the transport and properties of the non-equilibrium longitudinal optical phonons. We use a high quality, single-crystal wurtzite InN film grown on GaN and study the transport with picosecond/subpicosecond Raman spectroscopy. The built-in polarization and piezoelectric stress lead to an electric field of ~80 kV/cm in the sample, which is oriented in the growth direction. From the Raman data [1], we can determine not only the average velocities (the drift velocity), but also the distribution function of the carriers along the field direction. Individual electrons with velocities up to 2×10^8 cm/s and a transient electron drift velocity as high as $(7.5\pm0.5) \times 10^7$ cm/s have been observed at T = 300 K. As mentioned, these results are compared with ensemble Monte Carlo (EMC) simulations and reasonable agreement is obtained. While details of the exact conduction band structure and the intervalley coupling constants are not known, we can vary these to determine the sensitivity of the results to these factors. From these transient Raman studies, we have found that InN should provide considerably higher velocities, including overshoot velocity effects, than the comparable values in InGaAs. In addition, our experimental results on non-equilibrium longitudinal optical phonons strongly support a bandgap of around 0.8 eV in InN [2]. This also suggests that 0.8 eV-luminescence recently observed recently is due to a direct bandgap emission.

The sample studied in this work is a thick InN film grown on HVPE GaN template by a conventional MBE technique [3]. The excitation laser source is a DCM dye laser which is synchronously pumped by the output of the second harmonic of a cw mode-locked YAlG laser. The single-particle scattering (SPS) spectra were taken in the $Z(X,Y)\overline{Z}$ scattering configuration where $X = (100)$, $Y = (010)$, $Z = (001)$ so that only the SPS spectra associated with spin-density fluctuations were detected. The details of the sample growth and the measurement techniques are given in [4]. Raman pulse half-widths of 0.6 ps and 10 ps were used to compare the fast non-equilibrium behavior with the slower "steady-state" behavior.

To compare with the experiment, we have analyzed the velocity and the z-axis distribution function with an ensemble Monte Carlo process. Here, we use a multi-valley formalism for the conduction band [5], similar to that used in other semiconductors [6]. Unfortunately, we do not know many of the important parameters of InN, which complicates the analysis considerably. Nevertheless, we can say some things about the likely structure of the conduction band. We use the latest estimate of the effective mass of $0.045m_0$ and a band gap of 0.75 eV, although the computed results are not sensitive to this latter parameter. The carriers are optically injected into the conduction band from a photon source of 1.92 eV, and then evolve under the z-axis electric field of 80 kV/cm. One primary result of the calculations is the z-axis velocity distribution function, which measures the properties of the carriers in the Γ valley. Because of this latter point, we find that these results are not sensitive to the separation between the Γ valley and the subsidiary valleys of the conduction band. We take a value of 2.5 eV for this parameter.

In Fig. 1, we compare the measured z-velocity distribution function (VDF) with that calculated from the Monte Carlo process for two different values of the nonparabolicity factor α. Here, we use 0.4 eV^{-1} and 0.5 eV^{-1}, which are thought to be the most likely value [1]. There is little difference between the two, although the smaller value of α leads to a higher velocity in the VDF. In both cases, the main peak lies higher than that seen experimentally. On the other hand, the estimated average velocities of the carriers in the Γ valley compare well with those determined from the experiment, as is shown in Table I.

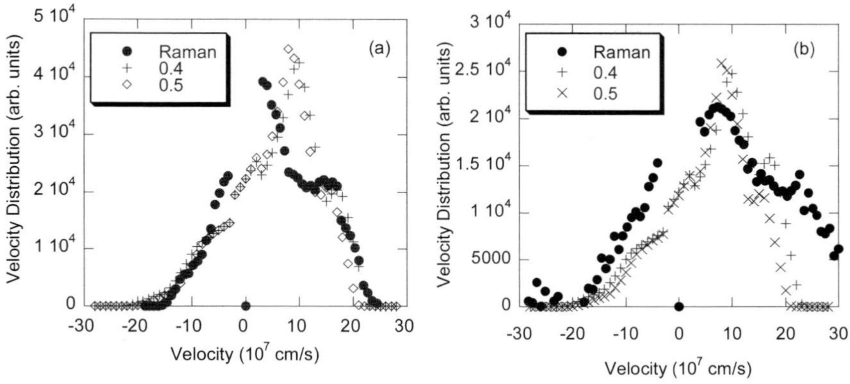

Fig. 1. Comparison of the calculated velocity distribution function with the experiment for the 10 ps pulse (a) and the 0.6 ps pulse (b).

Table I. Estimated Γ-Valley Velocities

Measure	10 ps pulse	0.6 ps pulse
Raman data	$5.1 \pm 0.8 \times 10^7$ cm/s	$7.5 \pm 0.5 \times 10^7$ cm/s
EMC ($\alpha = 0.4$)	$6.13 \pm 0.23 \times 10^7$ cm/s	$7.07 \pm 0.79 \times 10^7$ cm/s
EMC ($\alpha = 0.5$)	$5.26 \pm 0.26 \times 10^7$ cm/s	$6.06 \pm 0.62 \times 10^7$ cm/s

From the above data, it seems clear that value of a = 0.4 seems to give better agreement, particularly with the short pulse data. The uncertainty in the long pulse data could support either estimate for the velocity. In either case, it is clear that very high velocities can be obtained by the carriers in the Γ valley of InN, which suggests that this material is a good candidate for high frequency applications.

The difference between the 0.6 ps and 10 ps pulse data lies in the overshoot velocity that the carriers, *within* the Γ valley, experience in the high electric field. This occurs even with the dominant E_1 and A_1 polar mode intravalley scattering. But, for this to be seen in the data, this scattering cannot be too efficient. We also note that there are significant numbers of carriers above 2×10^8 cm/s in both distributions. Carriers, whose velocities lie in the range of 2-3 $\times 10^8$ cm/s, have energies of 0.6–1.6 eV for $\alpha = 0.4$ eV^{-1}, but the upper velocities are unobtainable for larger values of this parameter. This, of course, assumes that the **k•p**-derived hyperbolic band model applies throughout the energy range, which is unlikely. On the other hand, were an inflection point in the energy band present (as predicted for GaN [7]), above which the velocity began to decrease with increasing energy, these high velocities also would not be observable, since this represents a dramatically increasing nonparabolicity parameter. The very high velocities seen in the short pulse Raman data remain a poorly understood phenomena. On the other hand, the fact that the main peak is calculated to be too high in the long pulse data is likely due to the neglect of the intravalley deformation potential scattering due to the B_1 and E_2 phonons, as these are normally thought to be much weaker than the polar mode scattering. In addition, we have neglected the carrier-carrier scattering which broadens the distributions.

Finally, we note that varying either the position of the satellite valleys or the coupling to the satellite valleys tends to change the amplitude of the velocity distributions, due to changes in the number of carriers in the Γ valley, but does not change the shape of the distribution. Hence, these measurements cannot shed any light on these parameters.

The work at ASU is supported by the National Science Foundation (DMR-0305147) and the Office of Naval Research. The work at Cornell is supported by the Office of Naval Research and the Lawrence Berkeley Laboratory.

References

1. Tsen, K. T. *et al.*: 'Observation of large electron drift velocities in InN by ultrafast Raman spectroscopy', *Appl. Phys. Lett.* **86**, 222103, 2005.
2. W. Liang et al.: 'Observation of nonequilibrium longitudinal optical phonons in InN and its implications', *Appl. Phys. Lett.* **84**, 3849, 2005.
3. Lu H *et al.*: 'Growth of thick InN by molecular beam epitaxy', *MRS Proc.*, **743**, L4.10, 200
4. Tsen, K. T. *et al.*: 'Optical studies of high-field carrier transport of InN thick film grown on GaN', *J. Cryst. Growth, in press.*
5. Bulatay, C., and Ridley, B. K.: 'Theoretical assessment of electronic transport in InN', *Superlatt. Microstruc.*, **36**, 465-471, 2004
6. Ferry, D. K.: *Transport in Semiconductors*, Taylor and Francis, 2000.
7. Bulutay, C., *et al.*: 'Full-band polar optical phonon scattering analysis and negative differential conductivity in wurtzite GaN', *Phys. Rev. B*, **62**, 15754-63, 2000.

Monte Carlo Investigation of Dynamic Transport in Nitrides

L. Reggiani (1), P. Shiktorov (2), E. Starikov (2), V. Gruzinskis (2), L. Varani (3), J.C. Vaissiere (3), J.P. Nougier (3)

(1) Dipartimento di Ingegneria dell' Innovazione, CNR-INFM National Nanotechnology Laboratory, Università di Lecce, Via Arnesano s/n, 73100 Lecce, Italy; (2) Semiconductor Physics Institute, A. Gostauto 11, LT 01108 Vilnius, Lithuania; (3) CEM2 - Centre d'Electronique et de Micro-optoelectronique de Montpellier (CNRS UMR 5507), Universite Montpellier II, 34095 Montpellier Cedex 5, France

Summary. It is shown that in nitrides the instabilities based on the transit-time resonance inside the optical-phonon sphere of momentum space as well as the intervalley transfer of electrons can be realized in the TeraHertz frequency region. In both cases the maximum frequency of the instability is proportional to the effective rate of polar optical phonon emission.

1 Introduction

Various applications of nitride-based devices in micro- and optoelectronics [1] demand evaluations of high-frequency (HF) behavior of hot carrier dynamics and transport. From the theoretical point of view, this can be done by considering the spectral behavior of the linear response (e.g. differential mobility of carriers, admittance and impedance of devices) and velocity/current/voltage noise in a wide frequency range. The aim of this work is to present Monte Carlo (MC) simulations of HF behavior of hot carriers in nitride-based bulk materials and structures favorable for TeraHertz (THz) applications. Since nitrides belong to the group of many-valley polar semiconductors, the main features of their HF behavior are determined by the effect of electron heating in applied electric fields.

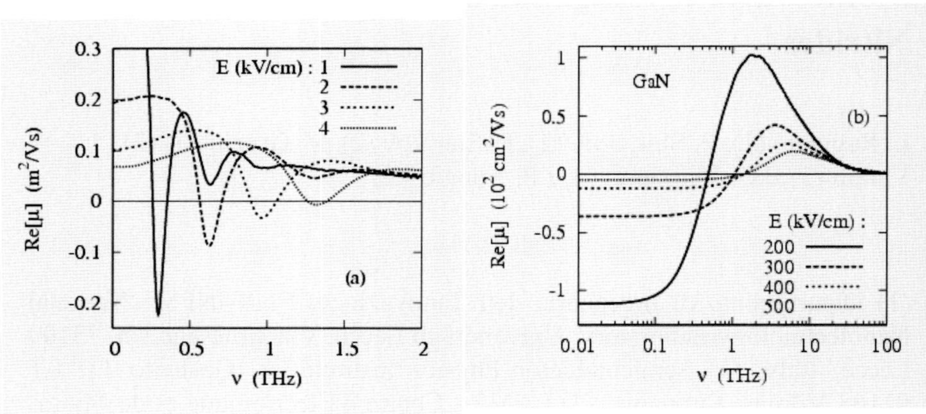

Fig. 1. Spectra of real part of the differential mobility as obtained by Monte Carlo simulations of: (a) optical-phonon transit-time resonance in InN at T=10 K, and (b) Ridley-Watkins-Hilsum-Gunn effect in GaN at T=300 K.

These effects can be subdivided into two main regimes: (i) heating controlled by polar optical phonon (POP) emission, and (ii) onset of intervalley transfer due to electron runaway. As shown below, both the regimes can be favorable for THz generation. In the former case, the HF instability can be realized due to the repeating ballistic-like motion of electrons in the passive area of momentum space where POP emission is absent. Such a dynamics leads to the appearance of negative differential mobility (NDM) in the restricted frequency range near the so called optical-phonon transit-time resonance (OPTTR) frequency [2]. In the latter case, electron transfer to upper valleys and associated NDM (the so called Gunn-instability) take place.

2 Monte Carlo Simulations

Numerical simulations are performed by the single-particle MC method as reported in [3]. All the models used here assume a conduction band consisting of three types of spherical, symmetric, and non-parabolic valleys. Furthermore, ionized impurity, polar optical phonon, piezoelectric and deformation acoustic phonon scatterings in each valley as well as all inter-valley transitions are taken into account. To compare the above regimes of instability, Figs. 1 and 2 show, respectively, the real part of the

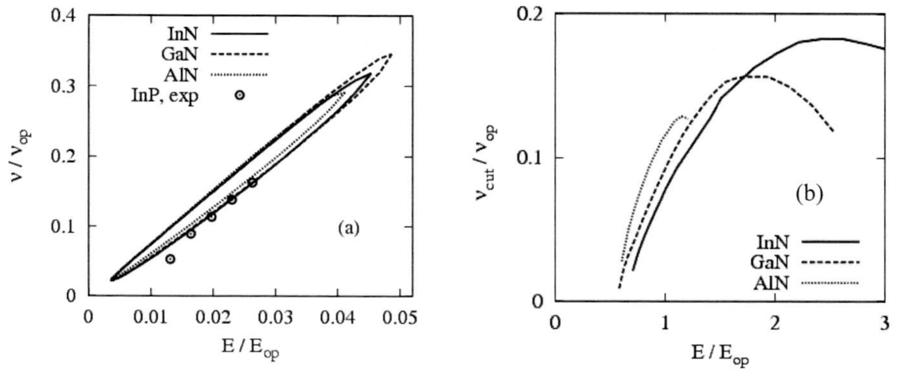

Fig.2. (a) Amplification band (area of dynamic NDM) due to OPTTR in bulk nitrides (T=10 K, n=10^{16} cm^{-3}), and (b) cutoff frequency of static NDM due to intervalley transfer (T=300 K, n=10^{17} cm^{-3}). The electric field and frequency are normalized to characteristic parameters of the POP emission: E_{op}= 99, 257 and 826 kV/cm and ν_{op}= 4.71, 8.96 and 17.86 THz for, respectively, for InN, GaN, and AlN. For comparison points show results of experimental observation of maser generation in InP at 10 K [4] (E_{op}=15.2 kV/cm, ν_{op} =1.23 THz).

differential mobility spectrum and the characteristic frequencies of the instability calculated for bulk nitrides. Figure 1 illustrates the difference between the dynamic and static NDMs, when, respectively, negative values of the differential mobility appear only near a separated frequency (OPTTR frequency in Fig. 1 (a)) or in the whole low-frequency region (Gunn instability, Fig. 1 (b)). The normalization of the applied electric field and frequency in Fig. 2 to the characteristic parameters of the POP emission, E_{op} and ν_{op} [4], is used to emphasize the similarity of these effects in nitrides. Here, the characteristic frequency ν_{op} is obtained from the POP emission rate $\lambda(\varepsilon)$ expressed as:

$$\lambda(\varepsilon) = \nu_{op}\sqrt{\varepsilon / \hbar\omega_{op} - 1} \quad (1)$$

where $\hbar\,\omega_{op}$ is the optical phonon energy and ε the carrier kinetic energy, and the field $E_{op}=\nu_{op}\, p_{op}/e$ is determined as the effective electric field in which an electron starting from the energy $\varepsilon = 0$ reaches the passive area boundary with momentum, $p_{op}=(2m^*\hbar\,\omega_{op})^{1/2}$, m^* being the electron effective mass, during the time ν_{op}^{-1}. The degree of maximum similarity is achieved in the OPTTR case, since in this case the characteristic features are determined by the POP emission parameters only. For the NDM associated with the Gunn-effect, a qualitative agreement among different nitrides still exists. In particular, the threshold field for the onset of the static NDM, $E_{th} \cong E_{op}/2$, is determined by the electron runaway from the POP emission. However, due to extra parameters related with the upper valleys, the difference between curves associated with different nitrides becomes more pronounced. As follows from Fig. 2, in nitrides a general rule can be formulated as following: for both instabilities the higher the ν_{op} the higher the generation frequencies (that is AlN for the case of 3D-bulk materials). It is worthwhile to underline, that in the case of the OPTTR, the maximum generation frequency can be increased further (up to five times with respect to the 3D case) by using the advantages of lateral transport in 2D nitrides-based structures (see, e.g., [6,7]). Another possibility is related with the vertical transport in n^+nn^+-like structures where the self-consistent electric field can improve the realization of instabilities. Such a situation has been found to occur in submicron and overmicron n^+nn^+ InN structures [8-10]. In conclusion, nitrides are proven to be promising materials to obtain devices working in the THz frequency range because of the high energy and large electron coupling of their polar optical phonons. In particular, both low and high temperature conditions can be favorably exploited to this extent.

Acknowledgment

This work is supported by NATO grant PST.EAP.CLG 980629. and the project "Noise models and measurements in nanostructures" of MIUR.

References

1. Foutz B.E. et al., J. Appl. Phys. **85**, 7727 1999.
2. Starikov E. et al., J. Phys. C : Condens. Matter **13**, 7159 2001.
3. Starikov E. et al., J. Appl. Phys. **89**, 1161 2001.
4. Sokolov V.N. et al., Appl. Phys. Lett., **84**, 3630 2004.
5. Vorob'ev L.E. et al., Pis'ma , ZETF **73** 253 2001.
6. Kim K.W. et al., J. Appl. Phys. **96,** 6488 2004.
7. Lu J.T.and Cao J.C., Semicond. Sci. Technol. **20**, 829 2005.
8. Bannov N. et al., Solid State Electron. **29**, 1207 1986.
9. Starikov E. et al., phys. stat. sol. (a), **190**, 287 2002.
10. Gruzinskis V. et al., Semicond. Sci. Technol. **19**, S173 2004.

High-Field Transport in Nitride Channels: a Hot-Phonon Bottleneck

A. Matulionis[1], L. F. Eastman[2], J. Liberis[1]

[1]Semiconductor Physics Institute, A.Goštauto 11, Vilnius, 01108 Lithuania
[2]Cornell University, 425 Philips Hall, Ithaca, NY 14853, USA

Summary. Experiments on channels with a high-density electron gas illustrate possibilities to control hot-phonon effect through variation of channel composition, growth conditions, and electron density.

1 Introduction

Two-dimensional electron gas (2DEG) channels support excellent high power performance of GaN HEMTs at microwave frequencies [1–3]. After a suitable deembedding procedure, the transistor cutoff frequency yields the electron drift velocity [1,4]. The resultant values at high electric fields ($1.3 \cdot 10^7$ cm/s [1], $1.5 \cdot 10^7$ cm/s [3], $1.2 \cdot 10^7$ cm/s [4]) are below the results of either traditional Monte Carlo simulation [5] or time-of-flight experiments [6]. Other techniques yield the velocities ranging from $1 \cdot 10^7$ cm/s [7] to $3 \cdot 10^7$ cm/s [8]. When channel self-heating is avoided [7], the non-equilibrium optical phonons (termed hot phonons) are expected to limit the drift velocity in a high-density 2DEG channel [9–11]. The paper considers how to reduce the excess friction caused by hot phonons.

2 Dependence on Electron Density

The dependence on electron density of the hot-phonon-induced friction has been calculated for electrons interacting with LO phonons in bulk GaN [10]. The saturated velocity increases as the electron density decreases. A similar dependence is obtained experimentally: the transistor cutoff frequency (and the drift velocity) increase when the electron density decreases as the channel pinchoff is approached [4]. However, the calculated velocity reaches $1.2 \cdot 10^7$ cm/s at $5 \cdot 10^{17}$ cm^{-3} [10] while the estimated electron densities are much higher in 2DEG channels (~10^{19} cm^{-3}). The misfit can be excluded if a shorter hot-phonon lifetime is assumed.

3 LO-Phonon Lifetime in a 2DEG Channel

Pump-probe Raman scattering is the standard way to measure hot-phonon lifetime [12]. The lifetime is ~3 ps in GaN at room temperature [13]. Since the Raman scattering has never been used to measure the lifetime in a 2DEG channel, the technique based on analysis of microwave noise is proposed [9]. Figure 1 shows that the lifetimes in 2DEG channels [9,14] are shorter than those in bulk samples (circles, diamond, star) [12,13,15].

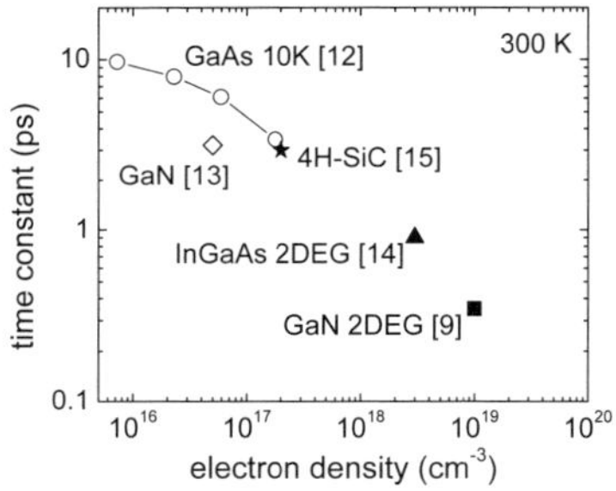

Fig. 1. Hot-phonon lifetimes in bulk and 2DEG samples: Raman [12,13] (open symbols) and noise [9,14,15] (closed symbols) data.

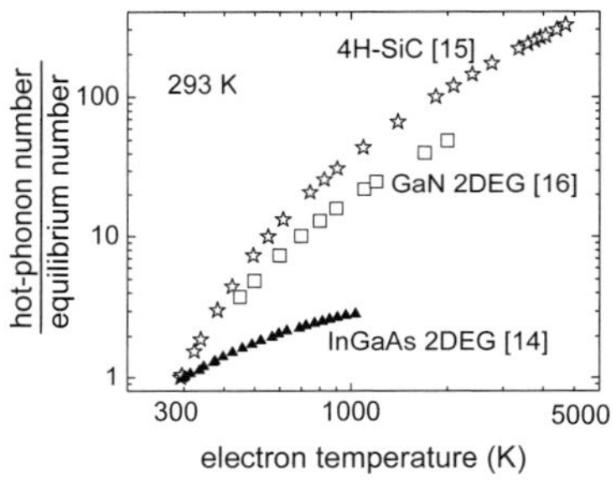

Fig. 2. Relative occupancy N_{eff}/N_0 of the involved hot-phonon states as a function of electron temperature for bulk and 2DEG channels [14–16].

4 Dependence on Channel Composition

The hot-phonon effect depends on the ratio N_{eff}/N_0 (N_0 is the equilibrium occupancy). The microwave noise experiments allow one to estimate the effective occupancy N_{eff} of the phonon states involved into the electron–LO-phonon interaction. The ratio N_{eff}/N_0 for AlInAs/GaInAs/AlGaAs [14], 4H-SiC [15], and AlGaN/AlN/GaN [16] is under comparison in Fig. 2. At a fixed electron temperature, say at $T_e = 1000$ K, the ratio N_{eff}/N_0 is large in SiC and small in InGaAs: thus, lighter channels demonstrate a stronger hot-phonon effect. An approximately linear function holds for the dependence of N_{eff}/N_0 on the reciprocal mean mass of atoms in the channel.

5 Dependence on Growth Conditions

Figure 3 shows the hot-phonon lifetime estimated from the noise experiments [17] on nominally "identical" heterostructures grown and processed at Cornell University. The shortest lifetime is obtained for the MBE-grown channel (squares). Triangles stand for different samples cut form the same MOCVD-grown wafer. Neighbor samples show similar lifetimes (up and down triangles), remote samples show different lifetimes (open and closed triangles). While the low-field current coincides for different samples (Fig.4), the high-field current is higher if the lifetime is shorter [18].

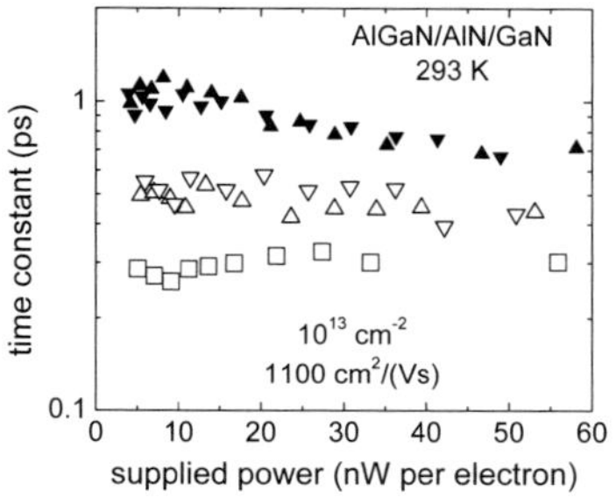

Fig. 3. Hot-phonon lifetimes deduced from microwave noise data [17].

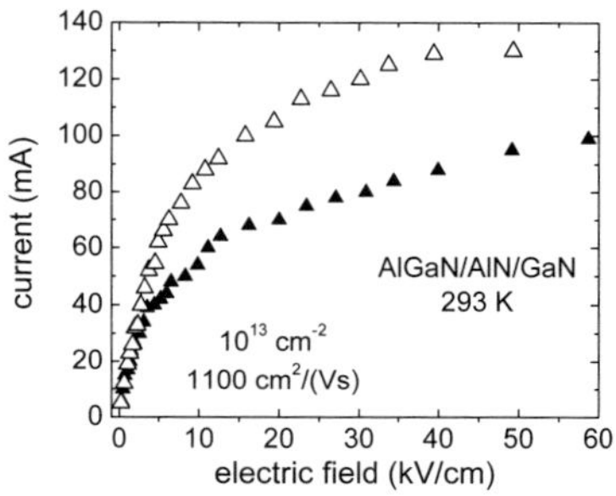

Fig. 4. Current–field dependence for the remote MOCVD-grown samples.

Acknowledgements. Support from the USA Office of Naval Research (Grant N00014-03-1-0558 monitored by Dr. Colin E.C. Wood and Grant N00014-05-1-4061) and the Lithuanian National Foundation for Science and Education (LVMSF Contract T-33/2005) is gratefully acknowledged.

References

1. Eastman, L.F., Tilak, V., Smart, J., Green, B.M., Chumbes, E.M., Dimitrov, R., Kim, H., Ambacher, O.S., Weimann, N., Prunty, T., Murphy, M., Schaff, W.J., and Shealy, J.R., 'Undoped AlGaN-GaN HEMTs for microwave power applications', *IEEE TED* **48**, 479–485, 2001.
2. Wu, Y.-F., Saxler, A., Moore, M., Smith, R.P., Sheppard, S., Chavakar, P.M., Wisleder, T., Mishra, U.K., Parikh, P., '30-W/mm GaN HEMTs by field plate optimization', *IEEE EDL* **25**, 117–119, 2004.
3. Inoue, T., Ando, Y., Miyamoto, H., Nakayama, T., Okamoto, Y., Hataya, K., Kuzuhara, M., '30-GHz-band over 5-W power performance of short channel AlGaN/GaN heterojunction FETs', *IEEE TMTT* **53**, 74–80, 2005.

4. Oxley, C.H., Uren, M.J., 'Measurements of unity gain cutoff frequency and saturation velocity of a GaN HEMT transistor', *IEEE TED* **52**, 165, 2005.
5. Yu, T.-H., Brennan, K., 'Monte Carlo calculation of two-dimensional electron dynamics in GaN-AlGaN heterostructures', *J.Appl.Phys.* **91**, 3730, 2002.
6. Wraback, M., Shen, H., Rudin, S., Bellotti, E., Goano, M., Carrano, J.C., Collins, C.J., Campbell, J.C., Dupuis, R.D., 'Direction-dependent band nonparabolicity effects on high-field transient electron transport in GaN', *Appl. Phys.Lett.* **82**, 3674–3676, 2003.
7. Ardaravicius, L., Ramonas, M., Kiprijanovic, O., Liberis, J., Matulionis, A., Eastman, L.F., Shealy, J.R., Chen, X., Sun, Y.J., 'Comparative analysis of hot-electron transport in AlGaN/GaN and AlGaN/AlN/GaN', *phys.stat. sol.* (a) **202**, 808–811, 2005.
8. Barker, J.M., Ferry, D.K., Koleske, D.D., Shul, R.J., 'Bulk GaN and AlGaN/GaN heterostructure drift velocity measurements and comparison to theoretical models', *J. Appl.Phys.* **97**, 063705, 2005.
9. Matulionis, A., Liberis, J., Matulionienė, I., Ramonas, M., Eastman, L.F., Shealy, J.R., Tilak, V., Vertiatchikh, A., 'Hot-phonon temperature and lifetime in a biased $Al_xGa_{1-x}N$/GaN channel estimated from noise analysis', *Phys.Rev.* B **68**, 035338 1–7, 2003.
10. Ridley, B.K., Schaff, W.J., Eastman, L.F., 'Hot-phonon-induced velocity saturation in GaN', *J. Appl.Phys.* **96**, 1499–1502, 2004.
11. Ramonas, M., Matulionis, A., Liberis, J., Eastman, L.F., Chen, X., Sun, Y.-J., 'Hot-phonon effect on power dissipation in a biased $Al_xGa_{1-x}N$ /AlN/GaN channel' *Phys.Rev.* B**71,** 075324 1-8, 2005.
12. Kash, J.A., and Tsang, J.C., in *Spectroscopy of nonequilibrium electrons and phonons*, North Holland, 1992, p. 151.
13. Tsen, K.T., Ferry, D.K., Botchkarev, A., Sverdlov, B., Salvador, A., Morkoc, H., ' Time-resolved Raman studies of the decay of the longitudinal optical phonons in wurtzite GaN', *Appl. Phys. Lett.* **72**, 2132–2134, 1998.
14. Aninkevičius, V., Matulionis, A., Matulionienė, I., 'Hot-phonon lifetime in a modulation-doped AlInAs/GaInAs/AlInAs/InP', *Semicond. Sci. Technol.* **20**, 109–114, 2005.
15. Matulionis, A., Liberis, J., Matulionienė, I., Cha, H.-Y., Eastman, L.F., Spencer, M.G., 'Hot-phonon temperature and lifetime in biased 4H-SiC', *J. Appl. Phys.* **96**, 6439–6444, 2004.
16. Matulionis, A., Liberis, J., Ramonas, M., Matulionienė, I., Eastman, L.F., Vertiatchikh, A., Chen, X., Sun, Y.J., 'Hot-electron microwave noise and power dissipation in AlGaN/AlN/GaN channels for HEMTs', *phys. stat. sol.* (c) **2**, 2585–2588, 2005.
17. Matulionis, A., Liberis, J., Eastman, L. F., Sun, Y.–J., 'Drift velocity saturation and hot-phonon disintegration in AlGaN/AlN/GaN channels', in *Digest 29th WOCSDICE 2005*, Cardiff University, 71–72, 2005.
18. Matulionis, A., Liberis, J. Eastman, L.F., Schaff, W.J., Shealy, J.R., Chen, X., Sun Y.J., 'Electron transport and microwave noise in MBE- and MOCVD-grown AlGaN/AlN/GaN', *Acta Phys.Polon.*A **107**, 361–364, 2005.

Quantum Transport and Spin Polarization in Strongly Biased Semiconductor Superlattices with Rashba Spin-Orbit Coupling

P. Kleinert and V.V. Bryksin

Paul-Drude-Institut für Festkörperelektronik, Hausvogteiplatz 5-7, 10117 Berlin, Germany
Physical-Technical-Institute, Politekhnicheskaya 26, 194021 St. Petersburg, Russia

Summary Based on the density matrix approach, the electric-field-induced spin polarization and depolarization is studied for lateral superlattices with spin-orbit coupling. We focus on nonlinear field effects. The striking analogy between the spin polarization and charge carrier transport is emphasized.

1 Introduction

Presently, much interest is devoted to studying device applications in the field of spintronics, which has stimulated the investigation of spin-orbit interaction in low-dimensional semiconductor heterostructures. Various mechanisms are under discussion that allow manipulations of the spin degree of freedom. One of the main objectives in this fast developing field of spin electronics is the treatment of all-electrical nonmagnetic mechanisms that allow coherent manipulation of electron or hole spins without the application of magnetic fields and magnetic materials. Two mechanisms are widely discussed in the literature, which permit the generation of a magnetization by exclusively applying an electric field to semiconductors with spin-orbit interaction: the spin-Hall effect [1] and the spin accumulation [2]. The spin-Hall effect gives rise to a transverse magnetization at the sample boundaries of a two-dimensional electron gas (2DEG) in the presence of an electric field. In contrast, the field-induced spin accumulation leads to a homogeneous in-plane magnetization. Another pure electronic mechanism is proposed in this paper, namely the spin depolarization due to field-induced resonant tunnelling.

To our knowledge, both the spin-Hall effect and the current-induced spin accumulation have been studied only in the linear response regime. Despite the exciting progress in this field, there are only few studies of nonlinear spin phenomena under nonequilibrium conditions. It is therefore the objective of this paper to consider nonlinear electric field effects in semiconductor heterostructures with spin-orbit coupling. Nonlinear field effects are most pronounced in superlattices (SLs) with a large lattice constant d and high Bloch frequencies $\Omega = eEd/\hbar$ (E denotes the electric field strength). To be more specific, we focus on biased lateral SLs with a strong potential modulation along the x axis as fabricated, e.g., by the cleaved-edge overgrowth technique.

2 Basic Theory and Numerical Results

Let us treat a tight-binding model for carriers moving in the x-y plane. They are subject to electric ($\boldsymbol{E}$) and magnetic ($\boldsymbol{B}$) fields as well as the spin-orbit interaction. The Hamiltonian of the biased lateral superlattice has the form

$$H_0 = \sum_{\boldsymbol{k},\sigma} \varepsilon_\sigma(\boldsymbol{k}) a^\dagger_{\boldsymbol{k}\sigma} a_{\boldsymbol{k}\sigma} + \sum_{\boldsymbol{k}} \sum_{\sigma,\sigma'} J_{\sigma\sigma'}(\boldsymbol{k}) a^\dagger_{\boldsymbol{k}\sigma} a_{\boldsymbol{k}\sigma'} - ie\boldsymbol{E} \sum_{\boldsymbol{k},\sigma} \nabla_\kappa \left(a^\dagger_{\boldsymbol{k}-\frac{\kappa}{2}\sigma} a_{\boldsymbol{k}+\frac{\kappa}{2}\sigma} \right)\Big|_{\kappa=0}, \tag{1}$$

where $\boldsymbol{k}=(k_x,k_y)$ denotes the in-plane quasimomentum. The tight-binding dispersion relation of the SL is given by

$$\varepsilon_{1,2}(\boldsymbol{k}) = \varepsilon(\boldsymbol{k}) \mp g^* \mu_B B_z, \quad \varepsilon(\boldsymbol{k}) = \frac{\Delta}{2} [1 - \cos(k_x d)] + \varepsilon(k_y), \tag{2}$$

where Δ, μ_B, and g^* denote the width of the lowest miniband, the Bohr magneton, and the Zeeman factor, respectively. The off-diagonal element $J(\boldsymbol{k})$ of the spin-orbit coupling

$$J(\boldsymbol{k}) = [m^* v_y(\boldsymbol{k})\alpha + b_x] + i\, [m^* v_x(\boldsymbol{k})\alpha - b_y], \tag{3}$$

is calculated from the components of the drift velocity. The Rashba coupling is denoted by α, and $b_{x,y} = g^* \mu_B B_{x,y}/2$. The Hamiltonian in Eq. (1) is used to derive the kinetic equations for the spin polarization vector $\mathrm{Tr}(f\sigma)$ with f being the density matrix and σ the vector of the spin matrices.

First, let us consider the spin polarization that results from the Zeeman splitting of a transverse magnetic field. The additional application of an electric field may lead to intersubband tunnelling that strongly mixes different spin states. The description of this tunnelling-induced spin depolarization requires a non-perturbative treatment of the electric field. Applying a canonical transformation, the Hamiltonian in Eq. (1) is exactly diagonalized. For weakly coupled SLs, the kinetic equations for the transformed density matrix are analytically solved. We focus on inelastic scattering on polar-optical phonons. Results for the out-of-plane magnetization f_z as a function of the electric field are shown in Fig. 1.

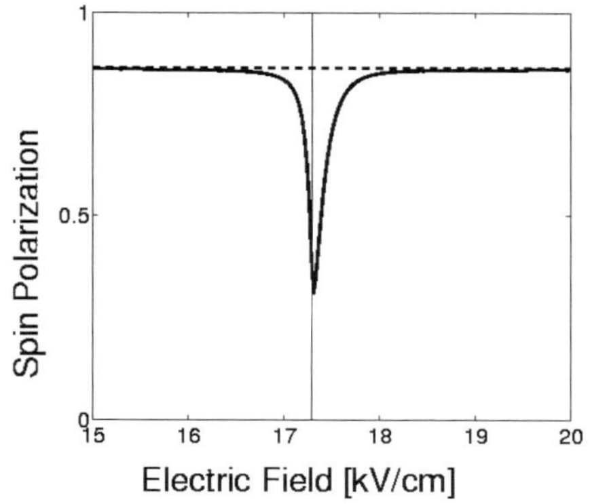

Fig. 1. Out-of-plane spin polarization f_z for Δ=0 (dashed line) and Δ=100 meV (solid line). Parameters used in the calculation are: B=10 T, T=77 K, d=10 nm, and $\alpha=10^{-9}$ eVcm. The cyclotron-Stark resonance is marked by a vertical line.

When the barriers are very thick so that tunnelling is absent ($\Delta\to 0$), the spin polarization becomes independent of the electric field and completely determined by the perpendicular magnetic field (dashed line). For SLs with a finite miniband width ($\Delta\neq 0$), resonant tunnelling can occur that leads to a strong depolarization at the cyclotron-Stark resonance (solid line). This antiresonance develops a sharp δ-like shape, when the Rashba coupling approaches zero. Furthermore, with increasing temperature T, the field-induced spin depolarization at the tunnelling resonance decreases [3].

An electric field cannot only destroy, but also induce a spin polarization. Let us treat nonlinear effects in the spin accumulation induced by dc and ac electric fields [$E(t)=E_{dc}+E_{ac}\cos(\omega t)$]. Again, we are going to carry out analytical calculations to a late stage by focusing on the main physics. We prefer to partly diagonalize the Hamiltonian in Eq. (1) and to solve the kinetic equations for the density matrix in the relaxation-time approximation [4]. Results for the current-induced in-plane magnetization are shown in Fig. 2.

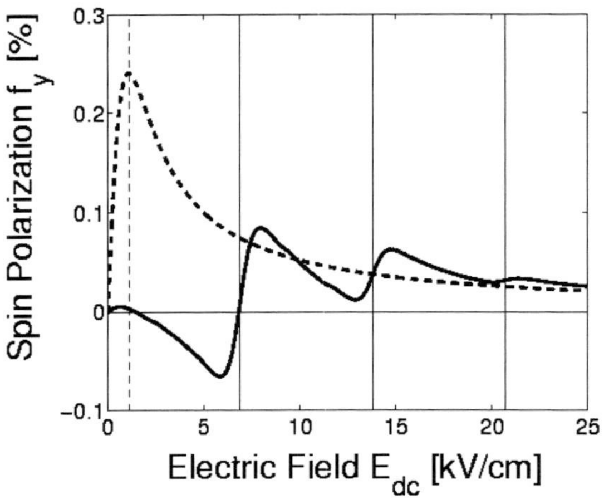

Fig. 2. Spin polarization f_y for E_{ac}=0 (dashed line) and $E_{ac}\neq 0$ (solid line, Ω_{ac}/ω=2). The positions of photon resonances are marked by vertical lines. Parameters are: Δ=100 meV, T=4 K, α=5 10^{-9} eVcm, ν_{ac}=1 THz, and a quasi Fermi energy of ε_F=100 meV.

A constant electric field induces a magnetization, the lineshape of which is very similar to the Esaki-Tsu current-voltage characteristics of conventional SLs. When the SL is exposed by an additional radiation field, photon resonances appear in the spin polarization. Furthermore, when the THz irradiation becomes sufficiently strong, a field-induced reorientation of the magnetization appears. This effect has its complete analogy in the field of spinless carrier transport by the appearance of absolute negative currents. This analogy extends also to dynamical localization and delocalization studied in the transport theory of SLs. Taking into account quantum corrections, tunnelling-induced depolarization is also observed in the field-induced spin accumulation. In the presence of combined Rashba and Dresselhaus spin-orbit couplings, the electric field induces also an out-of-plane magnetization.

References

1. J.E. Hirsch, *Phys. Rev. Lett.*, **83**, 1834 (1999).
2. V.M. Edelstein, *Solid State Commun.*, **73**, 233 (1990).
3. P. Kleinert, V.V. Bryksin, *J. Phys.: Condens. Matter*, **17**, 3865 (2005).
4. P. Kleinert, V.V. Bryksin, *Phys. Rev. B*, in press (2005).

Temperature Dependent Transport in Spin Valve Transistor Structures

R.Heer, J.Smoliner, J.Bornemeier, H.Brückl

ARCS - Nano-Systemtechnologien, Vienna, Austria; Institut für Festkörperelektronik, TU-Wien, Austria;Universität Bielefeld, Dept. of Thin Films and Nanostructures, Germany

Summary: In this work, temprature dependent ballistic electron transport through spin valve transistor structures is investigated by Ballistic Electron Emission Microscopy (BEEM). Co-Cu-Permalloy-Au layers sputtered onto n-type GaAs bulk substrates were studied between room temperature and T=10K. The magnetocurrent increases from 360% at room temperature to 790% at T=10K. The magnetocurrent was also investigated as a function of the electron energy, where we observe a saturation behavior at electron energies above 1.4 eV for all temperatures.

1 Introduction

Spin dependent hot electron transport in magnetic materials deposited on semiconductors has led to a new class of microelectronic devices including the spin valve transistor[1,2] (SVT). SVTs exhibit a huge magnetocurrent (MC), which is defined as the ratio of the collector current in the parallel ($I_{C,\uparrow\uparrow}$) and antiparallel ($I_{C,\uparrow\downarrow}$) magnetization configuration minus one ($MC = I_{C,\uparrow\uparrow} / I_{C,\uparrow\downarrow} - 1$). In SVTs, MC values up to 300% (see Ref.[2]) were already observerd at room temperature. As SVTs, however, are challenging to fabricate, spin-dependent hot electron transport was also studied by STM (scanning tunneling microscopy) methods. Ballistic Electron Emission Microscopy[3] (BEEM), is a three terminal extension of conventional STM, where hot electrons are injected from the tip of a STM into a semiconductor via a thin metal base layer. Using BEEM, the magnetic properties of Co-Cu thin films[4] and magnetic nanostructures[5,6] were studied recently. In the following, this technique is used to determine hot electron attenuation lengths in magnetic films[7], magnetic multilayers[8] and embedded ferromagnetic films[9]. In our previous work[10], we studied spin valve structures employed as base layers for BEEM, and we observed MC values up to 600% at room temperature. In this work, the temperature and energy dependence of the MC are investigated.

2 Experiment

For the present BEEM experiment, n-type (N_D=1 x 17cm^{-3}) GaAs [100] bulk wafers were used. A sputtered Au-film was employed as backside collector. The base layer consists of the following sequence of polycrystalline metal layers, which were sputtered onto the substrate: Co 4.5nm, Cu 4.2nm, Py 3.5nm, where Py denotes Permalloy, a NiFe alloy. The layers were deposited in the presence of a magnetic bias field along the [100] direction of the GaAs, which causes an uniaxial anisotropy in both layers. To establish proper tunneling conditions for STM operation, an additional layer of Au (4.0 nm) was sputtered on top.

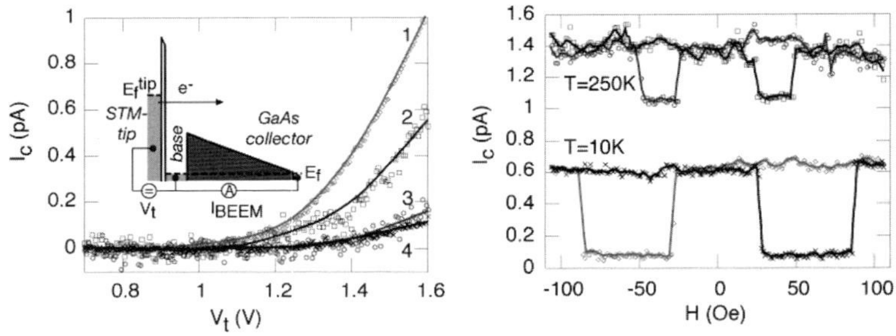

Figure 1: Left: Typical BEEM spectra recorded at T=10K, H=0Oe (curve 1), T=300K, H=0Oe (curve 2), T=300K, H=35Oe (curve 3), and T=10K, H=44Oe (curve 4). A schematic setup of the experiment is shown in the inset. The base represents the spin valve structure. Right : The collector current measured as a function of the magnetic field at T=10 K and T=250 K. For better viewing, the T=250 K curve is shifted by 1 pA.

Figure 1 (left) shows typical BEEM spectra taken at T=10K (curves 1,4) and T=300K (curves 2, 3) and under different magnetic fields. At zero magnetic field, (after the magnetization was driven into saturation), the onset is observed at approximately V_t=1.0V for both the 10K, and the 300K spectra. The current onset position is determined by the Schottky barrier height on the Co-GaAs interface at the specific temperature. Above the onset position, the collector current increases in a roughly quadratic dependence, as it is typical for BEEM signals on bulk GaAs samples. If the magnetization of the spin-valve is oriented in antiparallel configuration at an applied magnetic field of H=44Oe at 10K or H=35Oe at 300K, however, the ballistic collector current is considerably quenched. Interestingly, the spectra measured in antiparallel magnetization show only a weak dependence on temperature and the spectra taken at T=300K and T=10K (Figure 1, curve 3 and curve 4), only differ significantly at higher bias values (V_t > 1.4 V). In contrast to that, the spectra in parallel magnetization show a clear temperature dependence. In parallel magnetization the ballistic current increases approximately by a factor of two when the temperatures is lowered from T=300 K to T=10 K.

Figure 1 (right) shows the ballistic current measured as a function of magnetic field along the easy magnetization axis at a fixed value of V_t. As tunneling parameters we chose V_t=1.5V and I_t=20nA and temperatures of T=10K and T=250K in order to obtain a good signal to noise ratio. Both curves show similar characteristics. At T=10K and a magnetic field of H= -105Oe, the magnetic layers are in parallel configuration and in saturation magnetization. Running from H= –105Oe to H= +105Oe, the magnetic layers flip into the antiparallel configuration at H= +27Oe and the BEEM current is quenched. At H=+87Oe the hard magnetic layer changes its orientation too, and the spin filtering effect is switched off again. This behavior is symmetric on the H-field axis. The switching characteristics at T=250K is comparable to T=10K, only the switching fields of the soft magnetic Py and hard magnetic Co layer are shifted from H=27 Oe to H=25 Oe and H=87 Oe to H=40 Oe, respectively and the ratio between the currents in parallel and antiparallel magnetization configuration is reduced.

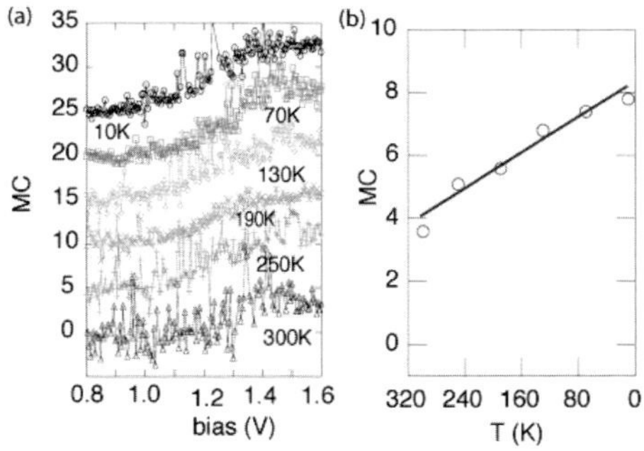

Figure 2 (a): Magnetocurrent MC plotted as a function of STM bias at T=10 K, 70 K, 130 K, 190 K, 250 K and T=300 K, respectively. An offset was added for better viewing. (b): Maximum of MC (V_t=1.4 V) as a function of temperature.

We finally focus on the spectral and temperature behavior of the magnetocurrent (MC). Figure 3(a) shows the MC behavior of our sample as a function of STM bias (which corresponds to the electron energy) and temperature. The shown MC curves were obtained from raw BEEM spectra by numerical division and did not undergo any curve fitting procedures. As one can see, the MC curves are zero below the Schottky barrier height (V_t<1V), then they increase and start to saturate at a bias of approximately 1.4 V. As a function of temperature, the maximum MC value increases linearly from approximately 400% (T=300 K) to 800% (T=10K) on the investigated sample. Figure 3(b) shows the corresponding data. Note that the maximum MC value also depends on the tip position and varies across the wafer. In our previous work, we reached values up to 600 % on other pieces of the sample[10] at a temperature of T=300 K.

3 Summary

In summary, we studied the temperature and energy characteristics of ballistic electrons transmitted through a spin valve structure. The magnetocurrent was found to be increased by a factor of two in the temperature regime between T=300K and T=10K and we also observed a saturation behavior of the magnetocurrent for electron energies above 1.4 eV. These observations indicate that the spin scattering rate in our system just weakly depends on temperature and dominates over all other scattering processes involved.

Acknowledgements : This work was sponsored by “FWF Austria", project No P16337-N08, and "Gesellschaft für Mikro- un Nanoelektronik (GMe)".

References

1 D.J.Monsma,J.C.Lodder, Th.Popma, B.Dieny, Phys. Rev. Lett.**74** 5260 (1995)
2 R.Jansen, Journal of Physics D, **36**, R289, (2003)
3 W.J.Kaiser and L.D.Bell, Phys. Rev. Lett. **60**, 1406 (1988)
4 W. H. Rippard and R. A. Buhrman Phys. Rev. Lett. **84**, 971 (2000)
5 W. H. Rippard, A. C. Perrella, P. Chalsani, F.J. Albert, J. A. Katine and R. A. Buhrman, Appl. Phys. Lett. **77**, 1357 (2000)
6 F. J. Albert, N. C. Emley, E. B. Myers, D. C. Ralph, and R. A. Buhrman Phys. Rev. Lett. **89**, 226802 (2002)
7 R.P.Lu,-B.A. Morgan, K.L Kavanagh, C.J. Powell, J.P. Chen,-F.G. Serpa, W.F. Egelhoff, Jr., J. Appl. Phys. **87**, 5164 (2000)
8 T. Kinno, K. Tanaka, and K. Mizushima, Phys Rev. B **56**, R4391,(1997)
9 T.Zhang, T.H.Shen, D.Greig, J.A.D.Matthew, M.Hopkinson, Journal of Physics: Condensed Matter **15**, 6485 (2003)
10 R.Heer, J.Smoliner, J.Bornemeier, H.Brückl, Appl. Phys. Lett.**85**, 4388 (2004)

Spin Filtering Effects in a Quantum Point Contact

R. Akis and D. K. Ferry

Center for Solid State Electronics Research and Department of Electrical Engineering,Arizona State University, Tempe AZ 85287-5706, USA

Summary. We simulate a quantum point contact (QPC). Using spin-density-functional theory (SDFT), we obtain a spin-dependent energy barrier structure in the QPC that may help explain anomalies observed in experiment.

Quantum point contacts can be formed in semiconductor heterostructures by depositing metal split-gate. Biasing the gates creates a quasi-1D channel, which separates the 2DEG (two dimensional electron gas) in the heterostructure into source and drain regions and through which current can flow. Typically, the conductance for such structures is quantized with plateaus at integer multiples of $G_0 = (2e^2/h)$ as function of gate voltage, a result readily explained by a single electron quantum mechanical theory. More recent experiments, however, have found additional non-integer features, most notably a $\sim 0.7\ G_0$ conductance anomaly [1]. Theoretically, while there is disagreement about the specifics[2,3] , it is generally believed that electron-electron interactions must be included to account for such effects. A typical way of doing this *approximately* is to incorporate spin-density-functional theory (SDFT) into the transport calculations, trusting that SDFT contains enough of the essential physics to accurately represent the effect of these interactions. This, however, is not guaranteed.

In this paper, we present our own transport calculations for a QPC using SDFT. For the initial 2D confining potential generated by a split gate structure, we have used the model potential suggested by Timp [4] :

$$V_{conf}(x,y) = f\left(\frac{2x-l}{2z_\circ}, \frac{2y+w}{2z_\circ}\right) - f\left(\frac{2x+l}{2z_\circ}, \frac{2y+w}{2z_\circ}\right) + f\left(\frac{2x-l}{2z_\circ}, \frac{-2y+w}{2z_\circ}\right) - f\left(\frac{2x+l}{2z_\circ}, \frac{-2y+w}{2z_\circ}\right) \quad (1)$$

where

$$f(u,v) = \frac{eV_g}{2\pi}\left[\frac{\pi}{2} - \tan^{-1}(u) - \tan^{-1}(u) + \tan^{-1}\left(\frac{uv}{\sqrt{1+u^2+v^2}}\right)\right]. \quad (2)$$

Here l and w are the lithographic width (350 nm) and gap (140 nm) between the electrodes, repectively, and V_g is the applied gate voltage. The vertical distance between the 2DEG and the gate, z_o, has been taken to be 70 nm. The inset in the lower right corner of Fig. 1(b) shows the contours of potential that arise for a V_g=-0.55 V in a domain at the very center of the QPC.

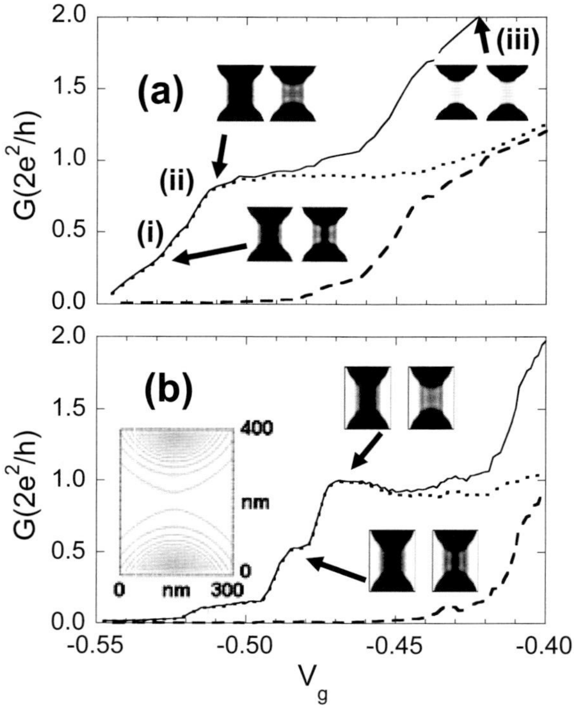

Fig. 1. QPC conductance for a one-dimensional density, $n_{1D} = 2.01 \times 10^6$ cm^{-1} (E_F = 13.5 meV) in (a) and $n_{1D} = 1.96 \times 10^6$ cm^{-1} (E_F = 13.4 meV) in (b). The solid line is the total conductance, while the dashed and dotted lines show the spin-up and spin-down contributions respectively. The self-consistent potentials for spin-down (right insets) and spin-up (left insets) at the indicated points are also shown as shaded contour plots.

Using a LSDT method applied earlier to quantum wires (see [5,6] and references therein for details), the total potential for spin σ is given by:

$$V_{tot}^{\sigma}(x,y) = V_{conf} + V_H + V_{exch}^{\sigma} + V_{cor}^{\sigma} \cdot \qquad (3)$$

where V_H, $V^{\sigma}{}_{cor}$, and $V^{\sigma}{}_{exch}$ are, respectively, the Hartree, correlation, and exchange potentials. Since there is strong confinement only in one direction, our approach was to break the QPC into a series of slices along the x axis, and solve for the self-consistent potentials of each slice individually. As a perturbation, a weak Zeeman term (~10^{-6} eV) is included to break the initial spin degeneracy [6] *(it should be noted using a Rashba term as the perturbation has a similar effect*). The conductance is computed using a lattice discretization of the single-particle Schrödinger equation. The discrete slices are translated across using a stable, iterative scattering matrix approach[7], yielding the transmission coefficients that enter the Landauer formula.

In Fig. 1, we plot the *total* conductance (the sum over the two spin channels) vs. V_g for two 1D electron densities, n_{1D} , set at the left boundary. This can be translated into a Fermi energy and a 2D density for a given QPC width. The trace (a) has a hump-like "plateau" at $0.7\text{-}0.8\ G_0$ and a point of inflection at ~$0.3G_0$. As is evident, the spin-down contribution (dotted) initially dominates the total conductance, even up to almost G_0. Thus, *two* spin-down modes can be almost fully transmitted through the QPC in this case before the first spin-up mode makes it through.

As shown in the insets, an additional potential barrier structure becomes superimposed on top of the QPC when self-consistency is introduced, a structure that depends on spin. This structure tends to weaken as more modes are allowed to pass through the QPC (note that the additional barriers are barely visible for $G\sim 2\ G_0$ at (iii) in Fig.1(a)). For (i), the spin-down and spin-up potentials line up at the center of the QPC. However, away from the center, they deviate, with spin-down dropping significantly below spin-up, splitting with the latter and developing "shoulders". Previously noted in the context of quantum wires[6], this splitting, *which can be larger than the level spacing of the modes*, is mainly the result of the exchange potential and oscillates as a function of the local density . The result is that spin-down sees a barrier of the same height, but is much narrower. It allows for the partial transmission of a single spin-down mode, but completely blocks the spin-up modes. For (ii), the spin-up barrier remains comparatively high, but the central spin-down barrier now collapses to the level of the "shoulders". With this collapse, the 1st spin-down mode is allowed to be fully transmitted and portions of a 2nd spin-down mode now also make it through, yielding the $0.7\text{-}0.8\ G_0$ hump. As V_g is reduced and the QPC is allowed to become more open, the spin-down and spin-up barriers gradually start moving back together again, and the spin-up modes finally start being transmitted through the QPC. By (iii), they have merged. The potential splitting is known to vanish when there are more modes[6], so this is not unexpected. However, the self-consistent potential still has an *extra* barrier on top of the initial QPC potential. *Its presence prevents modes from being cleanly transmitted through the QPC*. This is why the conductance only shows an inflection point at $2G_0$ rather than a fully formed plateau. To obtain a $2G_0$ plateau at this density, a wider initial QPC must be used. For the Timp potential, we find that these barriers are rather generic features. However, by changing n_{1D} one can change their *relative transparency*. As shown in Fig. 1(b), one can create a situation where the inflection point of (a) has evolved into a true plateau at ~0.5 G_0 and there is now a well-defined plateau at G_0. Büttiker [8] has shown in general that whether one obtains conductance plateaus, inflection points or simply a smooth curve is tied to the QPC barrier shape, in particular, to the ratio of its width to its thickness. Here, since the barriers evolve in a nontrivial fashion as a function of density, V_g and spin, we are able to obtain different characteristics in calculating a single conductance trace.

In conclusion, in agreement with recent experiments, our SDFT calculations yield additional structure besides the usual integer plateaus. They result from the formation of density and spin dependent barriers in the QPC region that act as spin filters. The partial transmission of modes leads to the appearance of features away from integer or half-integer values of G_0. That two spin-down modes can be transmitted before a spin-up is allowed implies a significant energy splitting, indicating that these effects should be robust at finite temperatures, which agrees with experimental observations [1].

References

1. Thomas K. J., *et al.* : 'Possible spin polarization in a one-dimensional electron gas', *Phys. Rev. Lett.* **77**, 135-138 (1996).
2. Hirose K., Meir Y. and Wingreen N.S. : 'Local moment formation in quantum point contacts', *Phys. Rev. Lett.* **90**, 026804 (2003)
3. Berggren K.-F. and Yakimenko I. I.: 'Effects of exchange and electron correlation on conductance and nanomagnetism in ballistic semiconductor quantum point contacts', *Phys. Rev.B* **66,** 085323 (2002).
4. Timp G. : 'When does a wire become an electron waveguide?', Semiconductors and Semimetals, **35**, 113-190 (1992).
5. Sun Y. and Kirczenow G. : 'Spin Density-functional theory of the electronic structure of Coulomb-confined quantum wires', *Phys. Rev. B* **47,** 4413-4419 (1993).
6. Wang C.-K. and Berggren K.-F. : 'Spin splitting of subbands in quasi-one-dimensional electron quantum channels', *Phys. Rev. B* **66,** R14257–R14260 (1996).
7. Usuki T., *et al.* : 'Numerical analysis of electron-wave detection by a wedge-shaped point contact', *Phys. Rev.B* **50,** 7615-7625 (1994).
8. Büttiker M.: 'Quantized transmission of a saddle-point constriction', *Phys. Rev.B* **41,** 7906-7909 (1990).

Exchange Effects in the Wigner-Function Approach

E. Cancellieri, P. Bordone and C. Jacoboni

National Research Center S3, INFM-CNR
Dipartimento di Fisica, Università di Modena e Reggio Emilia, Via Campi 213/A, I. 41100 Modena, Italy

Summary. In this paper, an analysis of the Wigner function (WF) for identical fermions is presented. Three situations have been analyzed. i) A scattering process between two indistinguishable electrons in minimum uncertainty wavepackets showing the exchange and correlation hole in Wigner phase space. ii) An equilibrium ensemble of N electrons in a box showing that the WF integrated over space assumes the shape of a Fermi distribution even for very small N. iii) The reduced one-particle transport-equation for the WF in the case of interacting electrons showing the first contribution to the BBGKY hierarchy.

1 Introduction

The Wigner-function (WF) approach has proved to be very useful for studying quantum electron transport [1-3], owing to its strong analogy with the semiclassical picture since it explicitly refers to variables defined in an (r,p) Wigner "phase space", together with a rigorous description of electron dynamics in quantum terms. In this work we present an analysis of the WF for identical fermions. In particular, three situations will be analyzed: i) A scattering process between two indistinguishable electrons in minimum uncertainty wave packets, showing the exchange and correlation hole in Wigner phase space. ii) An equilibrium ensemble of N electrons in a box, showing that the WF integrated over space assumes the shape of a Fermi distribution even for very small N. iii) The transport equation for interacting electrons, showing the BBGKY hierarchy when the integral, over the degrees of freedom of all the particles but one, are performed.

2 Wigner Function for Many Identical Particles

The WF was introduced by Wigner in 1932 to study quantum corrections to classical statistical mechanics. Thus, from the very beginning this function was defined for N particles as:

$$f_W(r_1,p_1,...,t)=\int ds_1...ds_N e^{-\frac{i}{\hbar}\Sigma s_j p_j}\psi\left(r_1+\frac{s_1}{2},...,r_N+\frac{s_N}{2},t\right)\psi^*\left(r_1-\frac{s_1}{2},...,r_N-\frac{s_N}{2},t\right).$$

If the particles are identical fermions, the wave function changes sign if two position variables are exchanged. This implies that the WF remains unchanged if positions and Wigner momenta of two particles are exchanged. The same property holds if the particles are bosons. A reduced single-particle WF can then be defined as:

$$f_W^{(N)}(r,p,t)\propto\int dr_2...dr_N\int dp_2...dp_N f_W\left(r,p,r_2,p_2,...,r_N,p_N,t\right),$$

where the superscript *(N)* indicates that the reduced single-particle WF is defined in a system of N particles.

Example of two colliding electrons. A one-dimensional situation where two electrons collide with each other has been simulated. The Schrödinger equation was solved for two minimum-uncertainty wave packets interacting through Coulomb coupling (Fig.1). Since we are dealing with a 1D system, the two particles having opposite wave vectors are expected to decelerate, scatter, and then move far away from each other.

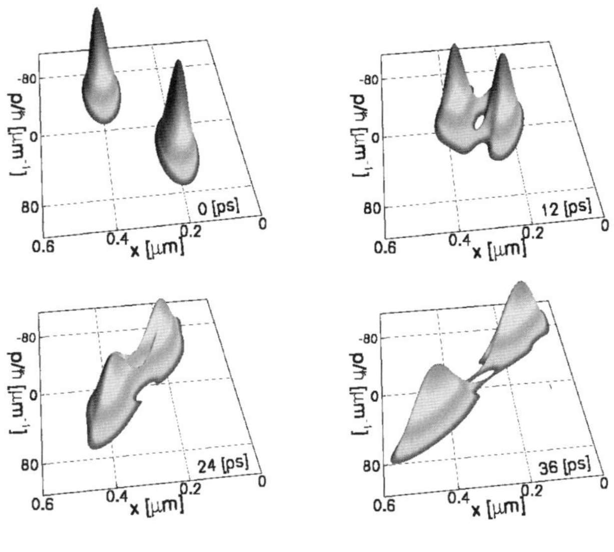

Fig. 1. Reduced single-particle WF of two interacting electrons at different times. The figure clearly shows the exchange hole due to the Pauli exclusion principle.

3 Equilibrium WF for Non-Interacting Particles in a Box

As a second example, we have studied the WF of an equilibrium distribution inside an infinite square potential well. To obtain the thermal distribution we explicitely write the WF in terms of the density matrix. To evalute this expression, the equilibrium density matrix at temperature T *exp(-H/K*$_B$*T)* and the many-particle wave functions are written using as basis of the Hilbert space the set of the eigenfunctions of the box. Thanks to the diagonality of the density matix at the equilibrium the non interacting single-particle WF results:

$$f_W^{(2)}(r,p) \propto \sum_l \sum_{n \neq l} e^{-\frac{\hbar^2 (k_l^2 + k_n^2)}{2 m k_B T}} f_{Wl}(r,p)$$

where $f_{Wl}(r,p)$ indicates the WF of the lth eigenstate of the infinite square potential well. Increasing the number of particles in the system, the integral over the remaining r variable of the reduced single-particle WF tends to the Fermi distribution (Fig.2).

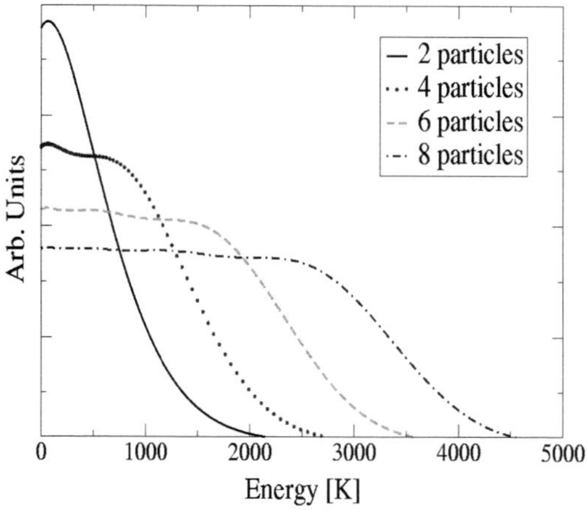

Fig. 2. Integral over the r variable of the reduced single-particle WF in an infinite square potential well. The width of the well has been adjusted in order to maintain a volumetric constant electron density equal to 10^{25} particles/m^3

4 Transport Equation

The transport equation for the WF in the case of e-e interaction with no phonons and no external potential is:

$$\frac{\partial}{\partial t} f_W(r_1, p_1, \ldots, r_N, p_N, t) = -\sum_l p_l \nabla_{r_l} f_W(r_1, p_1, \ldots, r_l, p_l, \ldots, r_N, p_N, t)$$

$$+\frac{1}{h^3} \sum_i \sum_j \int dp'_i \, dp'_j \, \delta(\Delta p_i + \Delta p_j) V_W(|r_i - r_j|, \Delta p_i - \Delta p_j)$$

$$\times f_W(r_1, p_1, \ldots, r_i, p'_i, \ldots, r_j, p'_j, \ldots, r_N, p_N, t)$$

where V_W is the standard potential kernel of the Wigner equation [3] and $\Delta p = p - p'$. By reducing the WF to the single-particle case, the BBGKY hierarchy appears:

$$\frac{\partial}{\partial t} f_W^{(N)}(r,p,t) = -p\nabla_r f_W^{(N)}(r,p,t) + \frac{1}{h^3}\int d\rho\, dp_\rho \int dp' V_w\left(|r-\rho|, 2\Delta p\right) f_W^N\left(r,p',\rho,p_\rho,t\right)$$

This equation shows that the transport equation for the reduced single-particle WF depends on the reduced two-particle WF. If, for simplicity, we consider the case of two particles in the system, the WF expanded over an anti-symmetric combination of wave packets $\psi_i(r)$ (i=1,2) it is given by:

$$\begin{aligned} f_W(r_1,p_1,r_2,p_2) &= f_{W1}(r_1,p_1) f_{W2}(r_2,p_2) + f_{W1}(r_2,p_2) f_{W2}(r_1,p_1) \\ &- \frac{1}{h^6}\int ds_1 ds_2 e^{-\frac{i}{\hbar}(s_1p_1+s_2p_2)}\left[\psi_1\left(r_1+\frac{s_1}{2}\right)\psi_1{}^*\left(r_2-\frac{s_2}{2}\right)\psi_2\left(r_2+\frac{s_2}{2}\right)\psi_2{}^*\left(r_1-\frac{s_1}{2}\right)\right. \\ &\left. +\psi_1\left(r_2+\frac{s_2}{2}\right)\psi^*{}_1\left(r_1-\frac{s_1}{2}\right)\psi_2\left(r_1+\frac{s_1}{2}\right)\psi^*{}_2\left(r_2-\frac{s_2}{2}\right)\right] \end{aligned} \quad (1)$$

If the wave packets are not overlapping the transport equation for the reduced single-particle WF in a system of N identical particles is:

$$\frac{\partial}{\partial t} f_W^{(N)}(r,p,t) = \frac{\partial}{\partial t}\sum_i f_{Wi}(r,p,t) = -p\nabla_r f_W^{(N)}(r,p,t) + \frac{1}{h^3}\sum_i \sum_{j;j\neq i}\int dp'\left[\int dr' V_w\left(|r-r'|, 2\Delta p\right)\left|\psi_j(r')\right|^2\right] f_{Wi}(r,p',t)$$

In conclusion, in the transport equation for the reduced single-particle WF, for non overlapping one-particle wave packets, beside a Liouvillian contribution, an interaction term appears where each one-particle contribution interacts with all the others as in the Hartree approximation. In the case of overlapping wave packets also the second contribution in Eq. (1) must be considered, and the exchange term of the Hartree-Fock approximation is restored.

This work has been partially supported by the U.S.Office of Naval Research (contract No. N00014-98-1-0777 / N00014-03-1-0289).

References

1. Frensley, W.: "Boundary conditions for open quantum systems driven far from equilibrium", Rev Mod. Phys., **62**, 745-791, 1990.
2. Nedjalkov, M., Kosina, H., Selberherr, S., and Ferry, D.: "Unified particle approach to Wigner-Boltzmann transport in small semiconductor devices", *Phys. Rev. B*, **70**, 115319-115335, 2004.
3. Jacoboni, C. and Bordone, P.: "The Wigner-function approach to non-equilibrium electron transport", *Rep. Prog. Phys.*, **67**, 1033-1071, 2004.

Few-Particle Quantum Transmitting Boundary Method: Scattering Resonances Through a Charged 1D Quantum Dot

A. Bertoni and G. Goldoni

National Research Center on "nanoStructures and bioSystems at Surfaces" (S3), INFM-CNR, via Campi 213/A, Modena, I-41100, Italy
and Dipartimento di Fisica, Università di Modena e Reggio Emilia, Italy

Summary. We present an exact approach for the inclusion of particle-particle correlation in the calculation of current-carrying states in open systems. The method, based on the *quantum transmitting boundary method* [C. Lent and D. Kirkner, J. App. Phys. **67**, 6353, 1990], is applied to compute the transmission amplitude of an electron crossing a 1D quantum dot with one or two other electrons in it.

1 Introduction

The quantum transmitting boundary method (QTBM) [1] can be considered as a generalization of the scattering matrix method and consists of a real-space solution (with a finite difference or finite elements approach) of the Schrödinger equation in an arbitrary domain, where a given single-particle potential is present. It has been widely used in the literature for the calculation of ballistic quantum transport characteristics in open devices, where it is often at the basis of the 1D or 2D single-particle scattering states calculation in a Schrödinger-Poisson self-consistent loop [2,3]. In the latter approach a mean-field approximation is implied and the quantum character of indistinguishable carriers only enters through the Fermi distribution function adopted to populate the scattering states. Consequently, it succeeds in explaining a large number of physical phenomena in micro- and nano-devices as long as the number of carriers in the active region (i.e. the simulation domain) is large, but fails to predict the coherent behavior of carriers when only few particles are present in the device. In fact a mean-field approach is inadequate to describe the correlated carriers dynamics and the whole class of purely quantum phenomena stemming from the carrier-carrier entanglement.

We propose a generalization of the QTBM allowing for the exact numerical calculation of the "few-particle scattering state" i.e. the wave function of N carriers, (N-1) of which are bound or quasi-bound in a local potential and an extra carrier incoming as a plane wave (with a given energy) from one of the leads and either reflected or transmitted through other leads. Both bosonic and fermionic particles can be described by our method using appropriate boundary conditions [4].

The basic ideas of the method and the boundary conditions adopted are sketched in Section 2, while, in Section 3, results on a model 1D system with one, two and three spinless electrons are presented. Finally conclusions are drawn in Section 4.

2 The Method

Let us consider an arbitrarily shaped *device* region connected to a number of leads which extend to infinity. We consider, in analogy with Ref. [1], a local coordinate system in each lead, with z direction orthogonal to the device boundary and we assume, as usual, that the potential defining each lead is independent from z. As a consequence a single-particle scattering state in a lead can be written as the product of a plane wave (incoming or outgoing) and a lead transversal mode. Only the amplitudes of the incoming components are usually known and the outgoing amplitudes, on different modes, are obtained through the QTBM that combines the partial knowledge of the boundary wavefunctions with the solution in the device region.

The generalization to a few-particle system is straightforward if we note that the N-spinless particle Schrödinger equation in D dimensions has the same form of the Schrödinger equation for a single particle in (N·D) dimensions. Now, in fact, the form of a N-particle scattering state with a particle in a lead is the product of a plane wave, a lead transversal mode χ and a bound eigenstate φ of the other (N-1) particles. As an example we explicitly report a component of the 3D few-particle scattering state with particle 1 in the lead, in the transversal mode p, with the other particles in the correlated state m:

$$\psi_{m,p}\left(\mathbf{r}_1,\ldots,\mathbf{r}_N\right)=\varphi_m\left(\mathbf{r}_2,\ldots,\mathbf{r}_N\right)\chi_p\left(x_1,y_1\right)\left(a_{m,p}e^{-ik_{m,p}z_1}+b_{m,p}e^{ik_{m,p}z_1}\right)$$

where, for the sake of simplicity, we neglected evanescent waves (although included in the numerical calculations). The general solution in the lead region consists of a linear combination of components with different p and m. The coefficients b, representing reflection/transmission amplitudes are unknowns of the problem. The antisymmetry of the wavefunction is automatically met by taking boundary conditions with different sign each time the coordinates of two particles are exchanged [4].

The region is discretized by a proper mesh and the equations for the leads are coupled to the Schrödinger equation for the internal points in analogy with the QTBM [1]. The resulting linear system is numerically solved and the transmission amplitudes are easily obtained from the boundary values of the wave function. We note that the entanglement between the bound few-particle system and the transmitted carrier can also be quantified, thus evaluating the degree of coherence in the transmission through an interacting system of indistinguishable particles [5].

3 Transmission Amplitudes of a Correlated 1D System

As an example of application we computed the transmission coefficients and transmission phases for the model 1D system depicted in Fig. 1A, representing an open quantum dot. In Fig. 2 the ballistic conductance (solid line) and the transmission phase (dashed line) are shown for a single (A), two (B) and three (C) electrons as a function of the dot potential. The initial state of the bound electrons is taken as the ground one. The Coulomb blockade moves the resonance peaks towards stronger confining energies of the dot while the transmission phase (that a mean-field approach would not be able to predict), though in agreement with the Breit-Wigner formula, shows a behavior substantially different from the single-particle case. Furthermore we note that, while Breit-Wigner resonances (appearing as large Lorentzian peaks) are present in the three cases, a number of sharp peaks, with typical asymmetric Fano lineshape, is present in the transmission spectra of the two- and three-particle systems (Fig. 1B): they are genuine few-particle effect brought about by electron-electron correlation.

4 Conclusions

The generalization of the QTBM to few-particle systems allows us to compute transmission amplitudes that include Coulomb correlation effects. We obtained, for a 1D system, Fano resonances that show an abrupt phase lapse of π that might be in relation with unexpected phase lapses found in the transmission spectra of correlated quantum dots [6] and not explained by mean-field approaches. We estimate that the method, based on a brute-force solution of the few-electron wave function, is scalable, using typical

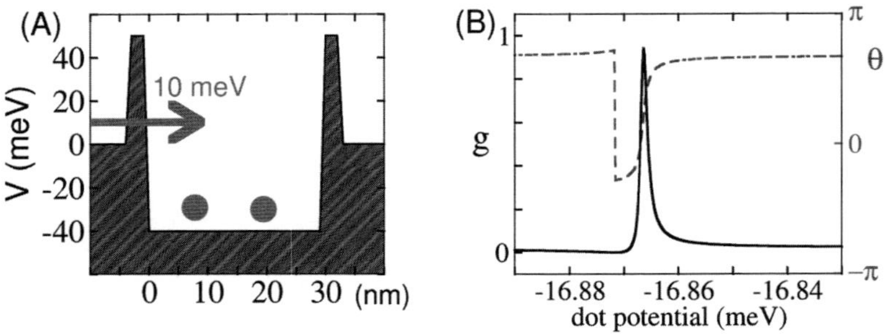

Fig. 1. (A) Potential profile of the simulated system. A 10 meV electron is scattered through a 1D dot with two other electrons initially in the ground two-particle state. Two 50 meV tunnel barriers mimic the quantum point contacts connecting the dot to the leads. (B) Transmission probability (solid curve) and phase (dashed curve) of the scattered electron as a function of the dot potential: a Fano resonance is shown, with typical asymmetric lineshape of the transmission probability and an abrupt phase lapse of π at transmission zero.

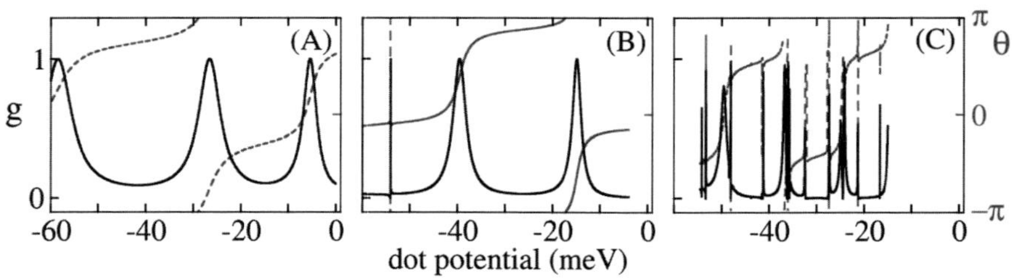

Fig. 2. Transmission probability (solid curve, left axis) and phase (dashed curve, right axis) for one (A) two (B) and three (C) electron system (see text) as a function of the bottom potential of the quantum dot.

computing facilities, to system of 4-5 particles in 1D or systems of 2-3 particles in 2D without introducing extra approximations.

References

1 C. Lent and D. Kirkner, J. App. Phys. **67**, 6353, 1990.
2 A. Abramo, Int. J. of High Speed Electronics and Systems **13**, 701, 2003.
3 S. E. Laux, A. Kumar, M. V. Fischetti, IEEE Trans. on Nanotechnology **1**, 255, 2002.
4 A. Bertoni and G. Goldoni, J. Comp. Electron., submitted, 2005.
5 A. Bertoni, J. Comput. Electron. **2**, 291, 2003.
6 A. Yacoby, M. Heiblum, D. Mahalu, H. Shtrikman, Phys. Rev. Lett **74**, 4047, 1995.

The R-Σ Approach to Tunnelling in Nanoscale Devices

M. Rudan, A. Marchi, R. Brunetti‡, S. Reggiani, E. Gnani

"E. De Castro" Advanced Research Center on Electronic Systems (ARCES)
and Department of Electronics, Computer Science and Systems (DEIS)
University of Bologna, Viale Risorgimento 2, I-40136 Bologna, Italy
Tel. +39-051-209-3016, mrudan@arces.unibo.it

‡Department of Physics and S^3, University of Modena
Via Campi 213/a, I-41100 Modena, Italy

Summary. The R-Σ method provides the time evolution of two dynamical variables extracted from a wave function, namely, the expectation value of the position and the dispersion. It overcomes the Ehrenfest approximation while keeping the Newtonian form of the equations, thus providing the basis for including quantum features into the description of the single-particle dynamics and for extending such features to the collective-transport case. Here the single-particle R-Σ equations are applied to the case of tunnelling, and the results are compared with a full-quantum calculation.

1 Introduction

In a recent paper, a theory has been proposed that leads to a set of two Newton-like equations describing the single-particle dynamics. The dynamical variables of such equations are the expectation value x of the wave function and its dispersion σ [1]. The equations inherently account for the Heisenberg uncertainty relation and exhibit a term proportional to $\hbar^2$. They will be termed in short "R-Σ equations". The theory has been devised for application to the modeling of nanoscale, solid-state devices, where it is necessary to consider the quantum effects. It overcomes some limitations of other quantum-correction methods (see, e.g., the discussion in [2,3]).

The key point in the derivation of the set of Newton equations is an approximate method to calculate the average of the force over the wave function. The method allows one to dispose of the Ehrenfest approximation without the need of completely determining the wave function itself. The outcome of the method of [1] is the set of equations

$$m\ddot{x}_i = -\frac{\partial}{\partial x_i}\left(V + \sum_{n=1}^{3}\frac{\sigma_n}{2}\frac{\partial^2 V}{\partial x_n^2}\right), \qquad m\ddot{\sigma}_i = \frac{\hbar^2}{2m\sigma_i} - \frac{\partial^2 V}{\partial x_i^2}\sigma_i\,, \quad (1)$$

where m is the particle (effective) mass and $V(\mathbf{x})$ the potential energy. In turn, $x_i = \langle \xi_i \rangle$ is the ith component of the expectation value of the particle's position and $\sigma_i = \langle (\xi_i - x_i)^2 \rangle > 0$ the ith component of the dispersion of the particle's position. In the derivation of (1) the wave function $\psi = \psi(\xi, t)$ is assumed to be of Gaussian form, $\psi = \psi_G$, and is normalized to 1. Eqs. (1) describe the dynamics of the expectation value and dispersion of the wave function for a given potential energy V. They provide an improved picture with respect to the standard one given by the expectation value alone, namely $m\ddot{x} = -\partial V/\partial x_i$. In particular, (1) incorporate a description of the wave function's dispersion, which is a typical quantum feature because it is related to the non-zero extension of the wave packet. Thus, it is of interest to compare the results derived from (1) with those obtained by solving the Schrödinger equation. The comparison is particularly significant as far as the dispersion is concerned. It can be shown that in the case of a free particle, or of a particle subject to a linear potential energy, or in the linear harmonic-oscillator case, and for time intervals in the range of the average time between collisions in a semiconductor, the time evolution of σ_i given by (1) is in good agreement with the full-quantum calculation (the details of such calculations will be given elsewhere). Another interesting comparison deals with the tunnel effect. It is addressed in this paper and is discussed in the following section.

2 Application to the Tunnel Effect

To discuss the application of the *R-Σ* method to the tunnel effect it is useful to focus onto some features of it, basing on a one-dimensional example. It may be argued that in devices the effective-mass approximation breaks down near a barrier; however, this drawback is not a failure of the *R-Σ* approach, it happens in other approaches using the effective-mass approximation, among which those discussed in [2,3]. Indicating the dynamical variables here with x and σ consider a normalized wave packet launched against a barrier. The barrier splits the packet into the reflected and trans-

mitted parts. Thus, when a sufficiently long time has elapsed, the two parts become essentially disjoint, and it is natural to think of them as two packets, ascribing an expectation value of position and a dispersion to each. The information about the reflection (P_r) and transmission (P_t) probability is given by the size of each part, due to the normalization constraint. However, one may also think of the two parts as a single packet of a more complicate form, to which only one expectation value of position and one dispersion are associated. In such a picture, the dynamics of x is qualitatively similar to the classical one, namely, the possible outcomes of the interaction are a full reflection, a full transmission, or a complete stop at the barrier as limiting case. The knowledge about reflection/transmission is partly recovered from the packet's dispersion (from the mathematical standpoint this is due to the fact that σ is a moment of $|\psi|^2$ of a higher order than x, so its inclusion improves the information about the shape of $|\psi|^2$).
The picture in which the two parts of the packet are treated as a whole lends itself better to the comparison with the R-Σ method. For this, consider again the one-dimensional example of a particle launched against a barrier with energy lower than the barrier top. As in the classical case, x reaches the turning-back position, say A, and bounces back. However, in the R-Σ approach the expectation value x pulls along with it the dispersion σ, whose value also changes with time according to (1). At a later time, say t', the particle reaches a new position B, to which a dispersion corresponds such that $\sqrt{\sigma}$ is some segment $D - C$ centered on x. Remembering that the R-Σ packet is Gaussian, from the segments $D - A$ and $A - C$ one easily reconstructs $P_t = \int_A^{\infty} |\psi_G|^2 \, d\xi$ and $P_r = 1 - P_t$ at t'. It may be argued that point D should not necessarily lie on the right of A. Indeed, the dynamics of σ depends on the initial conditions of (1) and on the barrier's form. This, however, is true also in the full-quantum analysis. Another issue is that in the full-quantum case the packet will eventually split into two disjoint packets, which in turn will make P_t and P_r independent of time. However, for this to happen the packets must typically move over distances larger than the mean free path in a semiconductor, which makes the comparison non realistic anyhow. A final remark is that, with the above provision, the method applies as well to the case in which the particle's energy is larger than the barrier top, and provides the reflection probability for a particle that according to classical mechanics would cross the barrier. Examples of the R-Σ and full-quantum calculation of x and σ are given in the figure, where the time evolution of x and $\sqrt{\sigma}$ is shown. Here the barrier is parabolic, centered at $x = 100$ nm with a 4 nm bottom width and a 150 meV height. The continuous lines show the R-Σ case, the circles show

the full-quantum case. The different pairs of curves correspond, respectively, to $x(t = 0) = 60, 80, 95$ nm. In all cases the injection energy is 140 meV, with $\sqrt{\sigma}\,(t = 0) = 4.66$ nm (about 20 times the interatomic distance in silicon). Note that the particle's energy has been chosen as almost equal to the barrier's height, to obtain the more difficult case where the wave function is forced to split into two parts of similar size. As expected, the *R*-*Σ* and full-quantum calculations agree for a short time. For the easier cases where the particle's energy is much lower or much higher than the barrier's height the agreement would last for a longer time. Apart from the above remarks, an aspect that makes the *R*-*Σ* equations appealing is that, following the standard prescriptions of the statistical mechanics, it is possible to derive from (1) a coherent extension of the transport equations still in a classical form [1]. In this way, the *R*-*Σ* equations provide the key to incorporating in a proper manner into the transport model the quantum information associated with the non-zero size of the wave function.

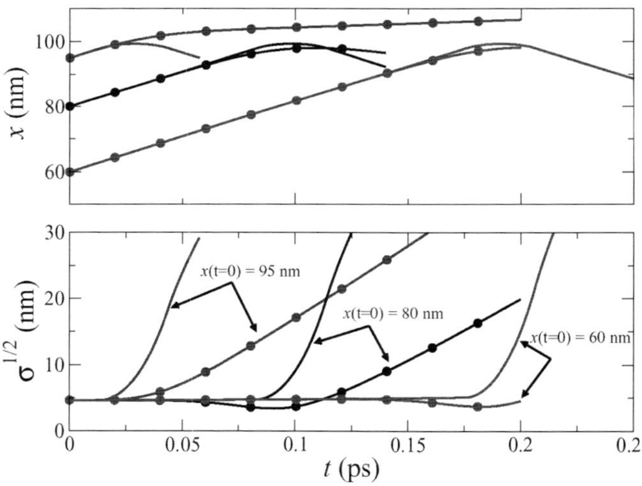

1. M. Rudan *et al.*, *A Coherent Extension of the Transport Equations in Semiconductors Incorporating the Quantum Correction. Part I – Single-Particle Dynamics; Part II – Collective Transport*, IEEE Tr. Nanotechnology **4**, no. 5, p. 495–502 and p. 503–509.
2. D. K. Ferry, J.-R. Zhou, *Form of the quantum potential for use in hydrodynamic equations for semiconductor device modeling*, Phys. Rev. B **48**, 1993, p. 7944.
3. M. Rudan *et al.*, *The Density-Gradient Correction as a Disguised Pilot Wave of de Broglie*, Proc. SISPAD 2003, Munich, G. Wachutka ed., 2003.

Monte Carlo Simulation of Solid-State Thermionic Energy Conversion Devices Based on Non-Planar Heterostructure Interfaces

Z. Bian and A. Shakouri

Electrical Engineering Department, University of California Santa Cruz, Santa Cruz, CA 95064

Summary. In this paper, electron emission from non-planar potential barrier structures is analyzed using a Monte Carlo electron transport model. Compared to the planar structures, about twice bigger emission current can be achieved for the non-planar tall barriers. The thermionic emission enhancement is attributed to combined effects of increased effective interface area and reduced probability of total internal reflection at the heterostructure interface.

1 Introduction

Heterostructure integrated thermionic devices are expected to offer larger thermoelectric power factor by selective emission of hot electrons while keeping similar electrical conductivity as the highly degenerate emitter material.[1, 2] However, it has been shown that the improvement in efficiency due to enhanced electronic transport properties is limited.[3] The main short-coming of planar barriers is that they only transmit "hot" electrons whose kinetic energy in the direction perpendicular to the barrier is large enough. In this paper, we show that it is possible to increase the number of electrons contributing to the electrical conductivity by using non-planar potential barriers. A schematic of a heterostructure thermionic device with zig-zagged interface is shown in Fig. 1. Electrons in a larger volume of the momentum space can be emitted due to multiple tilted directions of the barrier. In the real space, the effective interface area is increased for zig-zagged structures. However, an electron that crosses the interface may re-enter the emitter region in a rough heterostructure even without any scattering. On the other hand, an electron that is reflected from the barrier by total internal reflection may hit the next barrier surface with smaller angle with respect to normal. As it can be seen in Fig. 2, more electrons have a chance to pass over the barrier in a triangle region.

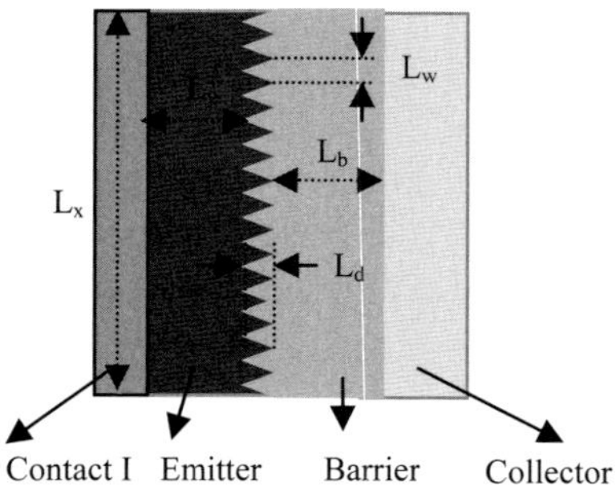

Fig. 1. A solid-state thermionic device with non-planar potential barrier.

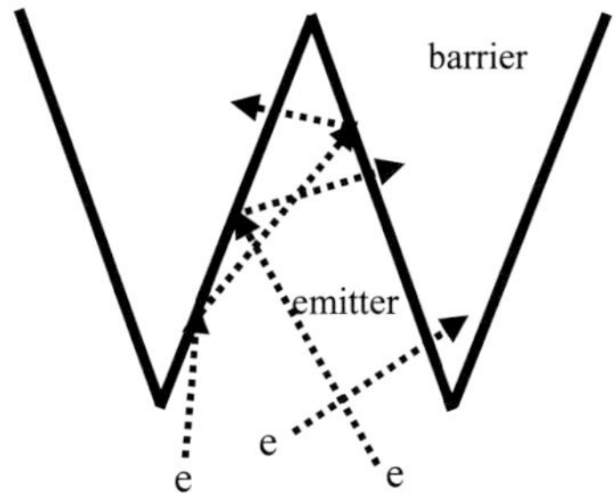

Fig. 2. Illustration of electron trajectories.

2 Monte Carlo Algorithms

We used a simplified ensemble Monte Carlo model to simulate the transport of a two-dimensional electron gas across a two-dimensional non-planar potential barrier. We included the random inelastic scattering in the Monte Carlo method which reassigns a random momentum to the scattered particle according to Fermi-Dirac statistics. In this way, the electron temperature was kept the same as the lattice temperature at the operation condition. The electron scattering was modeled with a constant relaxation time 88.5 fs for InGaAs material and the estimated electron mean-free-path was 0.188 μm for Fermi energy 526 meV. Since the mean-free-path is small at high doping densities and the electron wave generally loses coherence in the barrier, quantum mechanical interference and transmission are neglected. The simulation focuses on the effects of non-planar barrier; thus, a uniform barrier height of 500 meV was used, rather than a self-consistent band bending calculation. This will not change the results significantly because the emitter is much bigger than the interface region and energy distribution of electrons are mostly determined by the bulk emitter. A constant time step of 2 fs was used, which is much less than the scattering relaxation time. The carrier distribution at the quasi-equilibrium state is shown in Fig. 3. The zigzag interface can be clearly seen. Fig. 4 shows the energy distribution of the electrons along the structure. The hot electron filtering of the barrier structure can be clearly seen.

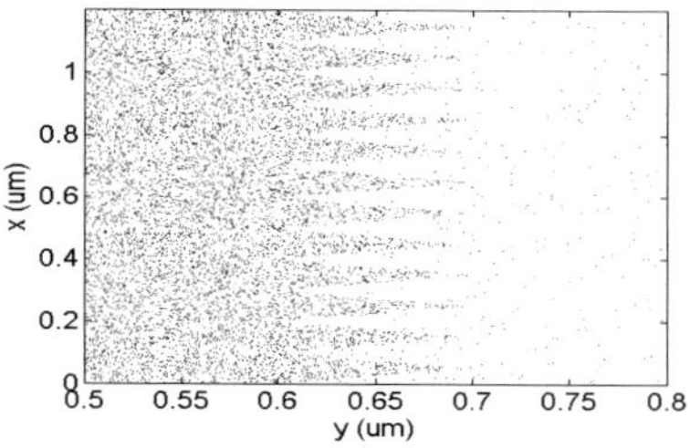

Fig. 3. Electron distribution in real space.

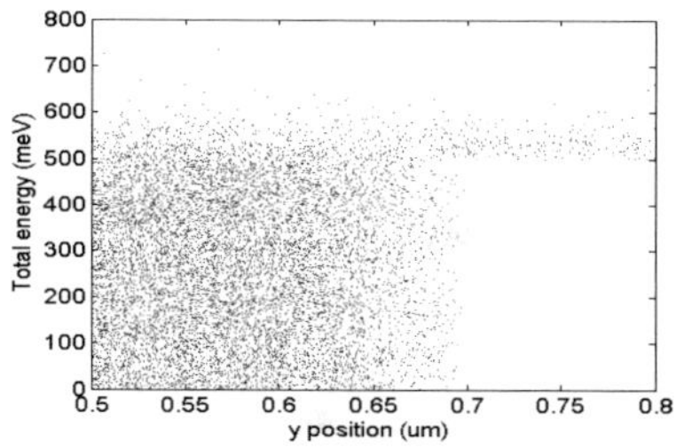

Fig. 4. Electron energy distribution along y direction

3 Simulation Results

Regardless of energy, the total current improvement for the zigzag non-planar barrier compared to that of the planar barrier (with width $L_b+L_d/2$) is shown in Fig. 5. It can be seen that the emitted current increases with the increase of depth L_d or the decrease of the period L_w. The dependence on period is easily understood since a larger period is related to a smaller effective interface area and the two regions in the momentum space have larger overlap. These two regions represent emitted electrons with enough kinetic energy perpendicular to each section of the barrier. An increase of the zigzag depth makes the affective interface area larger. However, when period L_w is small, emitted electrons have more chance to go back to the emitter region for a large zigzag depth. Thus, the improvement converges to an enhancement factor of 1.73 at small periods and large depths.

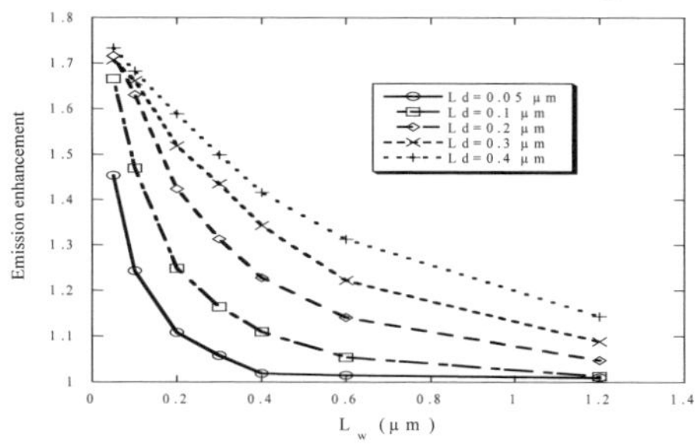

Fig. 5. The current enhancement as a function of zigzag dimensions.

The chance to have a larger total back-scattering and smaller transmission from a non-planar interface is small. One expects more current emission enhancement from more complex interface geometries. Fig. 6 shows a zig-zag interface with four tilted directions. The zigzag period L_w is divided evenly into four sections; and the zigzag depth is divided to two sections with the ratio of 1:2. The Monte Carlo simulation results of geometry dependence are shown in Fig. 7. Similar dependences on the zigzag period and depth as for the two-direction zigzag case can be seen. A factor of 2 maximum improvement compared to planar barriers has been achieved for small periods and large depths. One should note that at very small zigzag periods, when the feature size is smaller than the electron de Broglie wavelength (~8 nm), electrons will see an "effective" barrier profile. In this case a more accurate analysis should use 2D Schrodinger equation and calculate the quantum mechanical transmission coefficient. The overall improvement in the number of emitted electrons will persist as long as a larger volume of electrons in the momentum space can participate in the thermionic emission. [4]

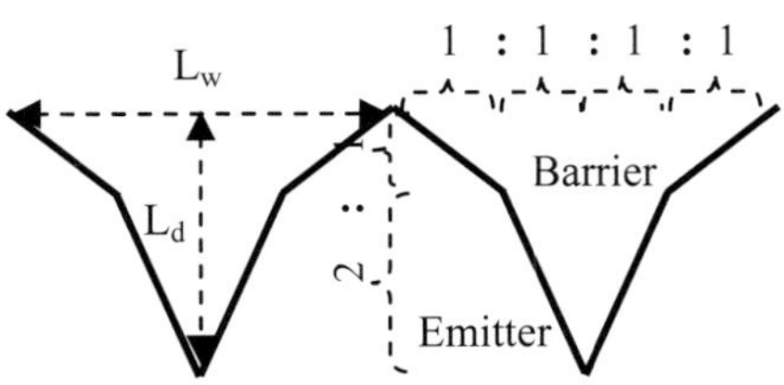

Fig. 6. Illustration of the 4-direction zigzagged interface

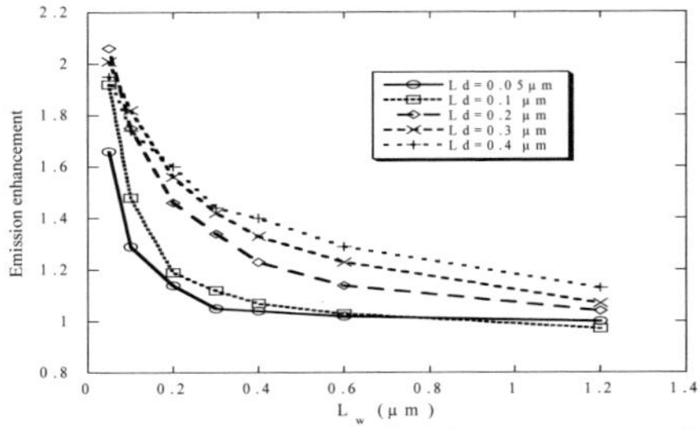

Fig. 7. The current enhancement of the 4-direction zigzag.

4 Conclusions

Non-planar heterostructure potential barriers can increase the number of electrons thermally emitted above the barrier. A factor of 2 of emission enhancement can be achieved with a 4-direction zigzagged barrier.

5 Acknowledgements

The authors thank Professor U. Ravaioli at UIUC for valuable suggestions.

References

1. A. Shakouri and J. Bowers, Appl. Phys. Lett. **71**, 1234 (1997).
2. G. D. Mahan and L. M. Woods, Phys. Rev. Lett. **80**, 4016 (1998).
3. M. D. Ulrich, P. A. Barnes, and C. B. Vining, J. Appl. Phys. **90**, 1625 (2001).
4. D. Vashaee and A. Shakouri, Phys. Rev. Lett. **92**, 106106 (2004).

Simulations of Inelastic Tunnelling in Molecular Bridges

A. Gagliardi, G. C. Solomon, A. Pecchia, A. Di Carlo, T. Frauenheim, J. R. Reimers, N. S. Hush

Institute for Theoretical Physics, University of Paderborn; PaSCO Graduate School, University of Paderborn, Germany; INFM- Dipartimento di Ingegneria Elettronica, Università di Roma Tor Vergata, Italy; School of Chemistry, The University of Sydney, NSW 2006, Australia.

Summary. We present results for a simulated inelastic electron tunneling spectra for octanedithiol chemisorbed on gold electrodes from calculations using the gDFTB code [1]. The geometric and electronic structure is obtained from calculations using a local basis density functional scheme and a non-equilibrium Green's function formalism is employed to deal with the transport aspect of the problem. The calculated spectra show good agreement with experimental results and suggest further details in the assignment and characterization of such spectra.

1 Introduction

The progress that has been made in molecular electronics since the first suggestion that single molecules could function as molecular components is remarkable [2]. Experimental methods have improved markedly from the first reports of single molecule conductivity [3] to highly sophisticated techniques to reproducibly measure molecular conductance [4]. Recently, two groups published results from inelastic electron tunnelling spectroscopy (IETS) measurements on molecules between metallic electrodes [5,6]. Indeed, these were the first measurements of molecular conductivity that conclusively showed the current flowed through the molecule in question confirming the mechanisms proposed for these configurations. IETS present some difficulties of interpretation. They may exhibit different features compared to more traditional spectroscopic techniques like infrared, Raman or HREELS, because of the absence of definite selection rules. Considerable theoretical work has been done in relation to IETS with varying strengths and weaknesses [7,8]. In this paper we present IETS simulations derived from coherent and incoherent tunnelling currents calculated using the non-equilibrium Green's function formalism for octanedithiol chemisorbed on gold. We investigate the correspondence between the experimental and theoretical IETS.

2 Method

We are interested in modelling the coherent and incoherent electron transport through a molecular wire bridging two metal contacts. While the electrons cross the system, they interact with the molecular ionic vibrations from which they can be inelastically scattered. The electronic system is described via a single-particle Tight-Binding Hamiltonian derived from Density Functional Theory (DFTB) [9,10]. The method has been recently extended to the non-equilibrium Green's function (NEGF) approach. This scheme allows to treat contacts and molecules to an equal footing including the open boundary and non-equilibrium conditions encountered in such transport problems. In order to study the electron-phonon coupling we expand the Tight-Binding Hamiltonian to first order in the atomic displacements [11]:

(1)
$$H_{el-ph} = \sum_{q,\mu,\nu} \gamma^{q}_{\mu\nu} c^{\dagger}_{\mu} c_{\nu} [a^{\dagger}_{q} + a_{q}]$$

where $c_\mu^\dagger$ ($a_q^\dagger$) and c_ν (a_q) are, respectively, the creation and annihilation operators of an electron (phonon quanta) in the local basis (vibrational mode *q*) and $\gamma_{\mu\nu}^q$ is the electron-phonon coupling which is related to the derivative of the Hamiltonian respect the normal coordinates of vibrations. The Hamiltonian of the decoupled mode oscillators are quantized by making use of the relationships between the position operator and the Bose field operator. Within the NEGF formalism the relevant quantities are the correlation functions, $G^<$ and $G^>$, representing the electron and the hole density respectively. The technicalities of the theory are mathematically rather involved and we refer the reader to specialized reviews which can be found, for example, in Ref. [12]. The relevant phonon self-energy is evaluated within the first order Born approximation, expressed as,

(2)
$$\Sigma^{<,>}_{el-ph}(E) = i\sum_{q} \gamma_q^2 \int \frac{dE}{2\pi} G^{<,>}(E-E') D^{<,>}_{0,q}(E')$$

where the $D_{0,q}^{<,>}(E)$ are the correlation functions related to the vibrational modes, which are assumed Einstein oscillators in thermal equilibrium with a bath. The current is computed using a generalized version of the Landauer formula [12] valid when sources of incoherent scattering are present:

(3) $$I = \frac{2e}{h}\int_{\mu_L}^{\mu_R} Tr[\Sigma_L^{<} G^{>} - \Sigma_L^{>} G^{<}]dE$$

where $\Sigma_L^{<,>}$ represents the inscattering of electrons (holes) trough the left contact of the device. The IETS is normally computed as the second derivative of the current, but in first approximation we use the first derivative of the trace in equation (3).

3 Results

The geometry of the molecule in the junctions is determined in two steps. First, an optimized geometry is obtained for octanethiol chemisorbed through the terminal sulfur to a single Au(111) surface. The geometry for the full electrode-molecule-electrode system was then generated by symmetrizing about a point of inversion between the C4-C5 bond to give octanedithiol bound to two co-facial Au(111). As shown in other works [11], gDFTB reproduces experimentally observed vibrational frequencies for octanedithiol chemisorbed on Au. In the case of a molecule bound to two metallic electrodes the modes of vibration are significantly perturbed and, as a consequence, the vibrational modes associated with the extremities of the molecule differ in frequency from modes of the same character associated with the central region. This is reflected in the calculated IETS for octanedithiol, showed in Figure 1 (hcp geometry) with the peaks assigned as showed in Table 1.

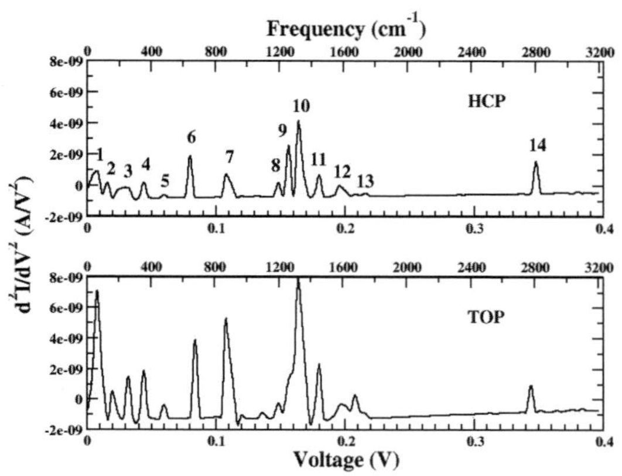

Table 1

1	0.008	C-C-C out-of-plane wag
2	0.017	S-C-C out-of-plane wag
3	0.033	Au - S stretch
4	0.044	S-C-C scissor
5	0.060	C-C-C scissor
6	0.083	C-S stretch
7	0.111	CH2 in-plane rock
8	0.150	CH2 in-plane rock
9	0.157	C-C stretch
10	0.164	C-C stretch
11	0.180	CH2 scissor
12	0.196	CH2 scissor
13	0.212	CH2 out-of-plane wag
14	0.348	C-H stretch sym

Figure 1: Simulation of the IETS for the octanedithiol (hcp and top bonding site). The numbers for peaks (hcp geometry) are related to vibrational modes in the table on the right.

The lack of definite selection rules means that, for a system of the complexity of octanedithiol, a complete assignment of the IETS from IR, Raman and HREELS is difficult to achieve. For example, in the experimental IETS for octanedithiol [5] there were a number of peaks attributed to the Si_3N_4 matrix. We have found that whilst the unassigned peaks at low frequency are likely to have come from the molecule (the simulated spectrum shows a number of peaks in this region) the peaks between 2000-2500 cm^{-1} probably arise from the Si_3N_4 matrix as there are no peaks in the simulated spectrum in this region. Another interesting result of our simulations, beside the interpretation of IETS, is the possibility of obtaining information about the geometry and the bonding site of the molecule between the electrodes. As showed in Figure 1, the IETS simulated for two different bonding sites of the octanedithiol on the gold surface (top and hcp) appear to be different, especially in the low frequency and in the 1000-1700 cm^{-1} regions.

4 Conclusions

IETS provides to be a very powerful tool to describe molecular devices due to the strong sensitivity to the molecule geometry and its environment. Nevertheless there are still many problems to solve due to the lack of definite selection rules. Simulations like ours can help experimentalists in interpreting and analyzing their data.

Acknowledgments: Australian Research Council and PaSCO for funding.

References

1. A. Pecchia and A. Di Carlo, *Rep. on Progresses in Phys.,* **67**, 1497, 2004.
2. A. Aviram and M. A. Ratner, *Chem. Phys. Lett.,* **29**, 277, 1974.
3. M. A. Reed et al., *Science (Washington D. C.)* **278** (5336), 252, 1997.
4. B. Xu and N. J. Tao, *Science (Washington D. C.)*, **301**, 1221, 2003.
5. W. Wang et al., *Nano Letters*, **4** (4), 643, 2004.
6. J. G. Kushmerik et al., *Nano Letters*, **4** (4), 639, 2004.
7. N. Lorente et al., *Phys. Rev. Lett.*, **86** (12), 2593, 2001.
8. N. Mingo and K. Makoshi, *Phys. Rev. Lett.*, **84** (16), 3694, 2000.
9. M. Elstner et al., *Phys. Rev. B*, **58**, 7260, 1998.
10. D. Porezag et al., *Phys. Rev. B*, **51**, 12947, 1995.
11. A. Pecchia et al., *Nano Lett.*, **4** (11), 2109, 2004.
12. H. Haug and A. P. Jauho: *Quantum Kinetics in Transport and Optics of Semiconductors*, Springer Series in Sol. State Sci., 1993.

Phonon Effects in Nanotubes: Phase Space Reduction and Electron Conductance

A. Raichura, M. Dutta and M. A. Stroscio

Electrical and Computer Engineering Department, University of Illinois at Chicago (AR, MD, and MAS), Physics Department, University of Illinois at Chicago (MD and MAS), Bioengineering Department, University of Illinois at Chicago (MAS)

Summary. Electron phonon effects are studied in nanotubes with fixed ends using the elastic continuum model for phonons. The dispersion relations are used to identify phonon phase-space-reduction effects in short nanotubes. The electron-phonon coupling constant is calculated and the change in conductance is estimated for limiting cases.

1 Introduction

Transport in nanotubes has been a field of active research ever since their discovery by Ijima in 1991 [1]. Various theories and experiments suggest ballistic transport in short nanotubes [2] which have shown promising applications in field effect transistors [3], infrared detectors [4] and other electronic devices. We study phonon effects in nanotubes using the elastic continuum model for phonons [5,6]. The dispersion relation obtained for the full set of phonon modes is shown to vary with the length of the nanotube and is applied in calculating the phonon-assisted electronic transitions in the nanotubes with fixed ends. For low temperatures, we calculate the change in the temperature-dependent quantized conductance related to the electron-phonon coupling [7].

2 Phonon Phase Space Reduction

The (10,0) nanotube with length 8 nm is investigated in this paper. Applying the elastic continuum model, the phonon modes are calculated using the Donnell's equations [5,6]. The zone folding and the force constant models, available in the literature neglect the curvature of the nanotube altogether. The dispersion relations, plotted in Fig. 1 show the three branches of phonon modes, namely radial, torsional and axial.

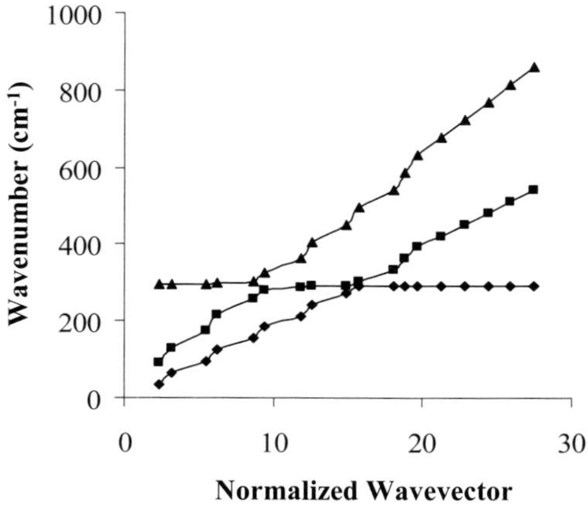

Fig. 1. Dispersion relation for (10,0) nanotube with length 8 nm.

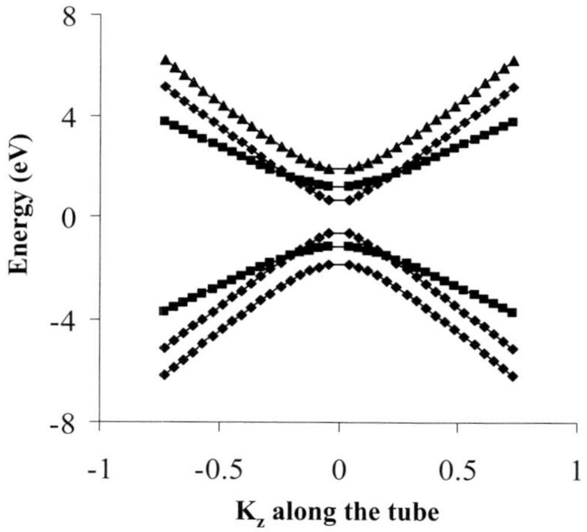

Fig. 2. Quantized electron energy levels for (10,0) nanotube.

The phonon modes are quantized as expected for short nanotubes with fixed nodes at its ends. Figure 2 shows the three lowest electronic subbands [8,9]. We investigate phonon-assisted electronic transitions for intravalley-intrasubband (1↔1, 2↔2 and 3↔3) and intravalley-intersubband (1↔2, 2↔3 and 1↔3) transitions which conserve both energy and momentum. For the intravalley-intrasubband transition, there are no phonons with the values of energy and momentum needed to cause transitions between any of the electronic states. For the intravalley-intersubband 1↔2 case, allowing for a 10% broadening of the transition energies, there are three allowed transitions involving an axial mode phonon with energy of 55 meV and a torsional mode phonon at 40 meV. There are two transitions allowed at 38 meV (torsional mode) and 50 meV (axial mode) for 3↔1, and four transitions allowed at 65 meV, 67 meV (torsional modes) and at 87 meV, 89 meV (axial modes) for 2↔3 intersubband case. These results imply a reduction in electron-phonon interactions in short nanotubes due to the quantized phonon energies. This phase-space-reduction effect is analogous to that underlying phonon bottleneck effects in quantum dots.

3 Electron Conductance

For a simple case of only one transmission channel in the nanotube [7], the change in the temperature-dependent quantized conductance due to phonon scattering is given by

$$G_{el-ph} = -\frac{e^2}{\pi\hbar}\eta\left(1-e^{-\frac{\Delta}{k_BT}}\right)^{-1}\left(1+e^{-\frac{(U-\Delta)}{k_BT}}\right)^{-1}, \text{with } \eta = \frac{D^2 l m^* k_f}{2\hbar\rho v_s E_f}$$

where D(deformation potential)=24 eV [5], l (length of nanotube)= 8nm, m*(relative mass of electron in first subband)= 6.887×10^{-32} kg [8], k_f(fermi wavevector)= 10^9 m^{-1}, ρ(mass per unit length)= 23.88×10^{-16} kg/m, v_s(fermi velocity)= 8×10^5 m/s, E_f(fermi energy)= 0.8 eV. The value of electron-phonon coupling constant, η, is calculated to be 0.006 for a (10,0) nanotube. For GaAs, η=0.03 [7]. Figure 3 shows the effect of temperature on electronic conductance. The conductance step is smeared with the increase in temperature. The low value of η suggests weak electron-phonon interactions. This, along with the phonon phase-space-reduction, suggests the enhancement of quasi ballistic transport in nanotube based devices [1,2,4].

4 Summary

The investigation of electron phonon interactions leads to phonon phase-space-reduction effects in nanotubes with short length. The variation of electronic conductance with temperature is studied under limiting cases. The calculated value of electron-phonon coupling constant indicates weak electron scattering due to phonons.

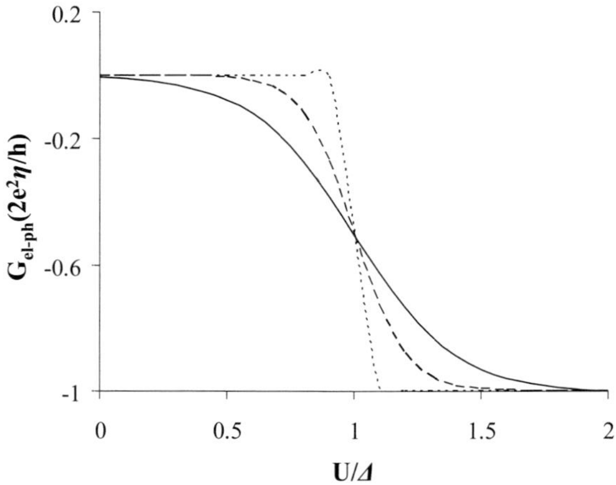

Fig. 3. Correction to electronic conduction due to electron-phonon interactions for $k_BT=0.2\Delta$ (solid curve), $k_BT=0.1\Delta$ (-- curve) and $k_BT=0.01\Delta$ (... curve).

References

1. Iijima, S.: 'Helical microtubules of graphitic carbon', *Nature*, **354**, 56, 1991.
2. Javey, A.; Guo, J.; Paulsson, M.; Wang, Q.; Mann, D.; Lundstrom, M.; and Dai, H. : 'High-field quasiballistic transport in short carbon nanotubes', *Phys. Rev. Lett*, **92**, 106804:1-106804:4, 2004.
3. Guo, J., Lundstrom, M., and Datta, S.: 'Performace projections for ballistic carbon nanotube field-effect transistors', *Appl. Phys. Lett.*, **80**, 3192-3194 , 2002.
4. Xu, J. M.: 'Highly ordered carbon nanotube arrays and IR detection', *J. Infrared Phys. and Tech.*, **42**, 485 -491, 2001.
5. Raichura, A., Dutta, M., and Stroscio, M.A.: 'Quantized acoustic vibrations of single-wall carbon nanotube', *J. Appl. Phys.*, **94**, 4060-4065, 2003.
6. Raichura, A., Dutta, M., and Stroscio, M.A.: 'Continuum model for acoustic phonons in nanotubes: phonon bottleneck', *Phys. Stat. Sol.*, **241**, 3448-3453, 2004.
7. Antonyuk, V.B., Malshukov, A.G., Larsson, M., and Chao, K.A: 'Effect of electron-phonon interaction on electron conductance in one-dimensional systems', *Phys. Rev. B*, **69**, 155308:1-155308-7, 2004.
8. Pennington, G., and Goldsman, N.: 'Semiclassical transport and phonon scattering of electrons in semiconducting carbon nanotubes', *Phys. Rev. B*, **68**, 045426:1-045426:11, 2003.
9. Raichura, A., Dutta, M., and Stroscio, M.A.: 'Accoustic phonons and phonon bottleneck in nanotubes', *J. Comp. Elec.*, **4**, 91-95, 2005.

Carbon Nanotubes Films for Sensing Applications: From Piezoresistive Sensor to Gas Sensing

M. Lucci[1], P. Regoliosi[1], F. Brunetti[1], A. Reale[1], A. Di Carlo[1], E.Tamburri[2], A. Fiori[2], S. Orlanducci[2], M. L. Terranova[2], P. Lugli[3]

[1]MINASlab and Dept. of El. Eng., Univ. Rome Tor Vergata

[2]MINASlab and Dept. Chem. Sc.Tech, Univ.Rome Tor Vergata

[3]Lehrstuhl für Nanoelektronik, TU München

Summary. In this work we present the study of the sensing properties of Carbon nanotube films. We show the realization of deformation sensors based on composite CNT-polymer films realized by electrochemical deposition starting from a dispersion of SWCNTs and EDOT monomer. The gauge factor (GF) of the sensors has been found to be up to 3-4 times higher than that of the commercial strain gauge. We show also that the CNT have gas sensing properties, that can be applied to different gases, included H_2O in gaseous form. Detection can be optimized by means of electrostatic polarization of the film substrate.

1 Introduction

It is known that most of the more important practical applications of SWCNTs come from systems or devices obtained by the controlled assembling of a large number of CNTs [1].

In particular, there has been much work involving the integration of SWCNTs with polymeric matrix to improve their mechanical, thermal and electrical properties [2-3], but especially to create a material which could be more easily processed from a technological application's point of view and which could have a more controlled manufacturing cost. In this work we present the realization of a CNT-polymer based device, which has shown to preserve the piezoresistive behaviour of SWCNT membranes [4,5].

In the meantime, CNT have proven to be a very efficient material for the realization of gas sensors, thanks to the extremely high effective area per unit mass (1580 m^2/g). It's demonstrated that the electrical conductance of the CNTs can change dramatically upon exposure to O_2, NO_2, or NH_3 gases [6].Experiments have shown that the nanotube based sensors can detect ppm or sub-ppm levels of gas molecules at room temperature. Chemicals induced perturbations on the resistance of nanotubes can give direct information, easy to read-out and to interface with conventional electronic architectures.

2 Piezoresistive Behaviour of CNT-Polimer Composites

For the preparation of our organic polymer/SWCNTs composite samples, we focused our attention on Poly(3,4-ethylenedioxythiophene) (PEDOT), one of the best-known π-conjugated conducting polymers.

We prepared composite film samples making use of an electrochemical oxidative polymerization on Au/Cr patterned anode of the EDOT monomer to form the PEDOT starting from a synthesis solution which also contains a fixed amount of pristine or chemically treated SWCNTs. The details concerning the polymerization could be found in [7]. After the polymerization, we transferred the patterned electrode to a polyimide support. After curing at 130 °C for 90 min, a flexible and tough polyimide film with thickness of 0.5 mm was peeled off from the anode surface [3]. In this way, we realized a completely plastic layer made of SWCNTs/PEDOT composite (see Figure 1a), with one conducting side and the other insulating.

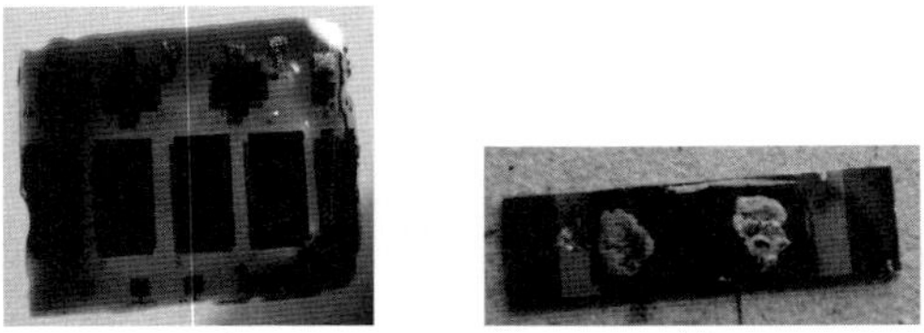

a) b)

Fig.1 : a) The complete plastic layer made of SWCNTs/PEDOT composite, after peeling off from the patterned anode; b) A test device, realized by sticking the composite film on a silicon bar substrate and forming ohmic contacts by deposition of silver paste

We compared the conductance modulation with either a sample formed by only PEDOT and either with a commercial strain gauge, with known GF, sticked on the same kind of substrate. As shown in Figure 2, the sensitivity to mechanical deformation of samples containing SWCNTs is quite higher than the one of polymer-only sample and the one of the strain gauge (these latter two give almost the same answer). Such a behaviour is therefore intrinsically due to the presence of SWCNTs, being independent from the treatment they underwent before. Since the substrates of the CNT/PEDOT composite and of the strain gauge are of the same kind, we estimated the gauge factor of our devices simply comparing their response curve in the figure: we obtained values in the range between

3 and 7, quite comparable with the values obtained for SWCNTs membranes [4].

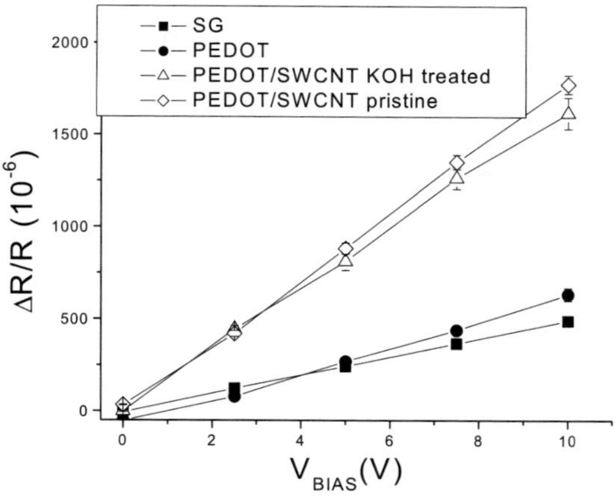

Fig.2: Piezoresistive response of electropolymerized samples.

4 Gas Sensing With CNT Films

The nanotube-based gas sensors have been prepared by depositing dispersions of SWCNT in Chloroform over Cr/Au interdigitated electrodes with 40 μm spacing. The electrodes were patterned over a SiO_2 insulating layer, grown on a p-doped Silicon substrate . To improve the efficiency of the sensor the SWCNT were aligned onto the multifinger electrode surface by a dielectrophoretic process following a protocol described in [6,9].
We demonstrate that SWCNT films can be used very efficiently for the detection of different gas species like NH_3, NO_x, H_2S, H_20, H_2 . The sensor response allows the detection of sub ppm concentration of the various species, and it is analyzed as a function of different parameters and conditions of operation. We also demonstrate that the gate voltage can improve the response time (detection/desorption) of the sensor, as shown in Fig. 3. Moreover, we checked that the gate voltage with only N_2 don't affect the response of the sensor.

Acknowledgments

The present work has been supported by MIUR PRIN 2004039399 and MIUR.FIRB RBNE019TMF.

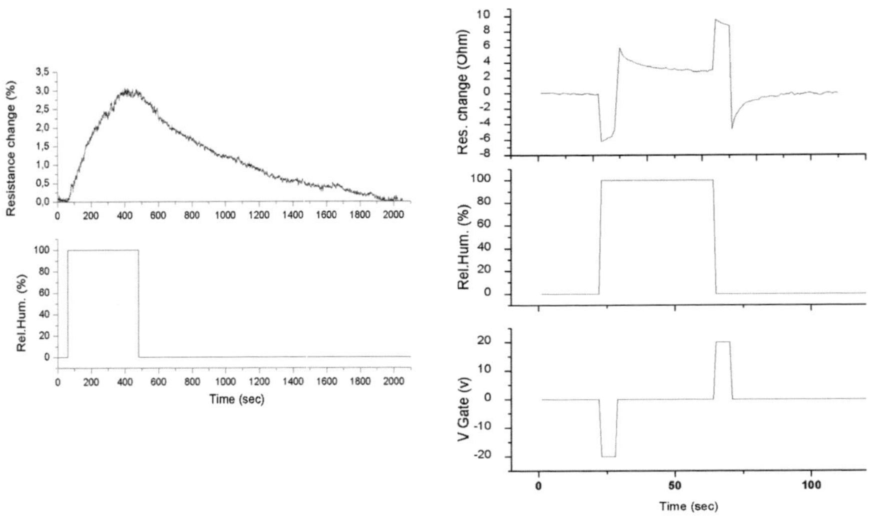

Fig.3: Gate bias applied to improve response time with H_2O vapour.

References

1) Baughman R. H., 'Carbon nanotubes – the route towards the applications', *Science* **297**, 787-792 (2002)
2) Cooper C. A., 'Distribution and alignment of carbon nanotubes and nanofibrils in a polymer matrix', *Composites Science and Technology* **62**, p.1105-1112, 2002
3) Brunetti F.,'Carbon nanotube/Conducting Polymer composites for electronic application: materials preparation and devices assembling', *Proceedings of IEEE Nano04*, 2004
4) Reale A., 'Evaluation of the gauge factor for membranes assembled by single-walled carbon nanotubes', *App. Phys. Lett.* **85**, p.2812-2814, 2004
5) Dharap P., 'Nanotube film based on single-wall carbon nanotubes for strain sensing', *Nanotechnology* **15**, p.379-382, 2004
6) M. Lucci, P. Regoliosi, A. Reale, A. Di Carlo, S. Orlanducci, E. Tamburri, M.L. Terranova, P. Lugli, C. Di Natale, A. D'Amico, R. Paolesse, accepted for publication on Sensors and Actuators B
7) Becker E., 'All-organic thin-film transistors patterned by means of selective electropolymerization', *App. Phys. Lett.* **83**, p.4044, 2003
8) Kong J, Franklin N R, Zhou C, Chapline M G, Peng S, Cho K and Dai H, Science 287 (2000) p. 622
9) M. Berliocchi, S. Orlanducci, A. Reale, P. Regoliosi, A. Di Carlo, P. Lugli, M.L. Terranova, F. Brunetti, G. Bruni, M. Cirillo, Synthetic Metals, 145, 171, (2004)

Electro-Thermal Transport in Silicon and Carbon Nanotube Devices

E. Pop,[1,2] D. Mann,[1] J. Rowlette,[2] K. Goodson[2] and H. Dai[1]

Dept. of [1]Chemistry and [2]Thermal Sciences, Stanford University, Stanford CA 94305, U.S.A. Contact: epop@alum.mit.edu

Summary. This work examines electro-thermal transport in silicon devices and in single-wall carbon nanotubes (SWNTs). Non-local transport is found to strongly affect heat generation in quasi-ballistic silicon devices. Under such conditions, Joule heat is mainly dissipated in the drain region, and increasing power densities may lead to phonon non-equilibrium. Significant current degradation is observed in suspended SWNTs, which is attributed to the presence of hot optical phonons and to a decrease in thermal conductivity (as ~$1/T$) at high temperature (T) under self-heating. The high temperature thermal conductivity can then be extracted by using the high bias characteristics of suspended SWNTs.

1 Introduction

Sharply increasing power densities are often considered the ultimate roadblock for the continued evolution of nanoscale electronics. As the dimensions and voltage of semiconductor devices are down-scaled linearly, the volume and area available for heat dissipation decrease cubically and quadratically. Three-dimensional integration and incorporating materials with lower thermal conductivities than silicon are accelerating such trends, while the thermal boundary resistance between materials also plays a role for designs with large surface-area-to-volume ratio [1]. Confined geometry devices (i.e. FinFET, ultra-thin body SOI, surround gate FETs, nanowires) will further limit the heat dissipation volume, leading to increased temperatures and the associated negative consequences on performance and reliability. Even carbon nanotubes, which have been shown to possess very high thermal conductivity [2], suffer from small thermal *conductance* owing to their small diameter. This combination may also yield power dissipation and self-heating issues under high bias current flow.

In this work, we examine the consequences of self-heating with emphasis on non-equilibrium effects in two types of nanometer scale devices. First, we describe Monte Carlo (MC) simulations which have been used to compute heat generation in bulk and strained silicon devices. We find that Joule heating occurs primarily in the drain of short devices when transport across the channel is quasi-ballistic. We show that non-equilibrium optical phonon (OP) effects may be important at power densities greater than 10^{12} W/cm^3, a range attainable in silicon devices with channel lengths below 20 nm under the current Technology Roadmap (ITRS) guidelines [3]. Second, we investigate electro-thermal transport in freely suspended single-wall carbon nanotubes (SWNTs), which represent the worst-case scenario of self-heating under high current flow. We find that the high bias electrical characteristics of suspended SWNTs exhibit Negative Differential Conductance (NDC) owed to non-equilibrium OPs and a decrease in thermal conductivity as ~$1/T$ at high temperatures ($T > 400$ K).

2 Joule Heating in Bulk and Strained Silicon Devices

The physical mechanism through which self-heating occurs in silicon is that of electron scattering with phonons, and thus only a simulation approach which deliberately incorporates all such scattering events can capture the microscopic, detailed picture. We have studied heat generation using special-purpose Monte Carlo (MC) simulations. The conduction band is modeled with the analytic nonparabolic approximation, and each phonon mode is reproduced with a quadratic dispersion [4]. The heat generation rate is computed as a sum of all phonon emission events minus all phonon absorption events. Hence, complete phonon generation spectra can be obtained both for bulk silicon samples (e.g., as a function of electric field) and in various device geometries [5]. In particular we find that the heat

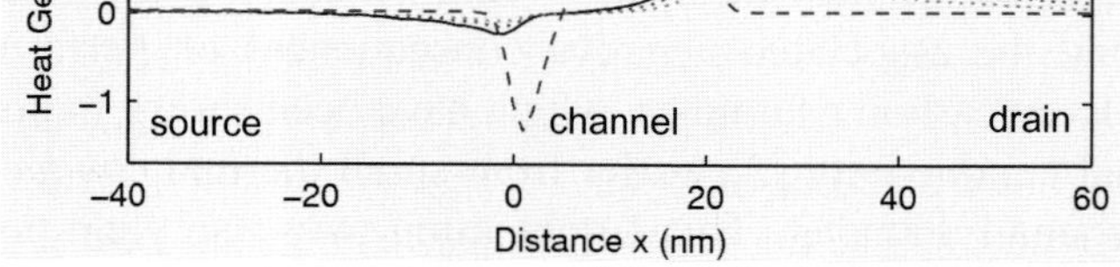

Fig. 1: Heat generation in a n+/n/n+ quasi-ballistic device with channel length $L = 20$ nm. The source and drain are doped to 10^{20} cm^{-3}, the applied voltage is 0.6 V. Unlike the classical (drift-diffusion) result, the MC simulation shows that heat is dissipated far into the device drain. The dotted lines represent the optical phonon (upper) and acoustic phonon (lower) heat generation profiles from the MC result.

generation region extends far into the drain of nanoscale devices (Fig. 1), since transport along the channel is quasi-ballistic. By contrast, the classical drift-diffusion approach is inadequate for computing heat generation within such scaled device geometries due to the highly non-local nature of transport. At the very least the hydrodynamic (carrier temperature dependent) approach ought to be used for calculating the location and magnitude of the heat generation term in TCAD device simulations [6].

In addition, we find that optical phonons (OPs) account for nearly two thirds of the heat generation in bulk silicon at most applied electric fields, and in strained silicon at high electric fields [4]. Dissipation with the acoustic phonon (AC) modes accounts for the remaining one third. This observation contradicts the long-held (simplifying) assumption that heat dissipation occurs predominantly with the OP modes [7]. Nevertheless we find that significant non-equilibrium OP populations may build up, particularly for the *g*-type longitudinal optical (LO) mode. The generation rates for the other phonon modes are either smaller or their density of states (DOS) is larger (the DOS is proportional to the square of the phonon wave vector, which is largest at the edge of the Brillouin zone) and non-equilibrium effects are less significant. Assuming a 10 ps phonon lifetime [8] we find the occupation number of the *g*-type LO phonon to exceed $N_{LO} > 0.1$ and become comparable to unity for power densities greater than 10^{12} W/cm^3 [5]. Such power densities are attainable in the drain of 20 nm (or shorter) channel length devices at operating voltages from the current ITRS guidelines (Fig. 1). Non-equilibrium phonon populations will increase electron scattering in the drain, leading (at the very least) to a magnification of the drain series resistance. The full ramifications of such behavior are still under investigation.

3 Electro-Thermal Transport in Suspended SWNTs

We have also recently explored electro-thermal transport in metallic single-wall carbon nanotubes (SWNTs). While such nanotubes on substrates are known to exhibit very high (> 20 μA) current-carrying capability, we have found that freely *suspended* nanotubes carry much lower currents due to significant self-heating. The suspended nanotube resistance at high bias in Fig. 2 is greater than expected near $T \sim 800$ K (the burning temperature of SWNTs in air), suggesting a lower lattice temperature and a higher non-equilibrium, hot OP population [9]. This observation is consistent with recent studies indicating much longer phonon lifetimes in suspended SWNTs [10]. This is attributed to the lack of intimate coupling with a substrate, which would otherwise provide additional phonon relaxation channels. We

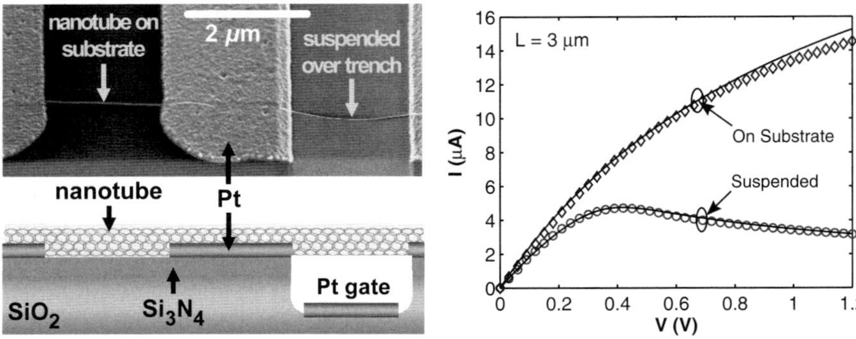

Fig. 2: SEM (top left) and diagram (bottom left) of a metallic SWNT with one portion grown across a substrate, the other suspended across a trench. The measured (symbols) and calculated (lines) *I-V* characteristics of the two nanotube segments are also plotted for a similar device with *L*~3 μm and *d*~2.4 nm (right) [9].

also observe Negative Differential Conductance (NDC) in our longest (10 μm) suspended SWNTs at much lower electric fields (~200 V/cm) than predicted by previous theoretical models which assume isothermal conditions (~5 kV/cm) [11]. This also indicates that the observed NDC is a thermal and not electrical (e.g. contact or field-related) effect.

A simple two-temperature (acoustic and optical) model is used to calculate the theoretical *I-V* characteristics (solid lines) in Fig. 2 [9]. The optical phonons are assumed to be stationary, and the acoustic phonons are solely responsible for thermal transport. The approach self-consistently computes the nanotube resistance, Joule heating and temperature along its length. A key feature is the temperature dependence of the SWNT thermal conductivity, which is found to be essential for reproducing the high bias behavior of the electrical characteristics. The $\sim 1/V$ shape of the suspended SWNT *I-V* characteristics at high bias is found to be a reflection of the $\sim 1/T$ dependence of the thermal conductivity at high temperatures due to Umklapp phonon scattering. This provides an indirect way to measure the thermal conductivity (k) of individual suspended SWNTs in the high temperature regime, and we find that $k \sim 3600(300/T)$ $\mathrm{Wm^{-1}K^{-1}}$ from $400 < T < 700$ K. This value is sensibly consistent with other recent measurements of the SWNT thermal conductivity [2].

4 Conclusions

This work briefly describes recent advances in the understanding of electro-thermal transport in nanoscale silicon devices and carbon nanotubes. Non-local transport is shown to strongly affect heat generation in quasi-ballistic silicon devices. Non-equilibrium phonon populations affect transport in suspended carbon nanotubes, which do not benefit from an intimate vibrational and thermal coupling with a substrate. A signature of the temperature dependence of the SWNT thermal conductivity ($\sim 1/T$ at high T) is observed in the measured high-bias current-voltage curves ($I \sim 1/V$). Aside from their fundamental importance, the "extreme" cases studied here may suggest ideas for the optimization of future nano-electronics (from SOI to nanotubes) through geometry, interface and materials design from an electro-thermal point of view.

References

1. E. Pop, C. O. Chui, S. Sinha, R. W. Dutton and K. E. Goodson, "Electro-thermal comparison and performance optimization of thin-body SOI and GOI MOSFETs," *Proc. IEDM*, p. 411, San Francisco CA (2004).
2. C. Yu, L. Shi, Z. Yao, D. Li and A. Majumdar, "Thermal conductance and thermopower of an individual single-wall carbon nanotube," *Nano Letters* **5**, 1842 (2005).
3. International Technology Roadmap for Semiconductors (ITRS). Online at http://public.itrs.net
4. E. Pop, R. W. Dutton and K. E. Goodson, "Monte Carlo simulation of Joule heating in bulk and strained silicon," *Appl. Phys. Lett.* **86**, 82101 (2005).
5. E. Pop, J. Rowlette, R. W. Dutton and K. E. Goodson, "Joule heating under quasi-ballistic transport conditions in bulk and strained silicon devices," *IEEE SISPAD*, Tokyo Japan (2005).
6. J.-H. Chun, B. Kim, Y. Liu, O. Tornblad, and R. W. Dutton, "Electro-thermal simulations of nanoscale transistors with optical and acoustic phonon heat conduction," *IEEE SISPAD*, Tokyo Japan (2005).
7. J. Lai and A. Majumdar, "Concurrent thermal and electrical modeling of sub-micrometer silicon devices," *J. Appl. Phys.* **79**, 7353 (1996).
8. S. Sinha, E. Pop and K. E. Goodson, "A split-flux model for phonon transport near hotspots," *Proc. IMECE*, Anaheim CA (2004) and *J. Heat Transfer* in press (2005).
9. E. Pop, D. Mann, J. Cao, Q. Wang, K. E. Goodson and H. Dai, "Negative differential conductance and hot phonons in suspended nanotube molecular wires," *Phys. Rev. Lett.* **95**, , 155505 (2005).
10. B. J. LeRoy, S. G. Lemay, J. Kong and C. Dekker, "Electrical generation and absorption of phonons in carbon nanotubes," *Nature* **432**, 371 (2004).
11. V. Perebeinos, J. Tersoff and P. Avouris, "Electron-phonon interaction and transport in semiconducting carbon nanotubes," *Phys. Rev. Lett.* **94**, 086802 (2005).

Silicon-Based Ion Channel Platforms

S. J. Wilk[1], L. Petrossian[1], M. Goryll[1], J. M. Tang[2], R. S. Eisenberg[2], M. Saraniti[3], S. M. Goodnick[1], T. J. Thornton[1]

[1]Arizona State University, Center for Solid State Electronics Research, Tempe, AZ 85287; [2]Rush Medical College, Department of Molecular Biophysics and Physiology, Chicago, IL 60612; [3]Illinois Institute of Technology, Department of Electrical and Computer Engineering, Chicago, IL 60616

Summary. We demonstrate that silicon substrates can be used as a universal platform for recording the electrical activity of ion channels inserted into suspended bilayers. The bilayers span narrow openings etched into silicon substrates using standard microelectronics processing techniques. Reversible Ag/AgCl electrodes are integrated around the circumference of the opening and provide long-term stable measurements of the ion channel currents. To demonstrate the utility of the silicon platform we have measured the electrical activity of OmpF porin ion channel proteins inserted into a lipid bilayer formed using the Montal – Mueller method. Systematic measurements of the lipid giga-seal characteristics are presented, including ac conductance measurements and statistical analysis in order to resolve the conductance of individual ion-channels.

1 Introduction

Cell membranes are made up of phospholipid bilayers that serve as high resistance impermeable barriers to the flow of charged ions and small molecules. Ion channels are proteins that form a pore across the cell membrane so that specific ions can pass through the cell wall. These proteins are of considerable interest because they are used to send signals throughout the human body. The patch-clamp was first demonstrated to measure single ion channel proteins in 1976 by Neher and Sakmann [1] and is able to form giga-ohm (GΩ) seals between a pipette tip and cell membrane by sucking the cell into the pipette. A stable high resistance seal is crucial for low noise measurements because it reduces electrical noise and isolates different sides of the membrane [2].

Recently, there has been considerable effort to planarize the patch-clamp setup into a high throughput system for applications in e.g. biosensing. Apertures in different substrates have been designed for both patch-clamp type experiments and ion channel reconstitution. We have recently reported on the use of Si as a substrate and then coated with polytetrafluoroethylene (PTFE, Teflon) to form stable high resistance, repeatable seals between painted lipid bilayers and the device [3], which is described in detail below.

2 Si Ion Channel Device Fabrication

Samples were prepared using double-sided polished Si (100) wafers. The aperture was designed to have a 150μm diameter similar to that currently used for Teflon and commercial devices. The substrates were patterned and then etched in a deep silicon reactive ion etcher using the Bosch process. After etching of the aperture, a thermal oxidation of 200 nm followed to produce an electrically insulating layer on the surface. The device was then coated with 75μm of SU-8 and patterned with conventional photolithography, so that resist entered the thinned region thus decreasing the overall capacitance of the device, as shown in Fig. 1a. Next, 800 nm of silver was evaporated onto the surface of both sides of the wafer. The silver layer was patterned using conventional lithography and then etched with a 1:1:20 mixture of sodium hydroxide, hydrogen peroxide and water. A Teflon layer was then chemically vapor deposited using the deep etcher and C_4F_8 as the gas source. Finally, the electrodes were chloridized in 5% NaOCl before experimental measurements were taken.

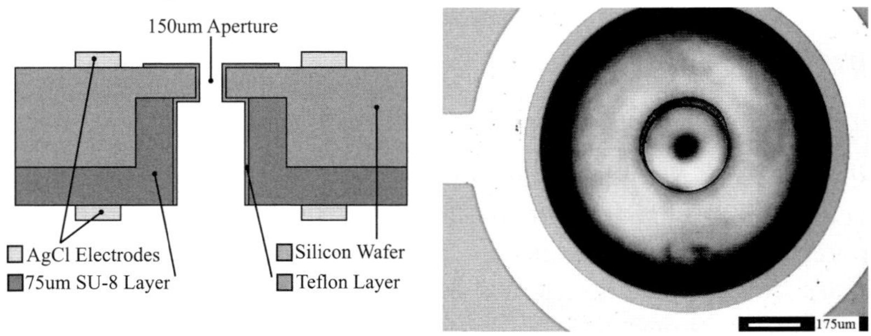

Fig. 1. a) Schematic diagram of the Si ion-channel sensor showing the capacitance reducing SU-8 layer and integrated silver/silver chloride electrodes; b) SEM micrograph showing the fabricated aperture from the SU-8 side of the device.

3 Current-Voltage Characteristics

Lipid bilayer experiments were performed using a Teflon bilayer chamber with a 5 mm diameter opening between two baths of electrolyte solution. Lipids were dissolved in n-Decane (10mg/ml) and used to form a bilayer with the techniques of Montal and Mueller [4]. Current and bilayer capacitance were measured using an Axon Instruments Axopatch amplifier. Ion channels were inserted into the membrane by adding OmpF porin to the trans (ground side) bath. OmpF is a relatively large polypeptide made of three monomers composed of 340 amino acids each.

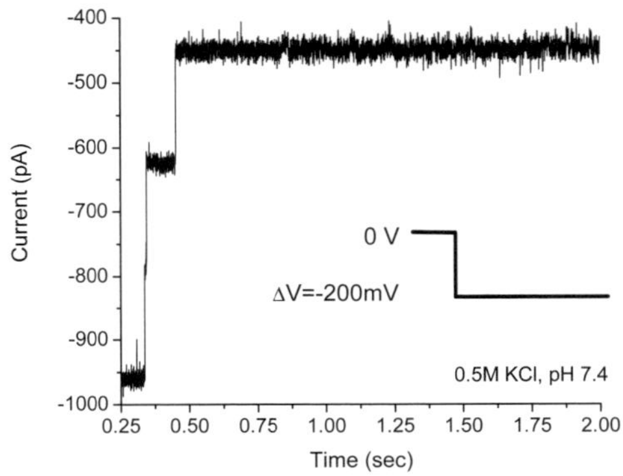

Fig. 2. Current recordings of an OmpF porin protein inserted into a lipid bilayer suspended on a microfabricated silicon device with -200mV applied potential.

Figure 2 shows typical data obtained for the current versus time when a -200 mV step voltage is applied across the membrane. Here a single protein has inserted, which is open prior to the voltage pulse. In response to the pulse, successive monomer channels within the trimer close stochastically in time, leading to the step-like reduction in conduction as shown in Fig. 2. This result shows that the planar integrated Si structure may be used to perform channel recordings with the same if not higher signal to noise than conventional patch clamp measurements.

We have recently investigated the use of lock-in amplifier measurements as a low noise alternative to the traditional Axopatch amplifier approach. Here, a summing amplifier is used to superimpose a 25Hz, 20mV rms, sine wave on the dc bias, and a phase sensitive detector used to measure the in-phase component of the current. A 100mV, 66% duty cycle, square wave voltage pulse is applied as an external bias, and the conductance measured as a function of time. We have performed a statistical analysis of the time dependent conductance, which is shown in Fig. 3 below. The sharp peak close to 0.5 nS corresponds to the conductance of the giga-seal, approximately 2 GΩ. Successive peaks correspond to integer multiples of a fundamental conductance, associated with that of single channels of the OmpF porin protein. We have performed such measurements for different KCl concentrations, which results in a roughly linear plot as shown in Fig. 3b. This statistical analysis of channel conductance provides a unique identification of the contributions of individual channels, and an unambiguous determination of the dependence of ion channel conductance on the electrolyte concentration.

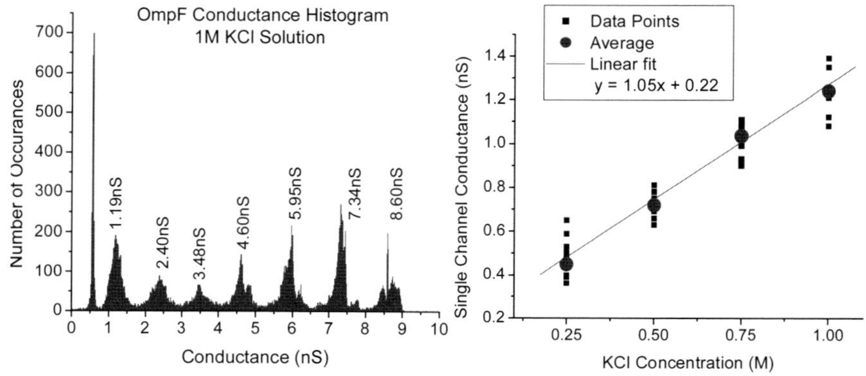

Fig. 3. a) Histogram of conductance samples measured using square wave excitation and lock-in amplifier measurements for a 1M KCl solution; b) The single channel conductance versus KCl concentration using this technique.

References

1. Neher, E. and Sakmann B.: 'Single-Channel Currents Recorded from Membrane of Denervated Frog Muscle-Fibers', *Nature*, **260**, 799-802, 1976.
2. Levis, R.A. and Rae, J.L.: 'Constructing a Patch Clamp Setup', *Methods in Enzymology*, **207**, 14-66, 1992.
3. Wilk, S.J., Goryll, M., Laws, G.M., Goodnick, S.M., Thornton, T.J., Saraniti, M., Tang, J., and Eisenberg, R.S.: 'Teflon (TM)-coated silicon apertures for supported lipid bilayer membranes', *Appl. Phys. Lett.* **85**, 3307-3309, 2004.
4. Montal, M. and Mueller, P.: 'Formation of Bimolecular Membranes from Lipid Monolayers and a Study of Their Electrical Properties', *Proc. of the Natl Academy of Science*, 69, 3561-3566, 1972.

Implicit Water Simulations of Non-Equilibrium Charge Transport in Ion Channels

U. Ravaioli, T. A. van der Straaten and G. Kathawala

Beckman Institute for Advanced Science and Technology, University of Illinois, 405 N Mathews Ave, Urbana, IL 61801, USA

Summary. Ion channels are natural nano-channels found in the membranes of all living cells, which exhibit a broad range of specific device-like functions to help regulate cell physiology. The study of charge transport in ion channels is imperative to understand how charge regulation is accomplished at the molecular level if one is to develop nanoscale artificial systems that mimic biological function and detection. Although Molecular Dynamics is the most popular approach to simulate ion channel behavior, the computational cost of representing all water molecules and ions in the system is prohibitive to study the timescales required to resolve ionic current and lead to structure design. A hierarchy of models of decreasing complexity is needed to address simulation of different time and space scales, similar to the set of models developed to study transport in semiconductors. This paper discusses the application of Monte Carlo and Drift-diffusion methods to simulate transport in ion channels, using the ompF porin channel as a prototype.

1 Introduction

Ion channels are protein macromolecules found in the cell membrane, that form nanoscopic aqueous pores puncturing the otherwise mostly impermeable membrane, allowing ions to enter and leave the cell. All channels carry a strong, sharply varying distribution of permanent charge which largely determines the channel's conductance. Most channels switch between conducting and non-conducting states in response to physiological triggers, thereby regulating the intracellular ion composition. Many channels will also selectively conduct a particular ion species while blocking the passage of others [1].

Evolution has generated an endless variety of channels with different sensing and actuating capabilities. The technique of mutagenesis also allows channel structure to be modified at the amino acid level, making it possible to engineer channels with specific conductances and selectivities.

As such the solid-state electronics community is paying increased attention to ion channels with potential device applications in mind. The detailed investigation of ion channel behavior should provide crucial information for future design of practical interfaces between solid-state and biological systems, or for the synthesis of artificial circuits with similar function.

In the last decade high resolution structural information has become available for several complete ion channels. At the same time, computational resources have advanced sufficiently that Molecular Dynamics (MD) simulations of realistic ion channels are possible, albeit over extremely short timescales [2]. While MD simulations can provide critical information on protein dynamics in atomic detail the computational power needed to simulate timescales of steady-state conduction far exceed the resources currently available. Over several decades, a comprehensive hierarchy of device simulation approaches has been developed to study charge transport in semiconductor devices, and there is a keen interest in adapting these to describe the behavior of ion channels from a device perspective. The goal of this paper is to focus on the application and adaptation of transport simulation for electronic devices, which may help broaden the range of accessible time and space scales in the simulation of ion channels.

2 Transport Monte Carlo Simulations

Ion channels have several features that lend themselves well to the application of the Monte Carlo methods used to study charge transport in solid state devices. The mobile ions form a natural ensemble of particles that interact with rest of the system (water, protein and membrane). The discrete nature of the ions is included naturally and their finite volume, which becomes increasingly important in the narrow confined spaces inside the channel, is readily handled using a pairwise repulsive potential. Interactions between ions and water molecules are treated as the scattering events that interrupt ion free flights. In addition, many-body interactions and localized phenomena can be incorporated into the model relatively easily. By modeling the protein and lipid atoms as static charges embedded in a dielectric background, and using the Poisson equation to evaluate local fields the effects of image charges induced at dielectric interfaces are also included naturally.

One of the main drawbacks of using particle methods to simulate ion channels is that successful ion traversals through the channel are rare events and the bulk of computation is performed to track ions that never approach the channel. Reliable estimates of current require simulations over microseconds, while ion-water scattering rates are $\sim 10^{13}$ s^{-1} necessitating trajectory integration time-steps of $\sim 10^{-14}$ s. Also, to ensure a proper treatment of boundary conditions large electrolyte baths must be included in the simulation domain. At the same time the domain must be discretized on a sufficiently fine mesh to resolve the complicated spatially varying protein boundary and charge distribution. This can lead to a very large number of grid points and a substantial computational overhead in the repeated solution of Poisson Equation which depends superlinearly on the number of grid points. For large channels the cost can be prohibitive.

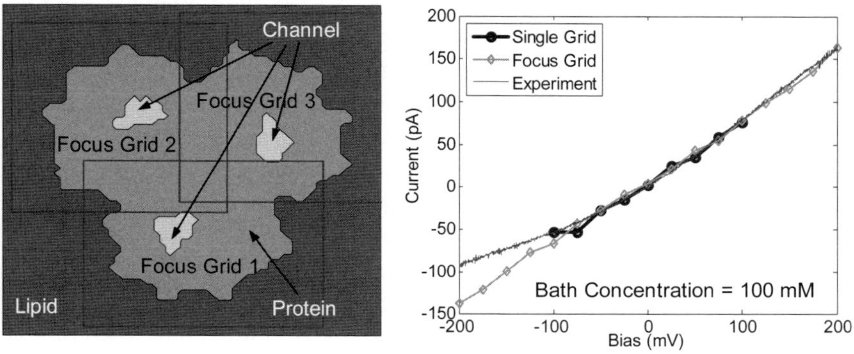

Fig. 1. (a) Representation of the focused grid domain for ompF porin. (b) Current-voltage curves computed for ompF porin immersed in 100mM KCl using both single and focused grid simulations, compared against experimental measurement.

With these constraints in mind we have developed the BioMOCA ion channel simulator, a 3-D Monte Carlo transport model for ion channels [3]. Simulations of bulk electrolyte have shown that using a coarser mesh in the bath regions in conjunction with the P^3M scheme [4] can speed up the computation considerably without compromising accuracy. A grid focusing scheme [5] has therefore been implemented in BioMOCA that allows the use of a coarse mesh spanning the entire domain and a finer mesh on a sub-domain containing the channel region simultaneously in a computationally efficient way. This methodology also allows multiple fine mesh sub-domains which can be useful for channels with multiple pores like ompF porin or an array of channels sharing common bath regions. Poisson's equation is first solved on the coarse mesh. Boundary conditions for the fine mesh are then obtained by interpolation from the coarse mesh solution along with some additional calculations to account for the short-range component of electrostatic potential. Figure 1 shows (a) a schematic representation of the grid focusing scheme applied to the ompF porin channel, and (b) current-voltage curves computed by BioMOCA for porin immersed in 100mM KCl using both a single grid and focused grid approach. Both results compare very well against experimentally measured data. The use of the focused grid scheme resulted in a 150% speed-up.

3 Drift-Diffusion Simulations

The next approximation in the hierarchy of transport models is the drift-diffusion model. Although this approach is straightforward to implement using well-established robust T-CAD packages [6] and yields complete current-voltage curves rapidly, the assumption of a charged fluid implies that the size and discreteness of the ions are neglected, which can lead to charge carriers entering very narrow regions of the channel where actual ions would be excluded, yielding ion densities that are higher than would be physically possible when the finite volume occupied by each ion is considered. This scenario is illustrated in Figure 2 which compares the 1-D ion occupancy profiles through the ompF porin ion channel, obtained from both Monte-Carlo and Drift-diffusion simulations. The ion occupancy is a measure of the average number of ions at a particular location along the channel. Although both models predict almost identical current-voltage curves that match experimental measurements remarkably well, the drift-diffusion simulation yields much higher ion densities inside the channel.

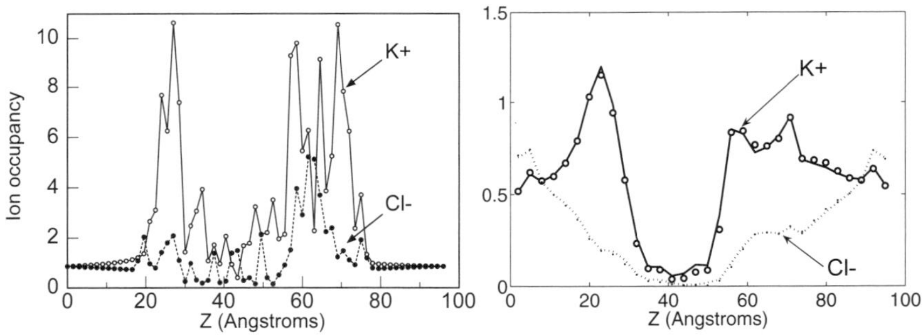

Fig. 2. Ion occupancy profiles for ompF porin in 100mM KCl, computed from Drift-diffusion (left) and Monte-Carlo (right) simulations.

A similar issue is well known in semiconductor devices [7]. Although electrons are well modeled by point particles, in narrow potential wells, such as at MOS interfaces and heterojunctions, a classical description of the charge yields a maximum of the density at the interface where the electric field is maximum. However, size quantization in the confined regions imparts an effective size to the electrons via the wave function, so that the maximum of density is offset from the interface by a measurable amount which is important for detailed simulation of devices. A possible approach is to account for this effect is the use of a correction potential in addition to the electrostatic one, thus giving a physical profile to the density of electrons which are still treated with a classical model. The equivalent approach for the classical ion fluid is to add a correction potential that considers the finite ion size (in terms of geometry and charge distribution) to prevent unphysical bunching of charges in restricted spaces, particularly if a strong fixed attractive field is present as in pores with heavily charged protein walls [8].

Acknowledgements

This work was supported by financial grants from the National Science Foundation – Network for Computational Nanotechnology (Grant No. EEC-0228390).

References

1. Hille, B.: *Ion Channels of Excitable Membranes*, Sinauer Associates, 1992
2. Tieleman, D.P., Biggin, P.C., Smith, G.R. and Sansom, M.: 'Simulation approaches to ion channel structure-function relationships', *Q. Rev. Biophys.*, **34**, 473-561, 2001.
3. van der Straaten, T.A, Kathawala, G., Trellakis, A., Eisenberg, R.S. and Ravaioli, U.: 'BioMOCA – A Boltzmann transport Monte Carlo model for ion channel simulation', *Molecular Simulation*, **31**, 151-171, 2005.
4. Hockney, R. and Eastwood, J.: *Computer Simulation Using Particles*, McGraw-Hill, 1981.
5. Baker, N.A., Sept, D., Holst, M.J. and McCammon, J.A.: 'The adaptive multilevel finite element solution of the Poisson-Boltzmann equation on massively parallel computers ', *IBM J. Res. Develop.,* **45**, 427-438, 2001.
6. van der Straaten, T.A., Tang, J.M., Ravaioli, U., Eisenberg, R.S. and Aluru, N.: 'Simulating ion permeation through the ompF porin channel using three-dimensional drift-diffusion theory', *J. Comp. Elect.*, **2**, 29-47, 2003.
7. Winstead, B. and Ravaioli, U.: 'A quantum correction based on Schrodinger equation applied to Monte Carlo device simulation', *IEEE Transactions on Electron Devices*, **50**, 440-446, 2003.
8. Gillespie, D., Nonner, W. and Eisenberg, R.S.: 'Coupling Poisson-Nernst-Planck and density functional theory to calculate ion flux', *J. Physics : Condensed Matter*, **14**, 12129-12145, 2002.

An Investigation of the Dependence of Ionic Conduction on the Dielectric Properties of Porin

S. J. Aboud[1], D. Marreiro[2] and M. Saraniti[2]

[1]Electrical and Computer Engineer. Dept., Worcester Polytechnic Institute
[2]Electrical and Computer Engineer. Dept., Illinois Institute of Technology

Summary. A previously validated P^3M force-field scheme, self-consistently coupled to a BD kernel, is used to investigate the influence of the protein dielectric constant on ion channel permeation in OmpF porin. The channel conductivity is 0.24nS for a protein dielectric constant of 4, and is in agreement with experimental measurements. Increased cation selectivity at low ionic concentrations is also observed in the simulations and appears to be dependent on the rings of aspartic acid residues around the mouths of the porin.

1 Introduction

The ion channel OmpF porin is a water filled trimer found in the outer membrane of *Escherichia coli*. Each monomer is a hollow barrel structure with a physical constriction near the center that reduces the width of the pore to approximately 6 Å. Highly charged residues line the inside of the pore constriction, generating a large electric field that facilitates the transport of ions through the channel. Because of the wide range of available experimental and simulation data, porin has become a valuable structure for investigating the atomic detail of charge transport in ion channels.

Brownian dynamics (BD) methods are attractive for modeling conduction in porin because they allow for simulation times that can capture the ion permeation process. The cost of simulating these systems for long times is an oversimplification of key physical features of the ion channel system, most notably, the polarization effects related to the solvent (water) and the protein are poorly represented by a stepwise constant dielectric constant. While the use of this model for the aqueous solution inside the permeation pore is arguably suitable because the ionic hydration shell remains intact (at least away from the central constriction), its validity is questionable when used to describe the polarization response of the protein. In literature, the values used for the dielectric constant of the protein can typically vary from ε_p=2 to ε_p=20, although values between ε_p=2 and ε_p=6 are generally the most common [1,2]. In this work, simulations are performed to investigate the impact of the value of the dielectric constant assigned to the protein on the conductivity of OmpF.

2 P^3M-Brownian Dynamics Simulation Approach

The simulation code is based on a BD kernel coupled to a 3D Poisson-P^3M force-field scheme that accounts for the electrostatic interactions, as described elsewhere [3]. The ion trajectories are modeled by the full Langevin equation, which is integrated using a third order scheme and a time step of 20fs. The long range electrostatic interactions, including the external boundary conditions and dielectric interfaces are accounted for self-consistently by a real-space Multigrid Poisson solver within the P^3M approach [4,5]. A third order scheme is used for both the charge assignment and force interpolation. Short-range van der Waals interactions are included using the standard Lennard-Jones potential. The time step used to update the solution of Poisson's equation during the simulation is 2ps. Values of the anion and cation diffusion coefficients and the Lennard-Jones parameters are taken from literature [6].

An illustration of the simulated system is shown in Fig. 1. The ion channel is modeled as a rigid structure embedded in a fixed dielectric slab

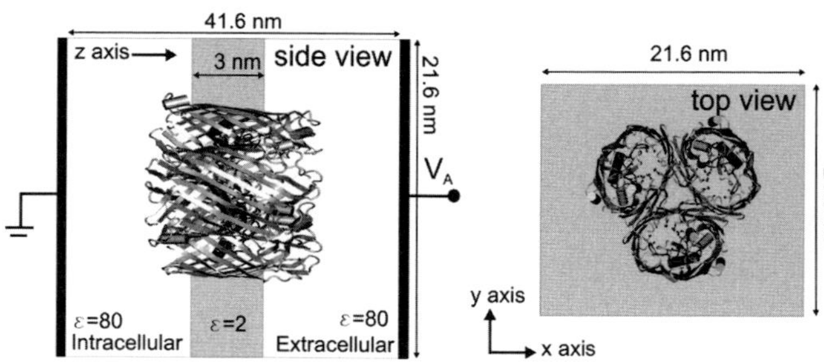

Fig. 1. Illustration of the simulated OmpF porin channel system.

3 nm wide, representing the membrane. The water in the electrolyte bath surrounding the channel/membrane system is also treated as a continuum dielectric. The dielectric constant of the membrane and electrolyte bath is set to 2 and 80, respectively. The charge distribution of the protein atoms is generated using Gromacs [6], starting from the atomic coordinates of the OmpF structure as determined by X-ray diffraction [7].The protonation states of the protein residues are taken at neutral pH, resulting in a net

charge of -30e for the entire trimer. The channel/membrane system is mapped onto a 30x30x50 inhomogeneous Cartesian grid with the permeation pathway of the channel aligned with the z-axis. The geometric center of the protein corresponds to the center of the simulation box. The mesh spacing of the grid varies from 1nm in the bulk electrolyte solution to 0.4nm in the interior of the channel, while a mesh expansion ratio less than 1.5 is maintained over the grid. Neumann conditions are applied at the boundaries of the x- and y-axis and Dirichlet conditions are used at the z-axis boundaries to model the externally applied bias. Ions are injected and ejected at the Dirichlet surfaces to maintain a constant concentration of the electrolyte and mimic the effect of far reversible electrodes. The external applied bias is defined by grounding the contact on the intracellular side of the channel and applying a voltage to the extracellular contact as depicted by Fig. 1.

3 Results and Discussion

The total charge flux through the contacts is shown in Fig. 2 for a KCl concentration of 0.1M at 1V and protein dielectric constants of 2, 4, and 6. The top and bottom contacts correspond to the extracellular and intracellular sides of the bath, respectively. The artificially high voltage was used to compensate for the small signal-to-noise ratio of the ionic flux, and does not affect the functionality of the simulated set-up, even if it would be

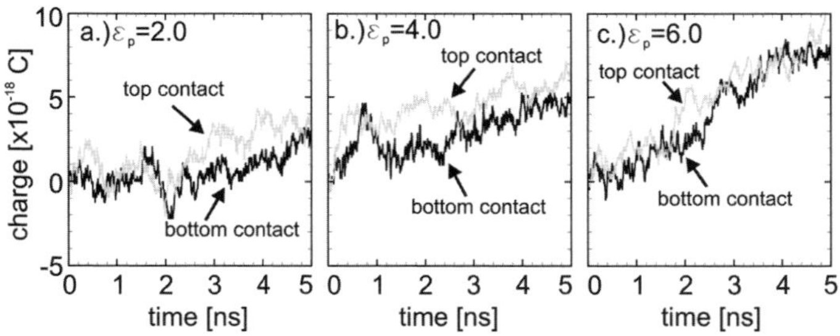

Fig. 2. The charge flux of ions passing through the Dirichlet contacts at 1V, 0.1M and protein dielectric constants of 2, 4, and 6.

unrealistic in a system with a flexible membrane and protein. From the charge fluxes the single pore conductivity is calculated to be 0.17nS at 0.1M and increases for larger protein dielectric values. When the protein dielectric constant set to 4 and 6, the calculated conductivity is 0.24nS and 0.58nS, respectively. A value of approximately 0.3nS is obtained from measurements of single pore conductivity in OmpF [9], corresponding to the simulation results obtained with a dielectric constant of 4. While this agreement with experiment is encouraging, care must be taken in interpreting these results. The simulations were run for only 5ns with averages taken over the last 2.5ns, and longer simulation should be performed in order to ensure that steady-state has been fully reached and to improve the statistics.

To further investigate how the dielectric constant assigned to the protein alters the passage of ions through the channel, the ionic concentration for different dielectric constants is shown in Fig. 3 along the z-axis at 0.1M KCl and a bias of 1.0V. The ion distribution is obtained by averaging along the x- and y-directions. The concentration of anions along the channel pore does not change significantly with protein dielectric while most of the variance in the average cation concentration is found on the region around the intracellular and extracellular mouths of the channel, with a higher number of ions found in simulations with lower dielectric constants. The higher cation density at the constriction region of the pore could be an artifact due to the use of a constant diffusion coefficient throughout the channel pore. The peaks in the cation concentration on both sides of the pore shown in Fig.2b are interesting since the transmembrane potential should act to deplete one side.

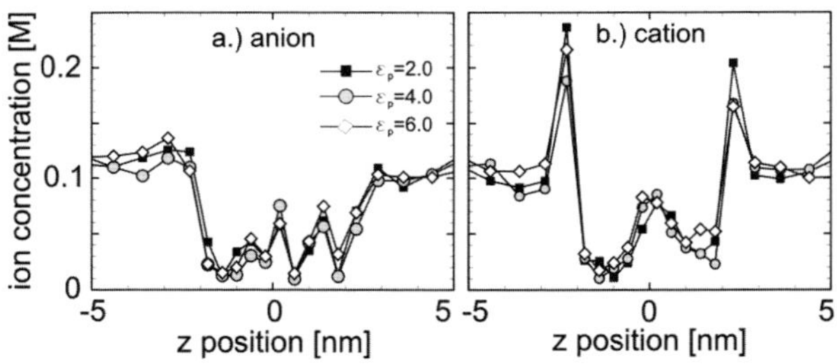

Fig. 3. The concentration of anions and cations along the porin channel axis plotted for several values of the protein dielectric constant. The electrolyte bath is 0.1M KCl and the external bias is 1.0V.

The OmpF structure shows rings of aspartic acid (Asp) residues in the xy plane, with each residue having a net charge of -e at the specific z position of the highest cation concentrations in Fig. 3b. This concentration of negative charge attracts the free positive charges in the surrounding bath, resulting in an increased cation population at the pore openings. OmpF is more cation selective at low ionic concentration, a fact that is often attributed to electrostatic screening in the pores interior [10]. Although the selectivity mechanism is still not well understood at the microscopic level

[6], one explanation may be found in the interaction of the negative Asp residue rings at the mouth of the channel. This electrostatic effect of the Asp residues is naturally screened by the surrounding bath at higher ionic concentration. In fact, as can be seen in Fig. 4, in a solution of 0.5M KCl there is no longer a peak in the cation density around the pore mouths.

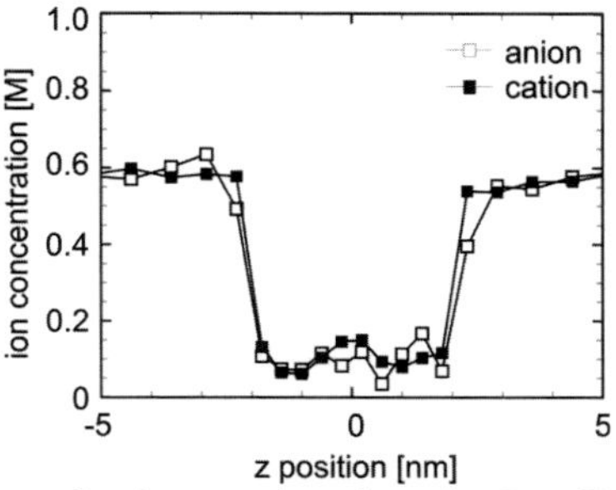

Fig. 4. The concentration of anions and cations in OmpF along the z-axis plotted for a protein dielectric constant of 2.0 at 0.5M KCl at 1.0V.

References

1. Varma, S. and Jakobsson, E. 'Ionization states of residues in OmpF and mutants: Effects of dielectric constant and interactions between residues', *Biophys. J.*, **86**, 690-704, 2004.
2. Bastug, T., Kuyucak, S., 'Role of the dielctric constants of membrane proteins and channel water in ion permeation', *Biohys. J.,* **84**, 2871-2882, 2003.
3. Aboud, S., Saraniti, M., and Eisenberg, R., 'Computational issues in modelling ion transport in biological channels: self-consistent particle-based simulations', *J. Comp. Elec.*, 2004.
4. Hockney, R. W., Eastwood, J. W., *Computer Simulation Using Particles, Adam Hilger*, 1988.
5. Wordelman, C. J., Ravaioli, U., '', *IEEE Trans. Elec. Dev.*, **47**, 410-, 2000.
6. Im, W., Roux, B., 'Ion Permeation and selectivity of OmpF porin: A theoretical study based on Molecular Dynamics, Brownian Dynamics, and continuum electrodiffusion theory', *J. Mol. Biol.*, **322**, 851-869, 2002.
7. Internet: http//www.gromacs.org, 2001.
8. Karshikoff, A., Spassov V., Cowan S. W., Ladenstein R., and Schirmer, T., 'Electrostatic properties of two porin channels from escherichia coli', *J. Mol. Biol.*, **240**, 372-, 1994.
9. Goodnick, S., personal communication.
10. Phale, P. S. Philippsen, A., Widmer, C. Phale, V. P., Rosenbusch, J. P., and Schirmer, T., 'Role of charge residues at the OmpF porin channel constriction probed by mutagenesis and simulation', *Biochemistry*, **40**, 6319, 2001.

Physical Mechanisms for Ion-Current Levelling Off in the KcsA Channel Through Combined Monte Carlo/Molecular Dynamics Simulations

E. Piccinini[2], F. Affinito[1], R. Brunetti[1], C. Jacoboni[1], and M. Rudan[2]

1 National Research Center S3, INFM-CNR and Dipartimento di Fisica, Università di Modena e Reggio Emilia, Italy

2 ARCES and Dipartimento di Ingegneria Elettronica, Informatica e Sistemistica, Alma Mater Università di Bologna, Italy

Summary. Conduction and noise properties of potassium ions in the *KcsA* membrane channel are analysed by means of a combined Molecular Dynamics-Monte Carlo numerical approach. The high-voltage part of the experimental *I(V)* characteristics shows a tendency to level off which is reproduced by computational results using a conduction model quite sensitive to the particular set of transition probabilities among the relevant ion occupancy configurations. Noise power spectra confirm the existence of correlation between consecutive ion exits from the channel.

1 Introduction

Ion channels are nanometric macromolecular pores in the cell membranes. Potassium channels have recently received considerable attention, especially since the publication of the atomic structure of the *KcsA* protein from *Streptomyces lividans* solved at *2.0 Å*[1]. Experiments on open single channels[2] exhibit a levelling-off effect at the highest potentials ($\pm 150 - 200\,mV$) at all tested potassium concentrations. The estimated electric field experienced by *K* ions inside the protein is of the order of $10^7\,V/m$.

Several Molecular Dynamics (MD) analyses[3] aimed at understanding the relevant functional features of the *KcsA* protein on an atomic basis confirm the hypothesis that the permeation process takes place as a single-file concerted motion of ions and they allow one to study the free-energy landscape between specific occupancy configurations. MD, however, is still unable to simulate the ionic flux due to the long time scale involved (ms). The aim of this work is to explore, by means of Monte Carlo (MC) simulations, the microscopic transport features which lead to the current levelling-off as a function of the applied potential and to evaluate the degree of correlation of ion motion from the analysis of the current noise spectrum.

2 Theoretical and Computational Framework

It has been established[3] that the *KcsA* channel exhibits six stable binding sites $(S_0, S_1, S_2, S_3, S_4, S_{cav})$ in which alternatively potassium ions and water molecules are found. MD estimates of the free-energy landscape between two given ion occupancy configurations are feasable, and the probability per unit time $k_{A \to B}$ to undergo a transition from channel configuration A to configuration B can be estimated combining MD with the transition-state theory. More details about our MD scheme can be found in [4].

The theoretical scheme described above is implemented in a MC simulation which generates random transitions between ion occupancy configurations, resulting in a net current through the channel evaluated as the net flux of incoming and outcoming ions at a channel boundary in a defined time interval Δt. Fluctuations arising from the discrete nature of current flow yield a "shot noise" analogous to that observed in electronic devices. The spectral density obtained from the classical theory of shot noise [5], in the hypothesis of Poissonian independent events, is given by the Schottky's formula: $S_{shot} = 2qI$, where q is the charge of the moving ion. Usually the spectral density is normalised to S_{shot}, and the Fano factor $F = S_I(\omega)/S_{shot}$ is taken as a measure of the degree of correlation in the ion motion. This correlation arises because of the single-file motion in presence of a strong Coulomb repulsion of ions in the channel. The current spectral density is obtained from MC by averaging the squared Fourier transform of the current fluctuations $\delta I(\omega)$.

3 Results and Discussion

The four-step mechanisms:

$(S_1,S_3,S_{cav}) \leftrightarrow (S_0,S_2,S_4) \leftrightarrow (S_2,S_4) \leftrightarrow (S_1,S_3) \leftrightarrow (S_1,S_3,S_{cav})$,

or, equivalently:

$(S_1,S_3,S_{cav}) \leftrightarrow (S_0,S_2,S_4) \leftrightarrow (S_2,S_4) \leftrightarrow (S_2,S_4,S_{cav}) \leftrightarrow (S_1,S_3,S_{cav})$.

are mainly responsible for the translocation of a single ion from the intracellular to the extracellular reservoir (and viceversa). After a complete cycle, the system changes its energy by an amount equal to qV, where V is the applied potential. Here we discuss how *I(V)* characteristics are sensitive to the fraction of the total electric potential energy associated with each of the steps above. We also analyse the influence of the input parameters of the MC code relevant for the permeation process (i.e., barrier heights and widths), keeping the experimental data in [2] as a reference. To this aim MD simulations provide an estimate of the free-energy barriers between different configurations in the absence of an external field, with an accuracy dependent on the choice of the selected force field and reaction coordinates, and the numerical techniques used to speed up computational convergence. Barrier heights used in our MC calculations usually range from 2.5 to 5.3 in units of KT (K being the Boltzmann constant and T the temperature); higher values (as sometimes MD simulations provide) lead to current saturation at voltages much higher than those suggested by the experiments. Furthermore, the existence of voltage-independent transition rates dominating the ion flow is suggested above 100mV to account for current levelling-off. The three-ion concerted motion $(S_1,S_3,S_{cav}) \leftrightarrow (S_0,S_2,S_4)$ is crucial for the ion flux: it exhibits the highest barriers at zero bias and it determines the order of magnitude of the current. In order to tune simulated to experimental data we found that most of the energy variation due to the external potential (up to 91%), is affecting this transition, being the residual energy distributed 1/3 to the transitions to/from cavity S_{cav} and 2/3 to the transitions from/to the outer mouth S_0. The two-ion motion is substantially unaltered. The resulting *I(V)* characteristics is reported in Fig. 1-a (continuous line; the dot-dashed line refers to the case in which all the potential drops on the transition involving the three-ion concerted motion). Alternatively, the main variation can be attributed to the two-ion motion, leaving unaltered the three-ion motion, but results are less satisfactory, expecially for forward bias (see dotted and dashed lines in Fig. 1-a).

Finally, the barrier width affects the pre-exponential factor appearing in the transition-state theory formula and takes into account the number and the type of reaction coordinates involved. The rates associated to the permeation path range from 1×10^{8} to $1\times10^{9}s^{-1}$.
An example of calculated noise power spectrum as a function of frequency is shown in Fig.1-b. White noise is found at low frequencies, until about $5x10^{6}$ Hz, which corresponds to $1/\langle T_{exit}\rangle$, being $\langle T_{exit}\rangle$ the average time between two successive ion exits. The obtained Fano factor is about 0.75, clearly indicating that correlation in the ion motion lowers the noise spectrum with respect to the Poissonian shot-noise value.

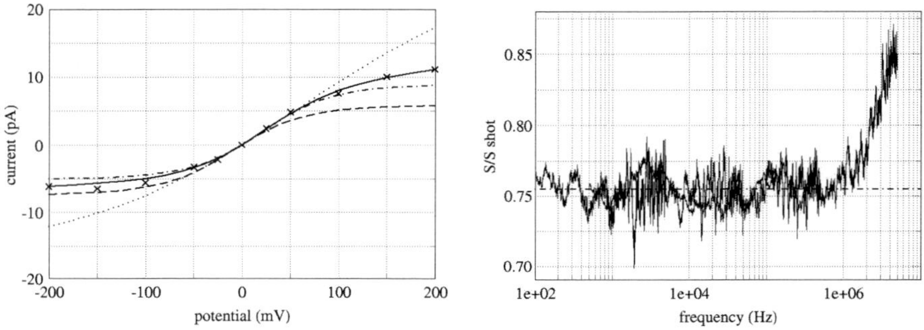

Fig. 1. a) Experimental and simulated *I(V)* characteristics (crosses refer to experiments in [2], curves refer to MC simulations (see text)). b) S/S_{shot} as a function of frequency at *V=100 mV* (see text). The horizontal line represents the average value of the Fano factor. A *K+* concentration of *100 mM* is assumed.

The results of the present computational study suggest that future research must be focused on MD simulations of the channel protein in the presence of an explicit electric field in order to explore the complex dependence of the barrier heights on the field strength.

References

1. Zhou Y., Morais-Cabral J., Kaufman K., and MacKinnon R., *Nature* **414**, 43-48 (2001).
2. LeMasurier M., Heginbotham, and C. Miller, *J. Gen. Physiol.,* **118**, 303-313 (2001).
3. For a review see, e.g.: Roux B., *Curr. Opin. Struct. Biol.* **12**, 182-189 (2002).
4. Affinito F., Brunetti R., Jacoboni C., Piccinini E., Rudan M., Bigiani A., and Carloni P., *J. Comp. Elect* **4**,171-174 (2005).
5. Shottky W. *Ann. Phys. (Leipzig)* **57**, 541-567 (1917).

Simulations of the Gramicidin A Channel by Using the TR-PNP Model

S. Hu and K. Hess

Beckman Institute, University of Illinois

Summary. We propose an extension (TR-PNP) of the traditional PNP model by explicitly taking into account the effects of ion trapping and release. Our solution to this problem has originated from the treatment introduced by Shockley, Read and Hall model in semiconductor theory. Simulation results are presented for the Gramicidin A channel by using both the TR-PNP and the PNP models and are compared with experiments.

1 Introduction

Living cells are surrounded by a thin membrane consisting of two layers of phospholipid molecules called lipid bilayer. Ion channels, a class of specialized protein molecules across the membrane, form water-filled nanopores through which ions can propagate. Many channels have the ability to selectively conduct or block the flow of particular ion types. Some of them demonstrate switching properties, which regulate the opening and closing of the ionic currents.

A large variety of device-like properties exhibited by ion channels has recently attracted the interest of the engineering community. As a matter of fact, the theories and techniques previously widely used in solid-state devices are playing an increasingly important role in treating problems of modeling and simulating ion channels. For instance, the Poisson and drift-diffusion model (usually named the Poisson-Nernst-Planck (PNP) theory by the biophysics community) has been utilized to predict ion currents through the channel [1] and this approach has achieved some satisfactory results. In addition, a partial differential equation solver initially intended for device simulation, PROPEHT [2], has been employed for solving 3-D PNP system of equations with high efficiency and robustness [3]. In the present paper, we introduce another well established theory of semiconductor devices into the arena of ion channel simulations with the hope to expand the understanding of this area.

2 The TR-PNP Formalism Based on the Shockley-Read-Hall Model

The binding sites of ions and the associated ion trapping and release effects have been demonstrated to be crucial for a wide variety of ion channels. The traditional PNP model is not able to cover this specific particle like property. The proposed TR-PNP model [4] explicitly treats the dynamic processes of ion trapping and release based on the approach of the Shockley-Read-Hall (SRH) model that is extensively used in device simulation. In the TR-PNP model, three characteristic parameters are used to express the ion trapping rate and ion release rate. They are the ion trapping coefficient c, the ion release coefficient e and the total trap density N_T. From a theoretical perspective, the traditional PNP model is not complete because it neglects the recombination and generation terms. These terms may, however, also be important in ion channels as soon as chemical interactions of the ions with the channel walls become significant.

The three characteristic parameters can be approximated based on the size (diameter) and mass of the particular ion type, the grid size of mesh and most importantly, the binding site properties. In particular, the ion release rate e is related to the potential of mean force (PMF) calculated from molecular dynamics (MD) simulations in terms of the Erying rate theory.

3 The Gramicidin A Channel

The Gramicidin A (GA) channel is an antibiotic polypeptide that consists of 15 amino acid residues. It is approximately 24Å in length and 4Å in diameter. It selectively conducts monovalent cations, binds divalent cations and completely rejects all anions. Because of its relatively small and simple structure, the GA channel has been extensively studied both experimentally and theoretically. The results from the NMR measurement and the MD simulation agree qualitatively with each other and both suggest two binding sites at each end of the channel.

The detailed structural information of the GA channel is accessible from the Protein Data Bank (PDB ID: 1MAG) [5]. The original PDB file is transformed into a PQR file by using the PDB2PQR script, one of the tools in the Adaptive Poisson-Boltzmann Solver (APBS) package [6], combined with the CHARMM force field. The BioMOCA package [7] is then employed to generate the material map file. Finally the mesh file and other input files (such as the characteristic parameters and the protein charge profile) are obtained through our own developed code. We have used PROPHET [2] as our simulation tool.

4 Simulation Results

We have employed the TR-PNP model (and the traditional PNP model as well for comparison) to predict the macroscopic properties of the GA channel. The simulation results of the current-voltage (I-V) curves and current-concentration (I-C) curves are displayed together with the experimental data [8] in Fig. 1(a) and Fig. 1(b), respectively.

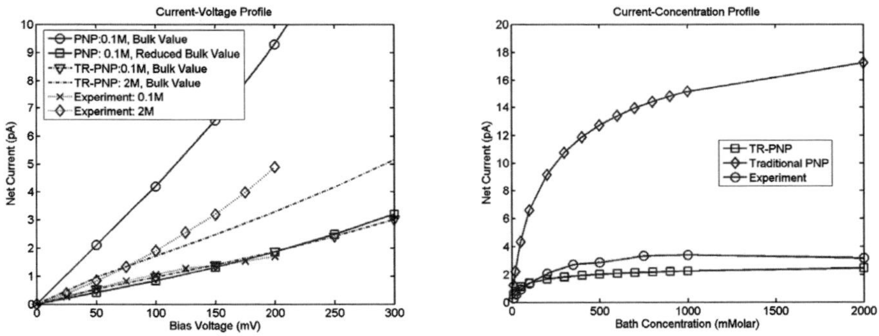

Fig. 1. (a) I-V curves with bath concentration fixed at 0.1M and 2M respectively, (b) I-C curves with bias voltage fixed at 150mV

The I-V curves predicted by the TR-PNP model agree very well with some measurements for (0.1M); only within a factor of 2 for others (2M). For the traditional PNP model, the diffusion coefficient has to be scaled down by a factor of 10 to achieve agreement with experiments. This, however, conflicts with the results of MD calculations. This necessary reduction of the effective diffusion coefficients [9] inside the channel pore is, in our opinion, necessitated by the lack of completeness of the PNP model and can be explained by the mechanism of ion trapping and release effects.

The I-C curves (fixed bias voltage at 150mV) predicted by the TR-PNP model agree with measurement within a factor of 1.5. The saturation effect is clearly reproduced and the half-saturation concentration is very close to the experimental value. In contrast, the traditional PNP model does not reproduce the saturation satisfactorily.

5 Conclusion

In conclusion, the TR-PNP model is capable of accounting for some particle properties in the framework of non-equilibrium continuum models. It provides a bridge connecting the microscopic and macroscopic models and simulation methodologies. It is a promising candidate to being integrated into the multi-scale simulation hierarchy.

References

1. Chen, D., Lear J., and Eisenberg, B.: 'Permeation Through an Open Channel: Poisson-Nernst-Planck Theory of a Synthetic Ionic Channel', *Biophys. J.*, **72**, 97-116, 1997.
2. http://www-tcad.stanford.edu/~prophet/
3. Van der Straaten, T.A., et. al.: 'Simulating Ion Permeation Through the ompF Porin Channel Using Three-Dimensional Drift-Diffusion Theory', *J. Comp. Elec.*, **2**, 29-47, 2003.
4. Hu, S., Hess, K.: 'An Application of the Recombination and Generation Theory by Shockley, Read and Hall to Biological Ion Channels', *J. Comp. Elec.*, **4**, 153-156, 2005.
5. http://www.rcsb.org/pdb/
6. Baker, N. A., Sept, D., Joseph, S., Holst, M. J., and McCammon, J. A.: 'Electrostatics of Nanosystems: Application to Microtubules and the Ribosome', *Proc. Natl. Acad. Sci. USA*, **98**, 10037-10041, 2001
7. Van der Straaten, T.A., et. al.: 'BioMOCA – A Boltzmann Transport Monte Carlo Model for Ion Channel Simulation', *Molecular Simulation*, **31**, 151-171, 2005.
8. Busath, D. D., et. al.: 'Noncontact Dipole Effects on Channel Permeation. I. Experiments with (5F-Indole)Trp^{13} Gramicidin A Channels', *Biophys. J.*, **75**, 2830-2844, 1998
9. Chiu, S-W, Novotny J. A. and Jakobsson, E.: 'The Nature of Ion and Water Barrier Crossings in a Simulated Ion Channel', *Biophys. J.*, **64**, 98-108, 1998

Phonon Emission and Absorption by Holes in the HOMO Bands of Duplex DNA

T. Yamanaka, M. Dutta, T. Rajh and M. A. Stroscio

Electrical and Computer Engineering Department, University of Illinois at Chicago (TY, MD, and MAS), Physics Department, University of Illinois at Chicago (MD and MAS), Chemistry Division, Argonne National Laboratory, Argonne, Illinois (TR), Bioengineering Department, University of Illinois at Chicago (MAS)

Summary. This paper focuses on the role of phonon absorption and emission in hole transport in duplex DNA. In these phonon emission and absorption calculations, carrier-phonon optical-deformation potential interactions in a quantum wire are modeled for hole-phonon interactions in duplex DNA.

1 Introduction

Based on recent studies of phonon spectra in DNA [1,2] and these phonon absorption and emission rates, hole transport is partially understood in light recent experimental studies of hole transport in DNA [3-6]. These results are especially interesting in the case when the hole is injected from a quantum dot linked covalently to the DNA as in the case of Rajh et al. [3] where TiO_2 quantum dots with dopamine (DA) surface ligands are bound to DNA. In these phonon emission and absorption calculations, the models of Stroscio and Dutta [7] for phonon emission and absorption in a quantum wire with the carrier-phonon interaction mediated via the Fröhlich coupling have been adapted to the case of hole-phonon interactions through the optical-deformation potential. The scattering rates presented in this paper provide a basis for studies of hole transport in DNA; in particular, if hole has enough energy to emit a phonon it may fall in the band of states near the HOMO level and it may become trapped near a region of guanine-guanine pairs since guanine has the lowest ionization potential --- and this corresponds to the highest HOMO level --- of the four bases in DNA. These results also provide the basis for future studies of nanoscale semiconductors integrated with biostructures [8,9].

2 Scattering Rates

The transport of holes in DNA via phonon-assisted processes has been considered by a number of authors in connection with the formation of polaron by holes in the negatively charged DNA molecule and as a possible means of enhancing carrier transport [3-6]. The rates of phonon emission and absorption in DNA have not been studied in detail and such studies are essential since these rates may play a major role hole detrapping in regions of DNA where the DNA base sequences result in a maximum value of the HOMO level near guanine-rich regions of the DNA. Accordingly, this paper models these phonon emission and absorption rates in terms of an optical deformation potential interaction having the form of the carrier-phonon interaction in the Su-Schrieffer-Heeger (SSH) Hamiltonian [10]. Measurements and models of the phonon spectrum in DNA [1-2] indicate that the phonon spectrum of DNA extends from zero to over 200 meV. Based on the hole band model of Conwell and Rakhmanova [5] and the SSH model, the hole energy in DNA is reduced by 0.25 eV due to polaron formation and the emission/absorption rates of phonons by holes is given by [5, 7, 10]:

$$\frac{1}{\tau(E)}=\frac{D^2}{4\pi^2\hbar}\cdot\frac{\hbar}{2\rho\omega}\cdot\left\{(2\pi)^2 16\left(\frac{8}{3\pi}\right)^4\right\}\cdot\frac{1}{\sqrt{\frac{\hbar^2}{2m}}\sqrt{\hbar\omega}}\left(N_q+\frac{1}{2}+\frac{1}{2}\varepsilon\right)\frac{1}{\sqrt{\frac{E}{\hbar\omega}-\varepsilon}} \quad \text{(Eq. 1)}$$

where D is the hole-phonon interaction coefficient *which is 6 eV/nm* [5], $\hbar$ is Planck's constant divided by 2π, ρ is the mass per unit length of the DNA molecule – 1.3×10^{-15} kg/m, ω is the angular phonon frequency, N_q is the temperature-dependent phonon occupation number, E is the energy of the hole, $\varepsilon = -1$ for absorption, and $\varepsilon = +1$ for emission. For the case of phonon emission, the scattering rate is zero below the phonon emission threshold $E = E_{phonon}$ and the rate for $E > E_{phonon}$ follows Eq.1 with $\varepsilon = +1$. For the case of phonon absorption, the Eq.1 holds for $E > 0$ with $\varepsilon = -1$. In addition, m denotes the hole effective mass. From Eq. 1, the emission/absorption rates depend weakly – as $m^{1/2}$ - on the hole effective mass. Thus, m is simply taken as the electron mass in our subsequent numerical evaluations. Figure 1 illustrates the room-temperature phonon emission rates for phonons for the case where the hole mass is taken as unity; Figure 2 gives corresponding absorption rates. Total absorption rates may be larger since the result of Bykhovskaia et al. shows groups of tens of modes around specific phonon energies [11].

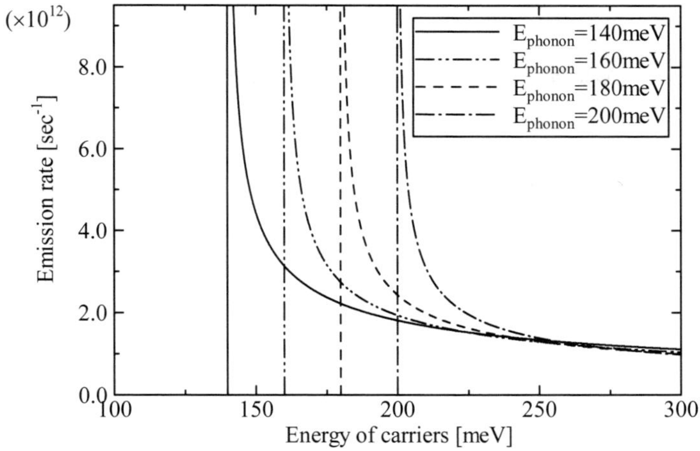

Fig. 1. Phonon emission rates with E_{phonon}=140, 160, 180, and 200 meV.

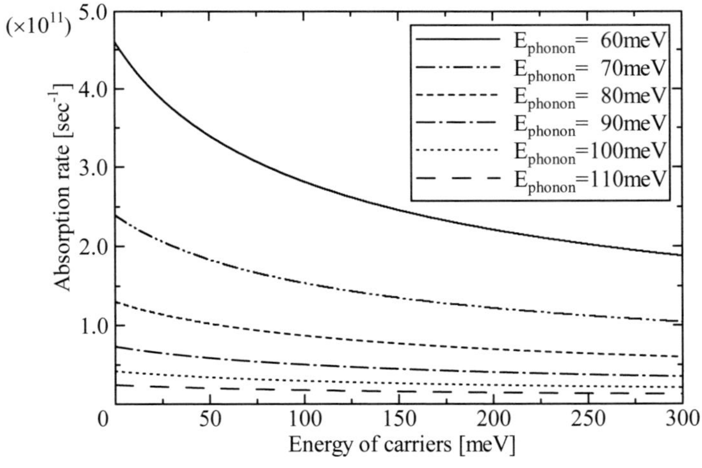

Fig. 2. Phonon absorption rates with E_{phonon}=60, 70, 80, 90, 100, and 110 meV.

3 Discussion

These calculations indicate the phonon emission rates for hole energies of 10 meV above the phonon emission threshold range from 4.5-2.5 x 10^{12} sec^{-1} for E_{phonon} in the range 140-200 meV. Moreover, phonon absorption rates for hole energies of 110 meV range from 2.8-0.18 x 10^{12} sec^{-1} for E_{phonon} in the range 60-100 meV. Thus, phonon emission occurs in a time of less than 0.1 ps for the indicated parameters, and phonon absorption occurs on a scale of very approximately a ps.

4 Summary

These results indicate that multiple-phonon absorption events of holes in DNA may take place on time scales of several ps for holes with energies below the phonon emission threshold.

References

1. Woolard, D. L., Globus, T. R., Gelmont, B. L., Bykhovskaia, M., Samuels, A. C., Cookmeyer, D., Hester, J. L., Crowe, T. W., Jensen, J. O., Jensen, J. L., and Loerop, W. R.: 'Submillimeter-wave phonon modes in DNA macromolecules', *Phys. Rev. E*, **65**, 051903-1-11, 2002.
2. Cao, Y., Jin, R., and Mirkin, C. A.: 'DNA-modified core shell Ag/Au nanoparticles', *J. Am. Chem. Soc.*, **123**, 7961-7962, 2001.
3. Rajh, T., Saponjic, Z., Liu, J., Dimitrijevic, N. M., Scherer, N. F., Vega-Arroyo, M., Zapol, P., Curtiss, L. A., and Thurnauer, M.: 'Charge transfer across the nanocrystal-DNA interface: Probing DNA recognition', *Nano Lett.*, **4**, 1017-1023, 2004; Rajh, T., Chen, L. X., Lukas, K., Liu, T., Thurnauer, M. C., and Tiede, D. M.: 'Surface restructuring of nanoparticles: An efficient route for ligand-metal oxide crosstalk', *J. Phys. Chem. B*, **106**, 10543-10552, 2002.
4. Schuster, G. B.: 'Long-range charge transfer in DNA: Transient structural distortions control the distance dependence', *Acc. Chem. Res.*, **33**, 253-260, 2000.
5. Conwell, E.: 'Polarons and transport in DNA', *Topics in Current Chemistry*, **237**, 73-101, 2004; Rakhmanova, S. V., and Conwell, E. M.: 'Polarons motion in DNA', *J. Phys. Chem.*, **105**, 2056-2061, 2001.
6. Schuster, G. B., and Landman, U.: 'The mechanism of long-distance radical cation transport in duplex DNA: Ion-gated hopping of polaron-like distortions', *Topics in Current Chemistry*, **236**, 139-161, 2004.
7. Stroscio, M. A., and Dutta, M.: Phonons in Nanostructures, Cambridge University Press, 2001; Stroscio, M. A.: 'Interaction between longitudinal-optical phonon modes of a rectangular quantum wire with charge carriers of a one-dimensional electron gas', *Phys. Rev. B*, **40**, 6428-6432, 1989; Stroscio, M. A., and Dutta, M.: 'Biologically-inspired chemically-directed self-assembly of semiconductor quantum-dot-based systems: phonon-hole scattering in DNA bound to DNA-quantum-dot complexes', International Journal of High Speed Electronics, in press.
8. Alexson, D., Li, Y., Ramadurai, D., Shi, P., George, L., George, L., Uddin, M., Thomas, P., Rufo, S., Dutta, M., and Stroscio, M. A.: 'Binding of semiconductor quantum dots to cellular integrins', *IEEE Trans. Nanotechnol.*, **3**, 86-92, 2004.
9. Stroscio, M. A., Dutta, M., Rufo, S., and Yang, J.: 'Dispersion and damping of acoustic phonons in quantum dots', *IEEE Trans. Nanotechnol.*, **3**, 32-36, 2004.
10. Su, W. P., Schrieffer, J. R., and Heeger, A. J.: 'Soliton excitations in polyacetylene', *Phys. Rev. B*, **22**, 2099-2111, 1980.
11. Bykhovskaia, M., Gelmont, B.,Globus, T., Woolard, D. L., Samuels, A. C., Duong, T. H., and Zakrzewska, K.: 'Prediction of DNA far-IR absorption spectra based on normal mode analysis', Theor. Chem. Acc., **106**, 22-27, 2001.

An Impedance Network Model for the Electrical Properties of a Single-Protein Nanodevice

V. Akimov[1], E. Alfinito[1], C. Pennetta[1], L. Reggiani[1], J. Minic[2], T. Gorojankina[2], E. Pajot-Augy[2] and R. Salesse[2]

(1) Dipartimento di Ingegneria dell'Innovazione, CNR-INFM NNL, Università di Lecce, 73100-Lecce, Italy, (2) Neurobiologie de l'Olfaction et de la Prise Alimentaire, Equipe Récepteurs et Communication Chimique, Domaine de Vilvert, 78352 Jouy en Josas Cedex, France.

Summary. A simple impedance network model to mimic the electrical properties of a single protein molecule nanodevice is presented. Within this model two sensing proteins of the GPCR family (bovine light-sensing rhodopsin and rat I7 olfactory receptor) in ground and activated states are studied. We predict a detectable impedance difference between ground and activated states of both proteins so proteins of this family are promising candidates as nanobiosensors.

1 Introduction

G protein-coupled receptors (GPCRs) constitute the largest family of membrane receptors, with functions going from transducing light to mediating the response to drugs, hormones, etc [1]. The challenge is to exploit the possibility to obtain an electrical signal from the conformational change of a GPCR associated with its sensing action. Here we present an impedance network (IN) which mimics the fundamental and activated states of two GPCRs and thus provides expectation of the corresponding change in their electrical properties.

2 Model and Results

By taking rhodopsin as the prototype of a GPCR, we have considered its spatial structure in contact with two metallic electrodes through which an external voltage sufficiently low to justify linear response is applied as sketched in Fig. 1. The device under test is modeled as an equivalent circuit consisting of an IN presented as a simple non-directed graph. The nodes of this graph correspond to the aminoacids of the protein and the links between any couple of nodes characterize some kind of interaction

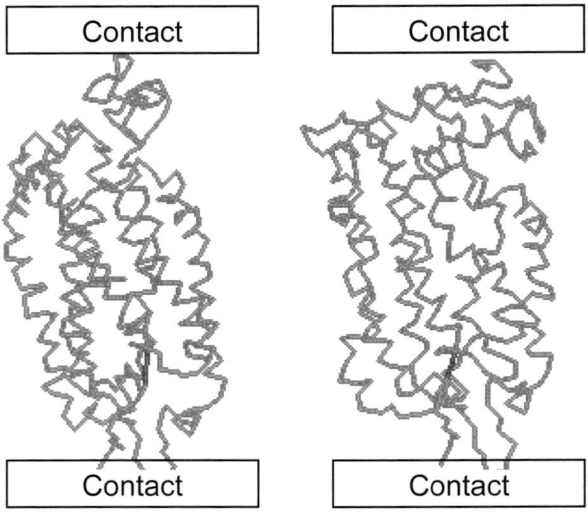

Fig. 1. Rhodopsin (left) and Metarhodopsin II (right) backbones in scale.

between aminoacids, which are neighboring in space within a given radius R_a. By associating an elemental RC impedance to each link between two neighboring amino acids we have constructed the corresponding IN which mimics the electrical response of the rhodopsin in its fundamental and light-activated state (Metarhodopsin II), as sketched in Fig. 2 (a) [2]. The network impedance depends on the radius of interaction R_a, which is chosen as the independent variable. Other parameters are chosen as follows: $\rho=10^6\ \Omega\times$m, $2\pi\omega=50$ Hz and ε of each link is determined from the intrinsic polarizability of the corresponding aminoacid [3]. Then the network has been analyzed in terms of its graph properties and the equivalent network circuit solved to determine its global impedance. The model predicts a detectable change of the global impedance (up to about 30%) associated with the conformational change due to the sensing action (see Fig. 3). The application to the case of I7 rat olfactory receptor has revealed similar results (see Fig. 3). The spatial structure of the core of I7 protein in both states was constructed with homologous modeling on the basis of rhodopsin, which is a standard technique and proven to be acceptable. Terminals of the proteins are removed from the network because their spatial structure is strongly uncertain. Contacts are taken as attached to the first and last aminoacids of the remaining backbone.

In conclusion, we have proposed an impedance network model to study the electrical properties of nanodevices embedding single protein between metal electrodes. Application to the sensing action of GPCRs shows

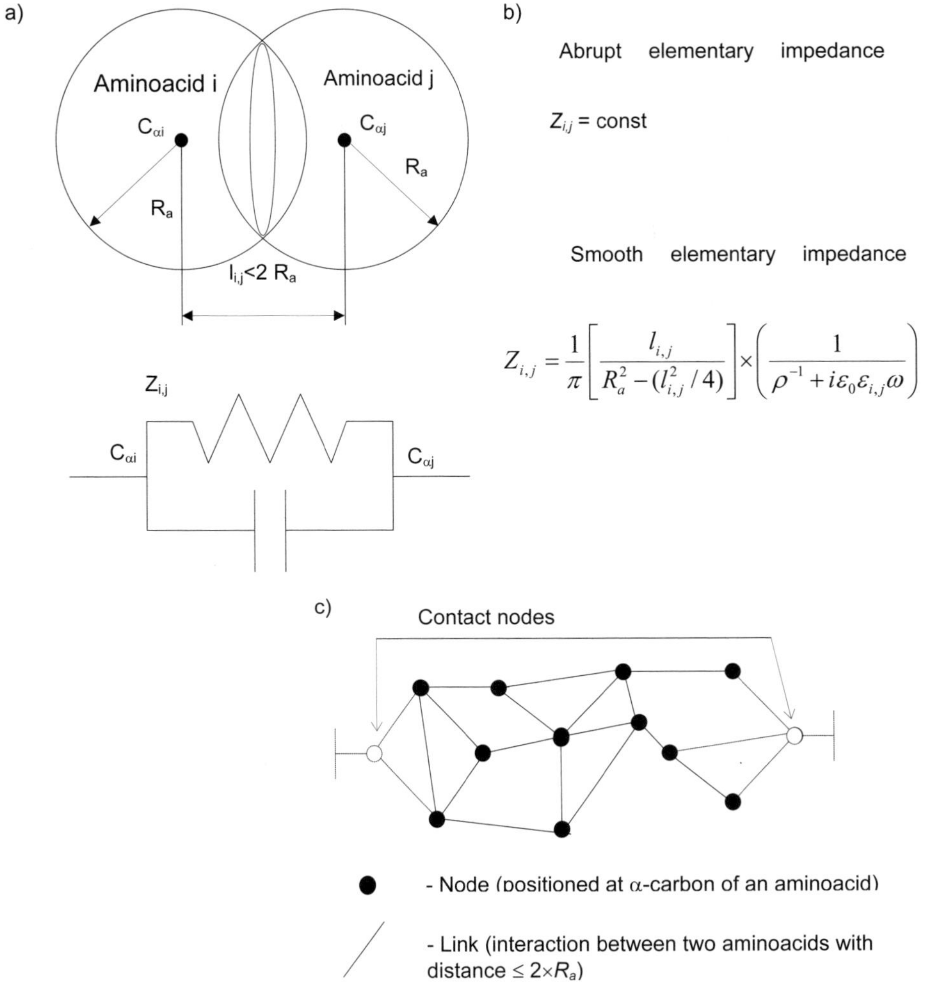

Fig. 2. (a) Overlap between two aminoacids and equivalent circuit element. $C_{\alpha i}$ identifies the center of the sphere corresponding to the alpha carbon atom of the *i*-th aminoacid. Parameter R_a is an interaction radius for the aminoacids. (b) Z_{ij} is an impedance of the link between i-th and j-th nodes. (c) Example graph for the equivalent circuit network.

promising possibilities for the development of electronic nanobiosensors, in particular based on olfactory receptors.

This research is supported by EC through the IST 2001 38899 SPOT-NOSED project.

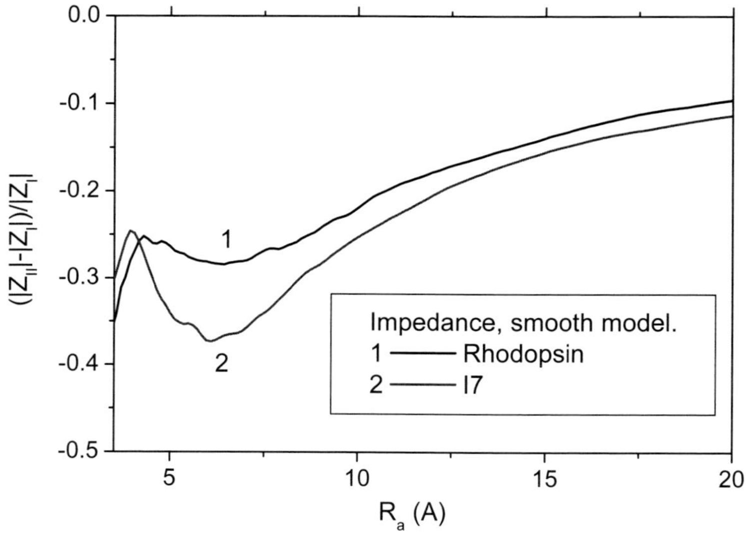

Fig. 3. Relative differences in impedance between activated and ground forms of Rhodopsin and I7 without terminals.

References

1. Lameh J. et al.: Structure and Function of G Protein Coupled Receptors, *Pharmaceutical Research*, **7** (12): 1213-1221, December 1990.
2. Pennetta C et al "Fluctuation of complex networks: Electrical properties of single nanodevice", *Spie Proc*, **5472**, p 172-184, (2004) eds. J.M. Smulko, Y. Blanter, M.I. Dikman, L.B. Kish.
3. Xueyu Song: "An inhomogeneous model of protein dielectric properties: Intrinsic polarizabilities of amino acids". *J. Chem. Phys.*, **116**, 21 (2002).

Field Effect Transistor Constructed of Novel Structure With Short-Period $(GaAs)_n/(AlAs)_m$ Superlattice

V.T. Trofimov[1], M.V. Valeiko[1], N.A. Volchkov[1], A.I. Toropov[2], K.S. Zhuravlev[2], E.V. Kiseleva[3], S.V. Obolenskii[3], M.A. Kitaev[4], V.A. Kozlov[5]

[1] P.N. Lebedev Physical Institute of RAS, 119991 Moscow, Leninskii pr. 53, Russia
[2] Institute of Semiconductor Physics, Novosibirsk, Russia
[3] N.I.Lobachevskii Nizhegorodskii University, Russia
[4] OSPP "Salujt", Nizhnii Novgorod, Russia
[5] Institute for Physics of Microstructure RAS, 603950, Nizhny Novgorod, Russia

Summary. A new type of field effect transistor is developed and realized. It is based on ballistic transport of hot electrons in a short GaAs channel. The channel is restricted by two short period superlattices. The gate has a V-grove shape. The transconductance of this FET exceeds 1 Sm/mm.

1 Introduction

The maximum of specific current in field effect transistors is achieved due to the overshoot effect when carrier motion under a high electric field becomes ballistic. To realize the ballistic motion it is necessary to accelerate an electron from the Γ-valley up to an extreme velocity as fast as possible without scattering to higher valleys. This requires a sharp increase of the high electric field inside a structure. For lateral transport it is difficult to get the high electric field with the large gradient and restrict the scattering of carriers. In this paper a novel heterostructure is described. It allows to realize a ballistic lateral transport of electrons near a heterointerface of undoped GaAs and a short-period $(GaAs)_n/(AlAs)_m$ superlattice.

2 Design of Heterostructure and Construction of the Device

The channel of the transistor with a length of 0.1 μm and a width of 50 μm were formed under a V-like groove breaking of a heavy-doped contact layer of the heterostructure (Fig. 1). A gate electrode was formed on the wall of the groove nearest to the source electrode. The base of the structure is the short-period $(GaAs)_n/(AlAs)_m$ superlattice (SL). It enables one to make a new field effect transistor (SLFet).

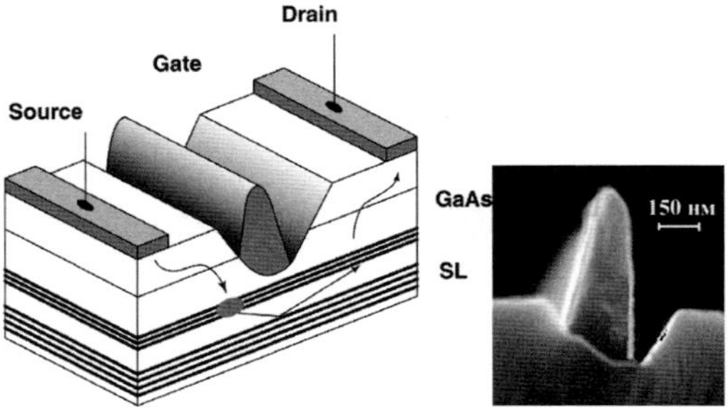

Fig. 1. Layout of $(GaAs)_n/(AlAs)_m$ short-period superlattice in SLFet transistor and SEM image of its gate (split off).

The SL period was optimized so to get a value of hot electrons reflection coefficient as high as 90%. At the top of the SL a 35 nm thick i-GaAs layer was grown containing in the upper part a δ-Si supply layer with sheet carrier concentration of $\sim 2.0 \cdot 10^{12}$ cm^{-2}, which serves as a conducting channel under a weak electric field. Under the high electric field the whole i-GaAs layer acts as the conducting channel because of the energy diagram distortion (Fig. 2).

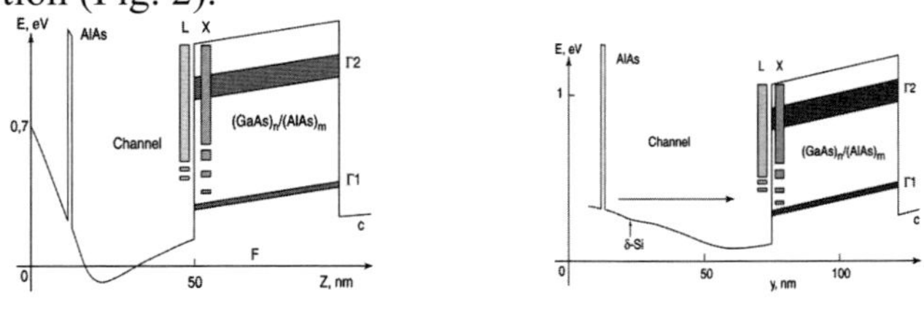

a) b)

Fig.2. SLFet energy diagram cross section in the vicinity of V-gate in the absence of electric field (a). SLFet energy diagram in high electric field along the electrons trajectories (b).

The carrier mobility in the GaAs layer was about 8000 and 65000 $cm^2/V \cdot s$ at 300 K and 77 K, respectively (comp. [1]). The conducting channel was covered by a 1.5 nm thick AlAs layer, which is used as an injector of hot electrons. The AlAs layer does not prevent penetration of electrons into the extractor zone located at the end of nanostructure close to the drain electrode.Effective scattering of hot electrons in the extractor zone leads to formation of a static negatively charged domain, which additionally increases the electric field in the ballistic zone. The structure was finally covered by a n+-GaAs<Si> contact layer. The proposed structure allows one to realize the quasiballistic regime with near the overshoot velocities over a length of more than 100 nm due to the following reasons.

1. The region of the high electric field is formed between the heavily doped injector and extractor zones where a depletion length is as low as 20 nm. It allows one realize a large gradient of the electric field. As the result electrons acquire the energy close to the X-valley edge at a time ~ 10^{-13} s which is comparable with the momentum relaxation time (Fig.3).

2. From the substrate side the electrons motion is confined by the reflector SL which forms an effective potential barrier of ~1 eV height both for cold and hot carriers of the Γ-valley and minimizes the length of electrons trajectories.

3. Electrons are injected into the high electric field region after tunneling through the 1.5 nm thick AlAs barrier, so their mean energy is much higher than the thermal one. The barrier brings an additional confinement of electrons in the injection and extraction zones.

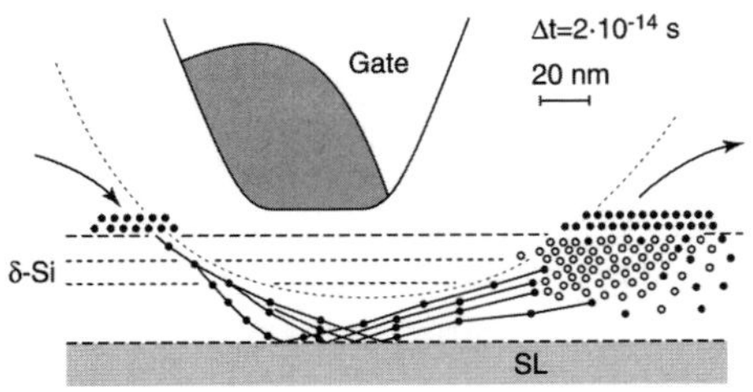

Fig.3. The electron trajectories in the vicinity of the V-gate. Dots denote the electron positions in time intervals of Δt=2·10-14 s: • - electrons localized in Γ-valley, o- electrons localized in L- and X-valleys. Dashed line shows the boundary of depletion space around the V-gate.

Transconductance (the derivative of drain current with respect to voltage on the gate electrode which controls a transversal electric field) of our FET reaches a value as high as 1 Sm/mm, which is rather high for the GaAs

FET and is comparable with the best values for p-HEMT structures with InGaAs channel (Fig. 4).

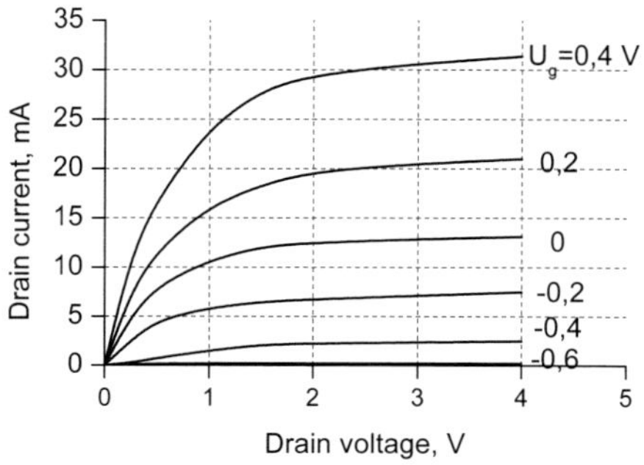

Fig.6. The dependence of drain current on source-drain voltage at T=300 K and various gate potentials, for gate length Lg~0.1 μm and its width Wg=50 μm. The value of transconductance exceeds 1 Sm/mm.

The measured capacitance of the gate electrode was ~ 30 fF. Value of the transconductance S allows one to estimate main transport properties of electron gas inside the channel. For a short-channel structure with length and width equal to L^* and W, respectively, the transconductance can be estimated as follows:

$$S = \frac{\Delta J_d}{\Delta U_g} = \frac{\Delta(e \cdot n_s \cdot v \cdot W)}{\Delta U_g} = \frac{\Delta Q}{\Delta U_g} \cdot \frac{v}{L^*} = \frac{C_0}{\tau}$$

where $\Delta Q=(e \cdot n_s \cdot L^* \cdot W)$ is the net charge of carriers in the structure, $\tau=L^*/v$ is the time of electron pass through the structure, e is the electron charge, n_s is the sheet carrier concentration and v is the mean electron velocity, C_0 is the active part of input electric capacitance which controls the dependence of electron concentration in the structure on transversal potential. The formula allows one to estimate the time of electron flight of $\tau \sim 2 \cdot 10^{-13}$s and the mean velocity of electron motion of $v \sim 5 \cdot 10^7$ cm/s.

This work was supported by RFBR Grant # 05-02-17121

References

1. Friedland K.J., et al: *Jpn. J. Appl. Phys.* **37**, 1340, 1997

Predominance of Geminate Process of Exciton Formation in AlGaAs Layers at Low Excitation

E.V. Kozhemyakina*, A.V. Efanov*, K.S. Zhuravlev*, J. Fuerst** and H. Pascher**

* Institute of Semiconductor Physics, Novosibirsk 630090, Russia
** Physikalisches Institut, Universitaet, Bayreuth, Germany

Abstract. The roles of different free exciton formation mechanisms in bulk AlGaAs were studied. The dependencies of the shape of Hanle curves on excitation power were analyzed. It was shown that geminate exciton formation mechanism dominates at low excitation intensities.

1 Introduction

At low temperatures the final stage of free carriers relaxation is coupling in excitons. In bulk III-V semiconductors the process of free exciton formation and relaxation is not well studied since purity of materials should be very high. Two processes of exciton formation are usually considered: direct (geminate) formation assisted by LO-phonons and indirect (bimolecular) formation from thermalized electron and hole. The rate of geminate formation is linear in the excitation intensity. The bimolecular formation rate is proportional to the product of electron and hole densities. It follows that for low pump intensities, the geminate process dominates [1].

In this paper we study the roles of different free exciton formation mechanisms in ultra-high purity bulk AlGaAs. For this, depolarization of photoluminescence (PL) in a transverse magnetic field (the Hanle effect) at different excitation intensities has been studied.

2 Samples and Experimental Techniques

In this study high quality $Al_{0.15}Ga_{0.75}As$ layers grown by molecular beam epitaxy were used. Details of growth were described in [2]. The samples were immersed in superfluid helium at 1.6 K. Stationary PL was excited by a laser diode with wavelength of 670 nm. The excitation power was varied from 140 μ W to 4.4 nW. The diameter of the laser spot was 200 μ m.

3 Experimental Results and Discussion

Fig. 1 shows Hanle curves taken at different excitation powers. At low excitation power they have a Lorentzian shape. An extremely high degree of circular polarization at zero magnetic field ρ = 35% was observed at wavelength of 710.7 nm, corresponding to an energy being 3.6 meV lower than the energy of the free exciton line. With increasing of excitation power, the degree of circular polarization decreases and the shape of the Hanle curves changes. A narrow contour appears and dominates at high excitation powers. The widths of the two contours differ approximately by a factor of 10.

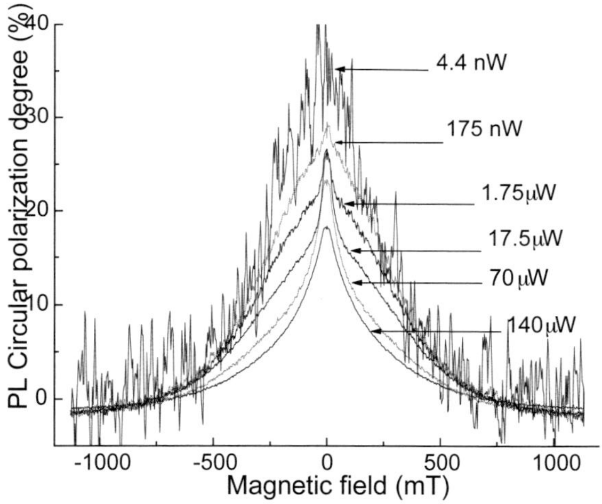

Fig. 1. Hanle curves taken at wave length of 710.7 nm.

The shape of the Hanle curves shows that there are two types of excitons that have different spin orientation degrees. We assume that the wide contour corresponds to the depolarization of geminate excitons. The narrow contour represents the two-cascade process of bimolecular exciton formation. The width of this contour is determined by free electron relaxation.

We fitted a sum of two Lorentz contours to the Hanle curves. The exciton lifetime as a function of the excitation power for both contours at different wavelengths within the PL-lines are shown in Fig. 2. The narrow contour should be described by the product of two Lorentz contours, but the correction is negligible.

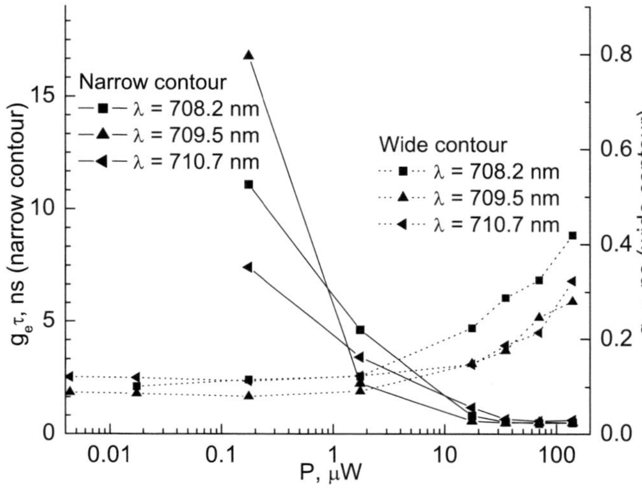

Fig. 2. Lifetime depending on excitation power defined for two contours ($g_e = 0.1$, $g_{ex} = 2.8$)

Our assumption about the origin of the two contours is confirmed by the following:

1) The lifetime defined for the narrow contour is 10 times longer than for the wide one. The lifetime of the geminate exciton should be shorter than the time of electron and hole binding in exciton.

2) Extremely high degree of circular polarization ($\rho = 35\%$) can be observed for recombination of geminate excitons only because the highest possible spin orientation degree of geminate excitons reaches 100%. The orientation of bimolecular excitons does not exceed 25%.

3) For the narrow contour the lifetime increases with decreasing excitation power. The time of electron and hole binding in exciton also increases with decreasing excitation.

Since the degree of circular polarization increases with decreasing excitation we can conclude that the role of geminate excitons dominates at low excitation power (P = 4.4 nW).

The very interesting result of our study is the short lifetime of excitons. It can be seen from Fig. 2 that it varies from 30 to 100 ps (for exciton g-factor g_{ex} = 2.8 [3]). This can be explained by trapping of excitons into localized states below the band gap characterized by short lifetime due to non-radiative recombination. This explanation is the most probable.

An alternative explanation one can give considering the exciton luminescence as Raman scattering of light by LO-phonons. In our experiment the energy difference $\Delta E = \hbar\omega - E_g = 123.8\ (meV)$ is approximately commensurable with the phonon energy $\hbar\omega_{LO} \approx 36\ (meV)$. The last stage of emission of LO-phonons may go via virtual intermediate state. In this case the outgoing photon corresponds to the energy below the free exciton band. Because of the fast transition via virtual state the observed short relaxation times in Hanle effect may be determined by relaxation of hot excitons.

4 Summary

In conclusion, a quantitative analysis of experimental Hanle curves was carried out. The lifetimes of excitons and the times of binding of electrons and holes into excitons were obtained. The dependencies of these times on the excitation intensity in different points of spectra were investigated. It was shown that geminate exciton formation mechanism dominates at low excitation intensities.

Acknowledgements

We acknowledge to the Russian Foundation for Basic Research (grant no. 04-02-16774).

References

1. C. Piermarocchi, F. Tassone, V. Savona, and A. Quattropani, Phys.Rev.B, **55**, 1333, 1997.
2. K. S. Zhuravlev, A. I. Toropov, T. S. Shamirzaev, and A. K. Bakarov, Appl. Phys. Lett. **76**, 1137, 2000.
3. K.S. Zhuravlev, A.V. Efanov, W. Kellner, H. Pascher, Physica B, **314**, 305–308, 2002.

Electron-Distribution Function for the Boltzmann Equation in Semiconductors

O. Muscato

Dipartimento di Matematica e Informatica, Università di Catania, Viale A. Doria 6, 95125 Catania (ITALY)

Summary. We prove the behavior of high-energy tails for the Boltzmann transport equation for silicon semiconductors, in the stationary and homogeneous regime, for parabolic band approximation, and acoustic and optical phonons scattering.

1 Introduction

The description of the high-energy tails for the electron energy distribution function (EED) is of paramount importance in the semiconductor environment. In this paper we shall tackle this phenomena in the Boltzmann Transport Equation (hereafter BTE) framework, and using an analytic technique, we shall prove the existence of maxwellian tails for the EED. The BTE for one conduction band is:

$$\frac{\partial f}{\partial t} + \mathbf{v}(\mathbf{k}) \cdot \nabla_{\mathbf{x}} f - \frac{q}{\hbar} \mathbf{E} \cdot \nabla_{\mathbf{k}} f = Q[f] \qquad (1.1)$$

where $f(t, \mathbf{x}, \mathbf{k})$ is the electron probability density, $\mathbf{k}$ the electron wave-vector, $\mathbf{v}$ the group velocity, $\mathbf{E}$ the electric field, $Q[f]$ the collision operator in the non degenerate case. In the sequel we shall consider the parabolic band approximation, and electron-phonon scattering (acoustic within the elastic approximation, and optical) for silicon at room temperature [1].

2 Exponential Tails

Since we expect that the behaviour of the steady-state solution of (1.1), for large values of $|\mathbf{v}|$ is

$$f \cong c \exp[-r\,|\,\mathbf{v}\,|^{S}] \qquad (2.1)$$

(where c, r and s are some real and positive constants), we introduce the following functionals [2]:

$$F_{r,s}(f) = \int f(\mathbf{k}) \exp[r\,|\,\mathbf{v}\,|^{s}] d\mathbf{k} \qquad (2.2)$$

and we study the values of r and s for which these functionals are positive and finite. In particular, we say that the function f has an *exponential tail of order s>0,* if the following supremum is positive and finite.

$$r_s^* = \sup\{r > 0 \,|\, F_{r,s}(f) < +\infty\}\,.$$

By using the *symmetric moments* of the distribution function, i.e.

$$m_p = \int f(\mathbf{k})\,|\,\mathbf{v}\,|^{2p}\, d\mathbf{k} \quad , \ p \geq 0 \text{ and real}$$

and by expanding the exponential function in (2.1) into Taylor series, we obtain (formally):

$$F_{r,s}(f) = \int f(\mathbf{k}) \sum_{n=0}^{\infty} \frac{|\,\mathbf{v}\,|^{sn}}{n!} r^n d\mathbf{k} = \sum_{n=0}^{\infty} \frac{m_{\frac{sn}{2}}}{n!} r^n \qquad (2.3)\,.$$

Then the value r_s^* can be interpreted as the radius of convergence of the series (2.3). In order to study the summability of the series (2.3), we take the symmetric moments of eq.(1.1) (in the stationary and homogeneous regime) and establish, for the sequence of the coefficients and for some s>0, the inequality [3]

$$\frac{m_{\frac{sn}{2}}}{n!} \leq \mathrm{CQ}^n \qquad (2.4)$$

(C and Q are positive constants depending also on the electric field **E**). Then from the Hadamard theorem we have

$$\lim_{n\to\infty} \sqrt[n]{\frac{m_{\frac{sn}{2}}}{n!}} = \frac{1}{r_s^*} \leq \mathrm{Q} \qquad (2.5)$$

and consequently the supremum r_s^* is strictly positive and bounded from below. That proves the existence of the tail (2.1), but no information is given about the order s.

3 Simulation Results

It has been widely proved by MC simulations [4,5], that the high-energy part of the EED for parabolic bands and electron-phonon interactions is maxwellian (i.e. s=2 in eq.(2.1)). By using a suitable logarithm transformation, we can rewrite eq.(2.1) into the following linear form,

$$y = \ln r + \frac{s}{2} x \qquad (3.1)$$

and then in figure 1 we compare this equation with s=2, and the corresponding MC data for bulk silicon (solid line). The asymptotic limit to the maxwellian is clearly indicated. Since the temperature of the maxwellian is related to the inverse of the radius of convergence r_s^*, in figure 2 we show this parameter as a function of the electric field. This figure proves that for moderate values of the electric field, the EED high-energy tails are maxwellians with the lattice temperature (dashed straight line), but for higher values this temperature increases. This behaviour can be justified as follows: r_s^* is the convergence radius of the power series (2.3), which is subject to the inequality (2.5), where Q is a function of the electric field **E**. For the future we shall try to extend this technique to the quasi-parabolic band approximation as well as to the electron-electron scattering.

References

1. Jacoboni, C. and Reggiani, L. : 'The Monte Carlo method for the solution of charge transport in semiconductors with applications to covalent materials', *Rev. Mod. Phys.*, **55**, 645-705,1983
2. Bobylev, A.V., Gamba, I.A. and Panverov, V. : 'Moment inequalities and high-energy tails for the Boltzmann equation with inelastic interactions', *J. Stat. Phys.*, **98**, 743-773,2004
3. Muscato O. : 'Moment inequalities and high-energy tails for the Boltzmann equation in semiconductors', preprint Università di Catania, 2005

4. Abramo, A. and Fiegna, C.: 'Electron energy distributions in silicon structures at low applied voltages and high electric fields ',*J. Appl. Phys.*, **80**, 889-893,1996
5. Leung, C.C.C. and Childs, P.A. : 'On the above supply voltage hot carrier distribution in semiconductors devices ', *App. Phys. Lett.*, **66**, 162-164, 1995.

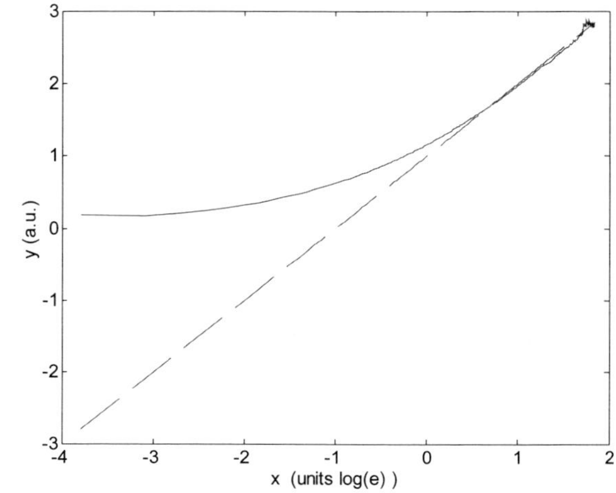

Fig. 1. Logarithmic plot of the EED obtained with MC simulation (solid curve), and with eq.(3.1) with s =2 (dashed straight line), $\mathbf{E}$ = 80 kV/cm.

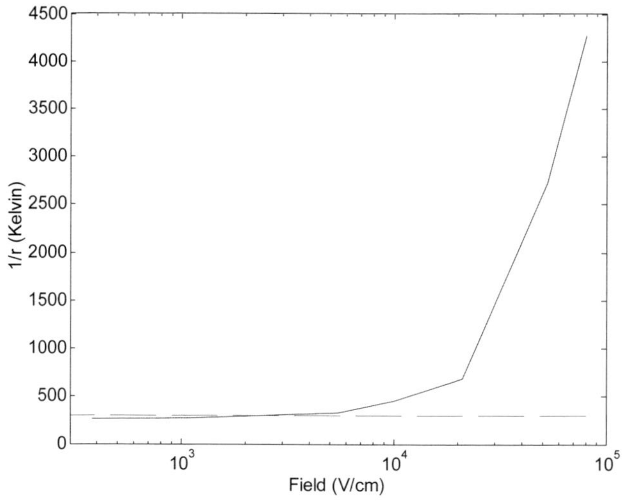

Fig. 2. The parameter $1/r$ in eq.(3.1) with s =2, as function of the electric field. The dashed straight line indicates the lattice temperature 300 K .

Giant Increase of Electron Saturated Drift Velocity in a MODFET Channel

V.G. Mokerov[1], J. Pozela[2], K. Pozela[2], V. Juciene[2]

[1]Institute of UHF Semiconductor Electronics, RAS, Moscow, Russia
[2]Semiconductor Physics Institute, Vilnius, Lithuania

Summary. The tenfold increase in electron saturated drift velocity at high electric fields in the AlGaAs/GaAs MODFET channel with InAs inserted layers containing quantum dots (QDs) in a GaAs quantum well is observed. The maximal drift velocity exceeding 10^8 cm/s is obtained. The maximal current density and transconductance of the MODFET with QDs increasing up to 35 A/cm and 10^3 mS/mm, respectively, are observed.

1 Introduction

Strong non-elastic electron-optical phonon scattering with phonon emission limits the increase of electron drift velocity at high electric fields. There are many attempts to enhance the saturated drift velocity in 2D structures. It has been shown that electron and polar optical (PO) phonon confinement in the GaAs quantum well (QW) decreases the non-elastic electron-phonon scattering rate [1]. Large decrease of the scattering rate is obtained when a GaAs QW is divided by a thin (a few monolayers) AlAs barrier. But this decrease is compensated by the increase of electron scattering rate by GaAs/AlAs interface (IF) PO phonons [2, 3]. As a result, noticeable increase of electron mobility and drift velocity is not observed in 2D structures.

With the aim to increase the saturated electron velocity we propose inserting thin (two monolayers) InAs barrier layers with high density of quantum dots (QDs) into a GaAs QW. We assume that chaotic destruction of the GaAs/InAs interface by QDs eliminates non-elastic electron scattering by the IF PO phonons.

The estimation of the maximal drift velocity $v_{\max}$ in the layer between PO phonon reflecting barriers in comparison with the saturated velocity v_S in the bulk material gives:

$$v_{\max}/v_S \approx 1+\left(L_{opt}/L_q\right)^2 , \tag{1}$$

where $L_{opt} = \pi\hbar/\sqrt{2m\hbar\omega_0}$, L_q is the width of a phonon QW, m is the effective electron mass and $\hbar\omega_0$ is the energy of a PO phonon.

In Γ-valley of GaAs, L_{opt} = 12.6 nm. At L_q = 4 nm, the maximal drift velocity in phonon QWs exceeds tenfold the saturated velocity in the bulk material.

2 The Giant Increase of Electron Drift Velocity

Figure 1 demonstrates schematically the modulation doped AlGaAs/GaAs/InAs/GaAs/InAs/GaAs heterostructures grown by molecular beam epitaxy on a GaAs semi-insulating substrate. In a thin InAs pseudomorfic layer, the self-assembled InAs quantum dots were formed. The width of the InAs/GaAs/InAs PO phonon QW of 3–5 nm corresponds, according to Eq. (1), to the tenfold increase of the saturated drift velocity. Two types of the structures were grown: S1 with high concentration of QDs of $3\cdot10^{10}$ cm^{-2} and S2 with low concentration of QDs of 10^{10} cm^{-2}.

The incorporation of InAs QD layer into the GaAs QW decreases the low field electron mobility. In structures S1, the measured electron mobility is $\mu = 3200$ cm^2/V·s, and the sheet electron concentration is $n_S = 6\cdot10^{10}$ cm^{-2}. Most part of free electrons in the GaAs layer is trapped on the QD levels and does not participate in conductivity. In structures S2 with lower QD concentration, the sheet electron concentration is much higher: $n_S = 8\cdot10^{11}$ cm^{-2}. Only a small part of electrons in GaAs layer is trapped by QDs. Electron mobility in structures S2 is $\mu_0 = 1000$ cm^2/V·s. Note that low-field mobility increases with increasing the number of QDs. It means that insertion of the InAs QD layer decreases electron – PO phonon scattering.

Figure 2 shows the measured *I-V* dependences of two types of structures. At low voltages, before ionization of QDs, the current through structure S1 is much less than that through structure S2 because of lower free electron concentration. At higher voltages, the increase of the current, due to ionization of QDs at high electric field, in structure S1 exceeds that increase in structure S2. The current in structure S2 is saturated at the level with the drift velocity of less than 10^7 cm/s. The current in structure S1 is saturated at 13 A/cm or at $n_S\, v_{\max} = 0.8 \cdot 10^{20}$ $\text{cm}^{-1}\text{s}^{-1}$.

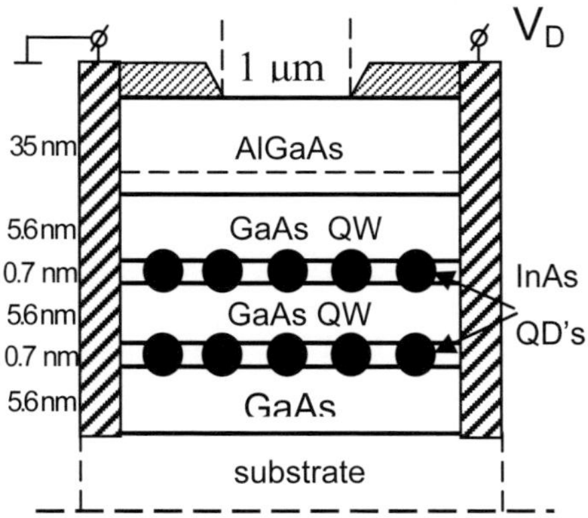

Fig. 1.

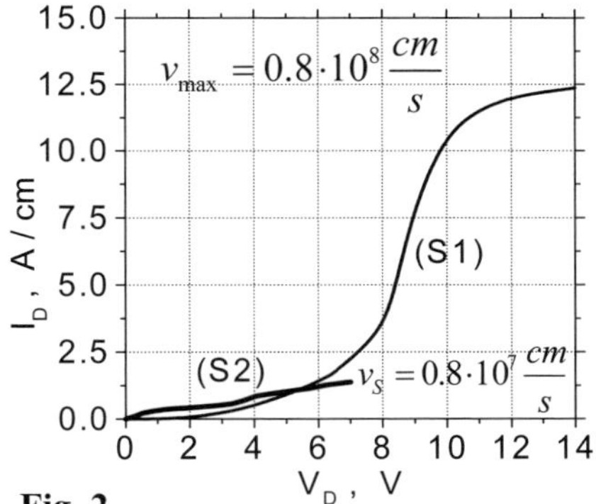

Fig. 2.

Fig. 1. A sequence of layers for AlGaAs/GaAs heterostructures with incorporated InAs layers having QDs. The dotted line shows a Si δ-doped layer ($2 \cdot 10^{12}$ cm^{-2}). **Fig. 2.** Current-voltage characteristics of the structures with high concentration, S1, and low concentration, S2, of InAs QDs.

Since the sheet electron concentration is not larger than 10^{12} cm^{-2}, we have to assume that the saturated drift velocity is $v_{\max} = 0.8 \cdot 10^8$ cm/s. This is tenfold larger than the saturated velocity $v_S = 0.8 \cdot 10^7$ cm/s in bulk GaAs at an electric field higher than 60 kV/cm.

We assume, that for so giant increase of the saturated drift velocity, are responsible both the confinement of electrons and PO-phonons in AlGaAs/GaAs/InAs/GaAs/InAs QWs and the suppression of the GaAs/InAs IF phonon mode by forming a large number of InAs QDs.

3 MODFET with QD Layers in a Channel

The structures AlGaAs/GaAs/InAs/GaAs/InAs/GaAs were used for creation of a new type of the FET with QDs incorporated into InAs layer (QD-MODFET) [4]. The *I–V* characteristics of the QD–MODFET based on the S2-type structure were similar to those of the conventional MODFET. The QD-MODFET with high density of QDs in the channel (S1-type structure) has unusual characteristics, quite different from those of the conventional FET (Fig. 3).

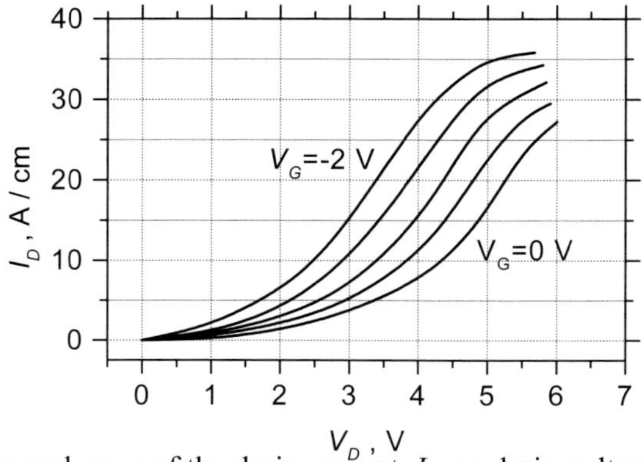

Fig. 3. The dependences of the drain current I_D on drain voltage V_D for the QD-MODFET with the submicron gate of $L_g = 0.4$ μm.

The transconductance of the QD-MODFET changes its sign depending on the drain voltage. The ionization of QDs at low drain voltages and the enhanced drift velocity saturation at high drain voltages are responsible for the specific *I–V* characteristics of the QD–MODFET.

The tenfold increase of the maximal drift velocity is responsible for the tenfold increase of the saturated current of up to 35 A/cm (Fig. 3) and for higher transconductance of $g_m = 1300$ mS/mm compared with those in the conventional FET. After ionization of QDs, the *I–V* characteristics of the QD-MODFET become similar to characteristics of the conventional FET with the tenfold increase of maximal drift velocity. We assume that the enhanced transconductance can exceed 10^4 mS/mm.

References

1. Pozela, J., Pozela, K., Juciene, V.: 'Electron mobility and electron scattering by polar optical phonons in heterostructure quantum wells', *Semiconductors,* **34**, 1011-1015, 2000.
2. Pozela, J., Namajunas, A., Pozela, K., Juciene, V.: 'Polar optical phonon confinement in quantum wells', *Physica E,* **5**, 108-116, 1999.
3. Bennett, C. R., Amato, M. A., Zakhleniuk, N. A., Ridley, B. K., Babiker M.: 'Effects of monolayer on the electron-phonon scattering rates in a quantum well: Dielectric continuum versus hybrid model', *J. Appl. Phys.,* **83**, 1499-1506, 1998.
4. Mokerov, V. G., Pozela, Yu. K., Fedorov, Yu. V.: 'Electron transport in unipolar heterostructure transistors with quantum dots in strong electric fields', *Semiconductors,* **37**, 1217-1221, 2003.

Technological Crossroads: Silicon or III-V for Future Generation Nanotransistors

M.J. Gilbert and D.K. Ferry

Department of Electrical Engineering and Center for Solid State Electronics Research, Arizona State University, Tempe, AZ 85287-5706

Summary. We present the results of a three-dimensional, self-consistent ballistic quantum mechanical simulation of an indium arsenide (InAs) quantum wire metal oxide semiconductor field effect transistor (MOSFET) with channel lengths of approximately 10 nm. We find that these devices exhibit exceptional I_{on}/I_{off} ratio, reasonable subthreshold swing and reduced threshold voltage variation. Finally, we compare the performance of the 10 nm InAs tri-gate device to a similar silicon device. We find that, when a suitable gate material is chosen, the InAs devices perform comparably to silicon devices in the ballistic limit.

While the prospects look promising for SNWTs, there are also other materials under consideration for next generation transistor applications. Within the last few years, there has been a renewed interest in the use of III-V materials instead of silicon for devices. This is mainly due to the fact that the III-V based transistors have the possibility of boasting much higher channel mobilities when compared to that of silicon. In this paper, we compare the performance of similar InAs and silicon nanowire transistors.

In Fig. 1, we display a schematic of the device geometry used for an InAs MOSFET. The thickness of the InAs layer is 9.09 nm. The source and drain of the device are *n*-type with a doping density of 6×10^{18} cm^{-3}, while the channel of the device is considered to be *p*-type, but undoped. The gate material is assumed to be platinum and the gate oxide on each side is composed of 1 nm of hafnium oxide (HfO_2). Underneath the device, we have assumed a generic insulating substrate. We model the device using recursive scattering matrices [1] where self-consistency is accelerated using the modified Broyden's method [2].

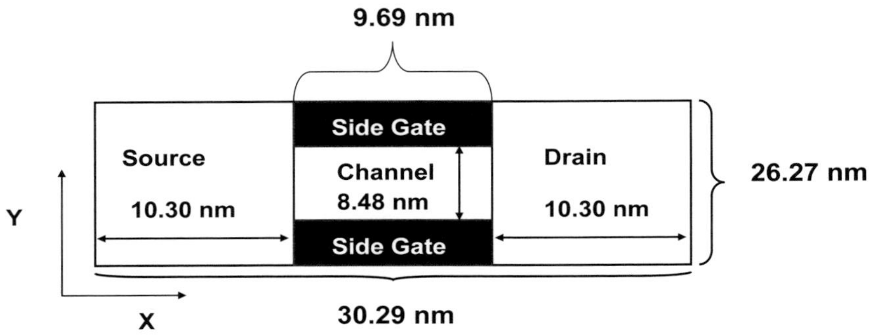

Fig. 1: Schematic of the device under consideration in the *xy* plane.

It should be noted that all of the simulations presented here have been performed at 300 K and ballistic transport is assumed throughout the paper.

In Fig. 2, we plot the I_d-V_g curves for the 10 nm InAs quantum wire MOSFET. The sub-threshold slope averages 169 mV/dec and the threshold voltage is 0.41±0.04 V. The I_{on}/I_{off} ratio is 4×10^3. Even at a gate length of 10 nm, these InAs devices perform well. The degraded sub-threshold slope, that is found in the 10 nm device, is due to the increased effect of the gate work function, enhanced source-to-drain tunneling, quantization, and a larger separation between successive subbands, as compared to similar silicon devices.

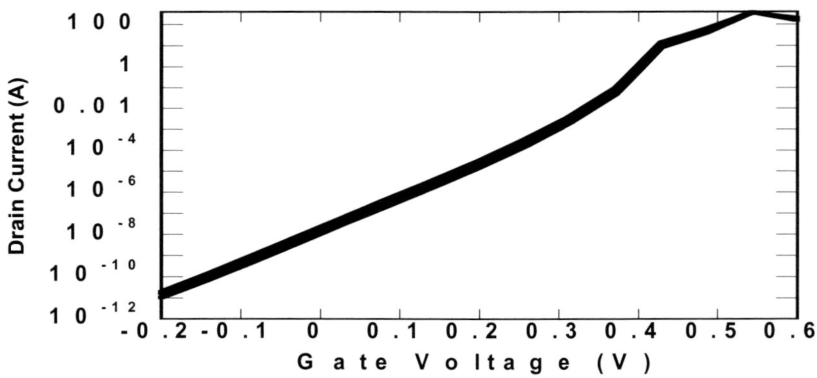

Fig. 2. I_d-V_g curves for 2 different 10 nm devices with $V_d = 0.6$ V.

We now compare the operation of two 10 nm devices, one in a silicon quantum wire and one device in InAs, of the exact same geometry as used to obtain the previous results. Nevertheless, some small modifications have to be made to the silicon device to account for the materials differences. In the silicon quantum wire tri-gate MOSFET, we use a silicon thickness of 6.51 nm and the length of the channel is set to be 9.77 nm. To account for the differences in thickness and lengths, we adjust the doping of the silicon device in the source and drain to 1×10^{20} cm^{-3}, while the channel of the device is again left undoped. Yet, we must deal with one additional issue before the devices can be compared, and this is the issue of the threshold voltage. For a device to be considered future CMOS applications, a threshold voltage of around 0.2 V is necessary. Clearly, by examining Fig. 2, we find that the threshold voltage for the InAs devices is far too high.

To reduce the threshold voltage, we use gate workfunction engineering. This is a desirable option, as we can adjust the threshold voltage without introducing additional scattering or creating further lithographical complications. The previous simulations were carried out using a generic metal that had a workfunction of $\varphi_{ms} = 5.5$ eV, which roughly corresponds to that of platinum. Here, we replace the platinum contact with a titanium nitride contact which has a workfunction of $\varphi_{ms} = 5.3$ eV.

In Fig. 3, we plot the I_d-V_g curves for the 10 nm InAs and Si quantum wire MOSFETs. Fig. 3(a) plots the I_d-V_g curves for InAs and Si quantum wire transistors on a linear scale, while Fig. 3(b) plots the same curves on a logarithmic scale. With the titanium nitride contact replacing the previous platinum contact, we now find that the threshold voltage for the device has shifted to 0.23 ± 0.04 V which is a significant improvement over the 0.41 ± 0.04 V we observed earlier. The subthreshold slope of the device averages to be 163.05 mV/dec and the I_{on}/I_{off} ratio is about 700. For the silicon device, we find that the threshold voltage for six devices averages to be 0.26 ± 0.12 V with an average subthreshold swing of 66.53 mV/dec. The I_{on}/I_{off} ratio averages to be about 900.

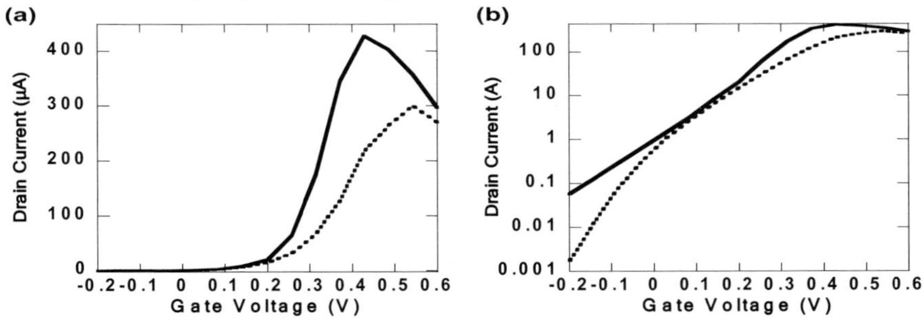

Fig.3. I_d-V_g curve for InAs (solid) and silicon (dashed) quantum wire tri-gate transistors plotted on a (a) linear scale and (b) logarithmic scale with $V_d = 0.6$ V.

The reason for the decrease in current after the gate voltage exceeds Vg = 0.43 V is due to tunneling through longitudinal barriers. In fig. 4, we plot the potential at a depth of 7 nm into the device with $V_g = V_d =$ 0.6. where we find a 1.25 k_bT barrier at the channel-drain interface.

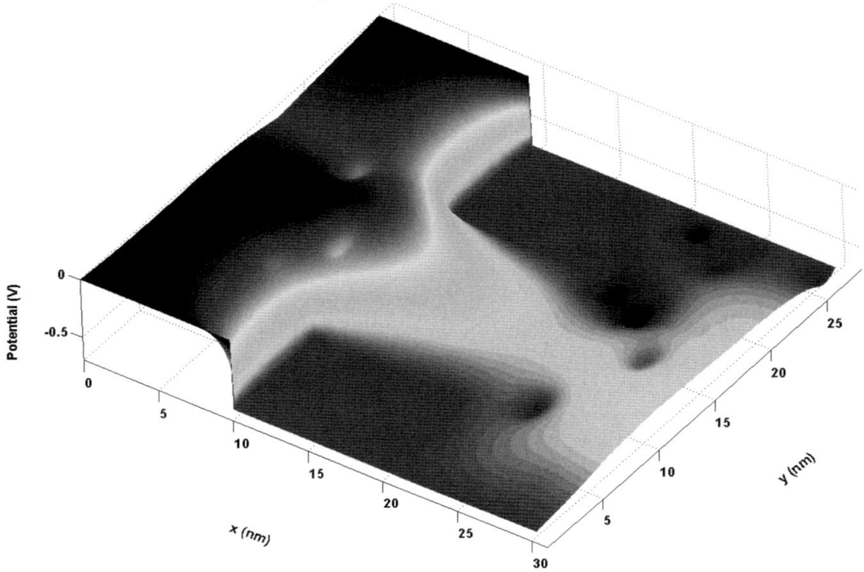

Fig.4. Potential profile at a depth of 7 nm into the device with $V_g = V_d = 0.6$ V.

Based on this data, we conclude that the InAs device is, at best, only marginally competitive with a comparable silicon based device. We find that even with the gate material adjustment, the I_{on}/I_{off} ratio fails to approach the value we find in the silicon devices. Furthermore, by changing the gate material, we find that now the longitudinal states in the channel begin to show up in the output characteristics. This was not the case in Fig. 2, where the effects of the longitudinal states were suppressed.

References

1. M.J. Gilbert and D.K. Ferry, "Efficient quantum three-dimensional modeling of fully depleted ballistic silicon-on-insulator metal-oxide-semiconductor-field-effect-transistors," *J. Appl. Phys.*, **95**, 7954-7960, 2004.
2. D.D. Johnson, "Modified broyden's method for accelerating convergence in self-consistent calculations," *Phys. Rev. B*, **38**, 12807-12813, 1988.

Optical Phonon Modes and Electron-Phonon Interaction in a Spheroidal Quantum Dot

M. Ishida[a)], M. Yamaguchi and N. Sawaki

Dept. of Electronics, Nagoya University, Chikusa-ku, Nagoya, 464-8603 Japan

Summary. Optical phonon modes in a semiconductor quantum dot were analyzed with exactly solvable model structure. A set of universal formulae for the electron and phonon wave functions was derived for spheroids. Using these wavefunctions, the intersubband scattering rates due to emission of phonons were investigated as a function of the size and the shape of the dot. It was shown that the maximum total scattering rate is slightly enhanced by reducing the height of the dot, which is attributed to the increase of the contribution of the interface modes.

1 Introduction

In a semiconductor quantum dot, the phonon modes, as well as the electronic states, are subject to the shape of the dot, which will determine the optical properties[1]. In quantum dots/wires, the contribution of the interface (IF) phonons might be dominant in the scattering phenomena as compared to those due to bulk like modes. But, several authors found that the total scattering rate is not changed and a sum rule holds[2,3]. Some assumed that the phonon energies involved are a constant given by those of bulk material. In a real structure, however, the phonon modes/energies show rather complicated behaviour depending on the size/shape of the dot[4,5]. In this study, we investigate the inter-subband scattering rate in a dot using a unified approach to analyze the electron and phonon modes.

2 Analytical Model and the Numerical Results

The self-organized quantum dot has often a shape of a truncated pyramid or cone, and the height (H) and width (B) will determine the electronic properties. In this paper, we study the spheroid as an exactly solvable model. It is not equivalent to a cone. Even though, by the virtue of the symmetry properties along z-axis, the results might represent the main issue of the overlap integral between electron and phonon wavefunctions that determines the electron phonon interaction.

The electronic states are analysed with the Schroedinger equation using the effective mass approximation, while the phonon modes are analysed assuming a dielectric continuum model. To do this, the general oblate spherical coordinates were transformed to the polar or Cartesian coordinates to reveal the general feature of the electron/phonon modes. By this transformation, we found that the results given by Klein et al[6] can be applicable to our cases. This is simply because the system is expressed by the Helmholtz equation. The modes were represented by a set of quantum numbers (*nlm*) in accordance with Klein et al[6]. The Froehlich interaction was analyzed and we found that the interaction Hamiltonian (scattering rate) in a spheroidal dot is transformed to that for a spherical dot using the ratio of the height (H) and the width (B). As the result, the interaction strength was found to be modified as a function of the ratio H/B.

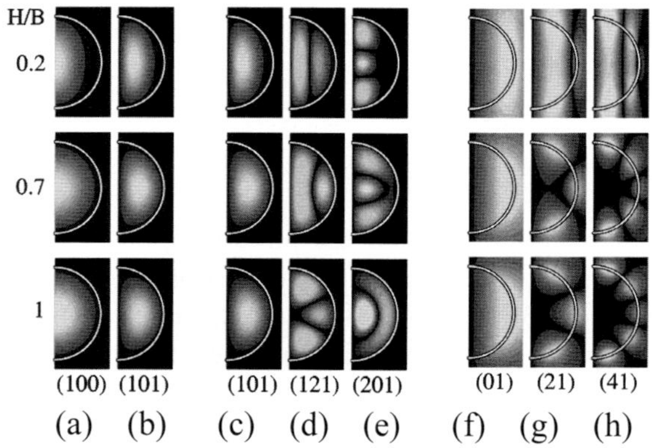

Fig. 1. The electron and phonon modes as a function of the height H normalized by the width B (The vertical axis is set to unity for comparison); (a-b) electron ground and the first excited state. (c-e) for confined (CF)phonons, and (f-h) for interface phonons (IF). The CF and IF modes are expressed by (*nl*m*) and *(l*m*), respectively, where *l*=l-m*.

In the light of the analytical results as above, the numerical analyses have been performed using differential calculus. The material parameters are assumed for an InAs dot embedded in a GaAs matrix. Typical mode structures are shown in Fig.1, where the images are shown in the cross section (x-z plane) and the vertical axis has been normalized for comparison use. Obviously, the modes with high indexes are more sensitive to the shape of the dot, e.g., the second peak in the modes labelled as (201) is along the surface of the sphere for H/B=1, while it is separated into two parts for the case of a thin disk (H/B=0.2). In Fig.1, the interface phonon modes are also shown. The shape effect is more enhanced for the interface modes. The relative amplitude of the interface phonon mode was enhanced by decreasing the ratio H/B (thin disks).

3 The Intersubband Scattering Rates

The intersubband scattering due to the Froelich interaction between the ground state (100) and the first excited state (101) was studied in detail, where the phonon modes with even l^* (=l-m) and m=1 have the dominant contribution by selection rule. Figure 2 shows the numerical results. The strong interaction is due to the confined (101) phonon, and the others are of the order of magnitude weaker. In the case of interface modes, by increasing the height H, the strength is enhanced for the (101) mode expressed as (01) in Fig.1, while it is decreased in other modes. This is in accordance with the variation of the phonon amplitude in the dot.

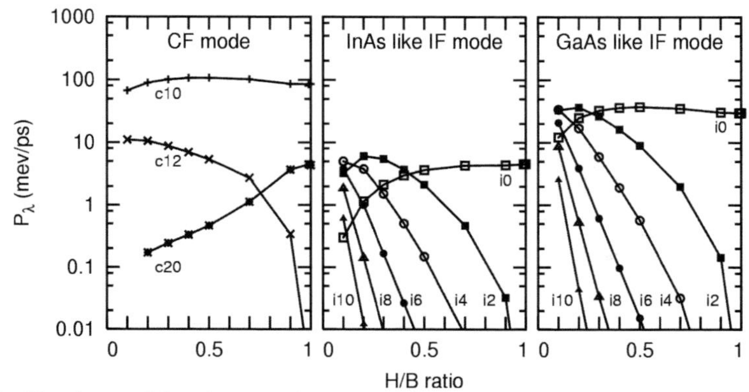

Fig. 2. The intersubband scattering strength due to each phonon mode. The numbers represent (nl^*) for the CF modes and (l^*) for the IF modes.

The total scattering rate given by summing each contribution is shown in Fig.3 as a function of the dot size. Because of the energy conservation law in the processes, we have a maximum scattering rate at a certain size where the energy difference coincides with the phonon energy. In case of a sphere (H/B=1), the scattering due to confined phonons gives the maximum contribution, while in case of a thin disk (H/B=0.1), the scattering with the GaAs like interface mode gives the maximum value. In all cases, the InAs like interface modes give small contribution. The main result in Fig.3 is that the total scattering rate is slightly increased by reducing the height of the dot. That is, the decrease in the confined modes cannot overcome the increase due to interface phonons.

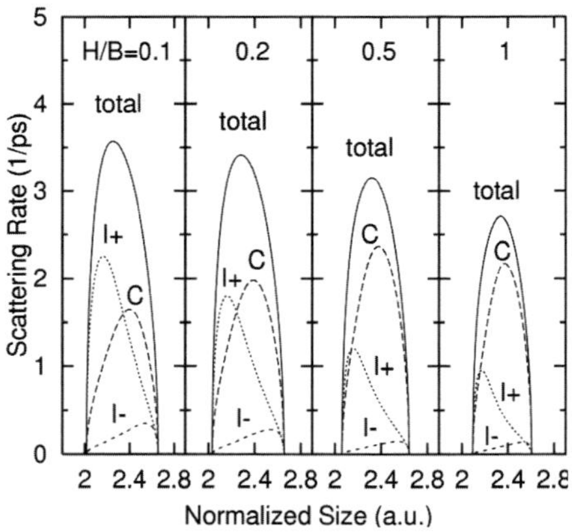

Fig. 3. Scattering rate due to confined (C), GaAs like (I+) interface, and InAs like(I-) interface mode.

4 Summary

The optical phonon modes in a model semiconductor quantum dot have been investigated as a function of the size as well as the shape. The interface phonon modes depended strongly on the shape of the dot. The intersubband scattering rate was enhanced by reducing the height of the dot due mainly to the contribution of the interface modes.

References

a. Present address; Nanoelectronics Collaborative Research Center, University of Tokyo, Tokyo 153-8505.
1. S.Moehl, F. Tinjod, K. Kheng, and H. Mariette, *Phys. Rev.* B**69**(2004) 245318.
2. N. Mori and T. Ando, *Phys. Rev.* B**40**(1989) 6175.
3. L.F. Register, *Phys. Rev.* B**45**(1992) 8756.
4. P.A. Knipp and T.L. Reinecke, *Phys. Rev.* B**46**(1992) 10310.
5. D.V. Melnikov and W.B. Fowler, *Phys. Rev.* B**64**(2001) 245320.
6. M.C.Klein, F.Hacke, D.Ricard, and C.Flytzanis, *Phys. Rev.* B**42** (1990) 11123.

Terahertz Negative Differential Conductivity in Heterostructures due to Population Inversion and Bunching of Ballistic Electrons

V.A. Kozlov, A.V. Nikolaev and V.A. Verbus

Institute for Physics of Microstructures RAS, 603950, Nizhny Novgorod, Russia

Summary. A simple quantum mechanical method for calculation of a high frequency differential conductivity of hot ballistic electrons in heterostructures is developed. Application of this method to the heterostructures with a special potential profile demonstrates that it is possible to obtain the negative differential conductivity at THz frequencies due to the interference and quantum bunching of ballistic electrons.

1 Introduction

The first successful attempt to develop a solid state source of terahertz (THz) radiation was made several decades ago when the p-Ge lasers [1] and masers [2] were developed. The creation of quantum cascade lasers [3] gave the possibility to make the solid state sources for high frequency part of THz band. Now it is necessary to get the solid state sources of THz radiation for the low frequency part of THz band. The dynamic negative differential conductivity (NDC) of hot ballistic electrons [4, 5] is very attractive for this purpose. At the low frequency part of THz region the energy of quantum $\varepsilon = \hbar\omega$ is comparable with the thermal energy $\varepsilon = kT$ (for temperatures 50 ÷ 300 K), so that there are many occupied electron states in the considered interval. The ordinary mechanisms for the creation of population inversion due to the overpopulation of a single selected level have failed and one has to consider the open quantum systems with continuous spectra. In this case the population inversion means that the distribution function f is satisfied the condition $\partial f / \partial \varepsilon > 0$. The ballistic transit time and tunnelling times [6] in short heterostructures lie in THz region, so it is possible to obtain the high frequency NDC in specially developed heterostructures. In this report we present a simple matrix method for the computation of differential conductivity. Using this method we calculated the dynamical NDC and demonstrated that this method is very useful for the development of heterostructures with ballistic NDC at the THz band.

2 Heterostructure Model and Perturbation Theory

The described method is based on the representation of a heterostructure as sequential set of layers (in *y-z* plane) with constant potential energy V and effective mass m^* in each layer. A boundary between adjacent layers is described by the Heavyside step function (Fig. 1). A natural thickness of the layers may be a lattice constant. The boundary conditions between layers are the following: $\psi_l = \psi_r$; $(m_l^*)^{-1}\,\partial\psi_l/\partial x = (m_r^*)^{-1}\,\partial\psi_r/\partial x$.

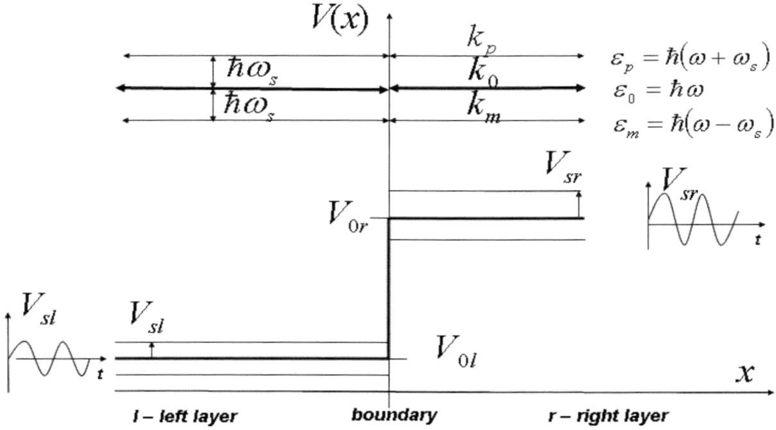

Fig. 1. Connection of two arbitrary adjacent layers of the heterostructure.

The applied external potential consists of dc and ac voltage bias and has the form: $V_{bias} = V_0 + V_s\left(e^{-i\omega_s t} + e^{i\omega_s t}\right)$ where $V_s << V_0$. Here V_0 is dc bias voltage and V_s is the amplitude of a small ac signal.

The first and the last layers are assumed to be heavily doped, so electric field and potential drop are absent in these layers. This means that we consider an open system with two reservoirs at the left and right ends. The applied voltage (which has a linear distribution along the x coordinate) changes the amplitudes of the step functions only. The left side potential is assumed to be zero, so that the amplitude of the ac voltage drop increases along in x direction. The wave function ψ of electrons in the heterostructure has the form: $\psi = \psi_0 e^{-i\omega_0 t} + \psi_p e^{-i(\omega_0+\omega_s)t} + \psi_m e^{-i(\omega_0-\omega_s)t}$, $(\psi_p, \psi_m << \psi_0)$. The wave function of electrons in the first layer of the system consists of an incident wave equal to $A_0 e^{-i\omega_0 t + ik_0 x}$ and the reflected wave. The amplitude A_0 defines the intensity of an incoming electron beam. This can be achieved by using the special injector with the resonant tunneling. We consider ballistic motion of electrons in short heterostructures, and take into account the quantum interference effect. It is convenient to introduce the complex gradient γ of the wave function ψ by the formula $\gamma = (\partial\psi / \partial x)/(im^*)$. This quantity is continuous at the boundary together with wave function ψ. The total current $j = j_0 + j_1$ consists of dc current j_0 and ac current j_1. The current j_0 is constant along the whole heterostructure. The ac current $j_1 = j_a e^{i\omega_s t} + j_a^* e^{-i\omega_s t}$ may vary with the time t and with the coordinate x due to an electron bunching. Our calculations show that the ac current amplitude j_a has a beating along x axis and may change sign. In this model the potential difference exists only at the potential steps, so the work done by the ac electric field on the electron beam takes place at the boundaries only. To obtain the total work of the ac electric field it is necessary to sum up these works over all boundaries. Studying the dependence of the current j_a and the signal voltage drop on the coordinate x one can find the conditions when the work of the ac field on the electron beam becomes negative. This implies the appearance of the dynamic NDC in the system.

3 The Results of Computation and Discussion

The result of the computation of ac current is shown in the Fig 2a. Hatching on the x axis indicates the variable potential profile of the heterostructure. The first and the last layers with constant potential have no hatching. The ac currents in these layers are the idler currents because they do not produce work. They are shown to demonstrate the quantum beating of the wave function. The small fast ripples in Fig 2a arise due to the step size approximation of the potential profile and they should be ignored. At the initial part of the heterostructure there is modulation of wave function by the small ac potential. During the propagation along the x direction the ac current changes its sign and begins to make the negative work. As a result negative differential conductivity arises. This behaviour corresponds to electron bunching. The real and imaginary parts of the high frequency conductivity are shown in Fig 2b. The real part of the conductivity has a dynamic NDC shape: it is positive at low frequencies and negative at high frequencies. It is very important for the stability of the system and it offers the possibility to connect these systems in a sequence. We have shown by the computer simulation that the heterostructures with the special form of potential profile have a sufficiently higher value of the dynamic negative differential conductivity at THz frequencies.

This work was supported by RFBR Grant # 05-02-17121

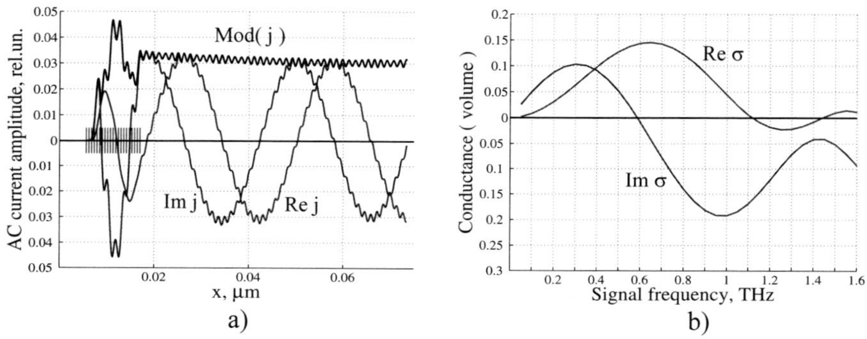

Fig. 2. The alternate current along heterostructure a) and the frequency dependence of conductivity b) for the heterostructure with a linear potential profile.

References

1. Andronov, A. A. et al: *Pisma v ZhETF,* **30**, 578, 1979; **40**, 69, 1984.
2 Kozlov, V. A. et al: *Pisma v ZhETF*, **37**, 142, 1983;
Ivanov, Yu. L. et al *Pisma v ZhTF,* **9**, 613, 1983.
3. Faist, J., Capasso, F., Sivco, D. L., Sitori, C. et al: *Science*, **264**, 553, 1994.
4. Kozlov, V. A. et al: *Mater. Sci. Forum,* **384,** 139, 2002;
Semic. Sci. Technol. **19**, S99, 2004.
5. Gribnicov, Z. S., Vagidov, N. Z., Mitin, V. V. et al: *Appl. Phys*, **9**, 5435, 2003.
6. Capasso, F. et al: *IEEE J. Quantum Electron.,* **QE-22**, 1853, 1986;
Price, P. J.: *Semic. Sci. Technol.*, **19**, S241, 2004.

Carrier Dynamics of Single ZnO Nanowires

L. Wischmeier, C. Bekeny and T. Voss

Institut für Festkörperphysik, Universität Bremen, P.O. Box 330440, D-28334 Bremen

Summary. The optical properties and carrier dynamics of an ensemble and of single nanowires with diameters < 200 nm are investigated. The optical properties of single nanowires are compared with the PL spectra of the as-grown ensemble. Furthermore, we present measurements at high excitation densities in which a sharp line appears resulting from resonator effects inside the nanowires.

Zincoxide (ZnO) is a direct wide band-gap semiconductor (E_{gap}=3.37 eV at room temperature) with an exciton binding-energy of 60 meV being large compared to the thermal energy of $k_BT \approx 25$ meV at room temperature. Due to these properties ZnO nanowires are promising building blocks for miniaturized optoelectronic devices operating in the blue to UV spectral region [1-3]. A detailed knowledge on the electronic processes which are involved in the generation of the photoluminescence (PL) or even stimulated emission and especially their dynamics is of fundamental importance for constructing and optimizing optical devices.
The ZnO nanowires with diameters < 200 nm are grown by a chemical vapor transport and condensation technique (vapor-liquid-solid) on (110)-sapphire substrates [2, 3]. In Fig. 1 scanning electron-microscopy (SEM) images of dispersed single nanowires are shown. The inset emphasizes the good crystalline quality of the nanowires. The dispersed single nanowires can be prepared either by tapping a new substrate onto the as-grown ensemble or by sonication in a dissolver bath.
For the investigation of the optical properties of single nanowires a micro-photoluminescence setup (μ-PL) and a HeCd laser at a wavelength of 325 nm are used. The laser beam is focused via a reflective microscope objective onto the sample which is kept in a temperature-variable cryostat. The PL of the wires is collected with the same objective. By use of a beamsplitter the collected intensity is detected spectrally resolved with a combination of a spectrometer and a CCD camera (spatial resolution ≈ 1.5μm).
As can be seen in Fig. 1 single nanowires of different shapes or even particles of ZnO were deposited on the substrate by the preparation process. The PL from undamaged nanowires shows strong and sharp excitonic features

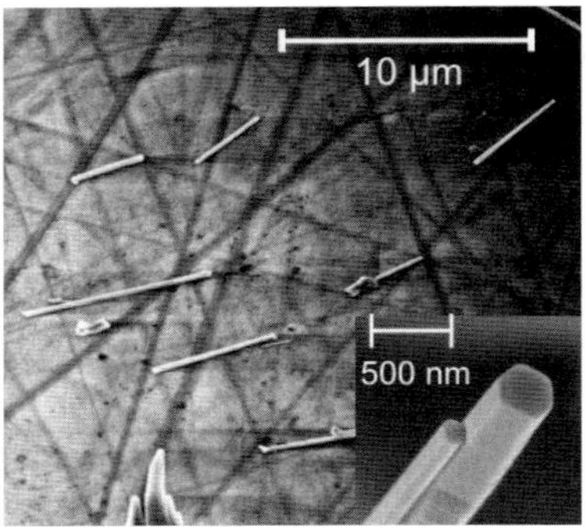

Fig. 1. SEM image of dispersed single ZnO nanowires. Perfect single nanowires, damaged wires, and particles can be found. Inset: Image of the tips of two as-grown ZnO nanowires.

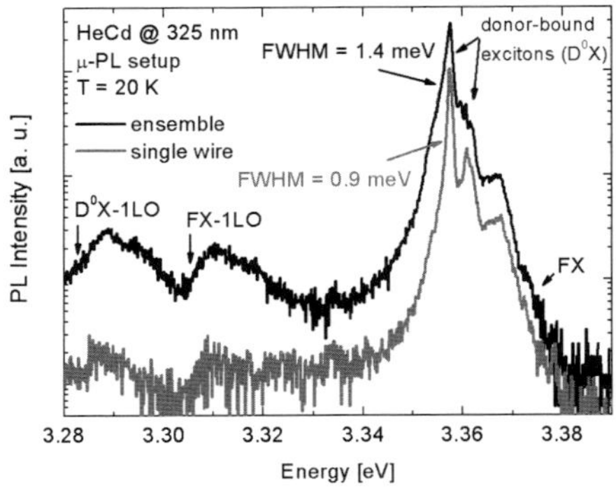

Fig. 2. Micro-Photoluminescence (µ-PL) of the nanowire ensemble and of one single nanowire are compared. The spectra show same excitonic features with different linewidths.

(Fig. 2) due to high crystalline quality which is comparable to that of bulk ZnO crystals [4]. Opposite to this, in the PL of the damaged nanowires and the ZnO particles (which is not shown here) the lines are broadened, and their relative intensities differ. In Fig. 2 the PL of an as-grown sample and of dispersed single nanowires are compared. Both PL spectra are dominated by the emission peaks at $\approx$ 3.36 eV originating from the radiative recombination of donor-bound excitons (D^{0}X). At an energy position of $\approx$ 3.375 eV a small shoulder due to the recombination of the free exciton (FX) can be observed. With increasing temperature the intensity of the free exciton PL increases and dominates the spectrum at room temperature. At $\approx$ 3.28 eV and $\approx$ 3.31 eV the phonon replica of the D^{0}X peaks (D^{0}X-1LO) and of the FX recombination (FX-1LO) are observed. The occurrence of the phonon replica proves the high crystalline quality of our samples. Apart from the width of the lines the spectra of single wires and of the ensemble are comparable. For the single nanowire and for the ensemble a linewidth of 0.9 meV and of 1.4 meV is obtained respectively, the latter due to an inhomogeneous broadening of the ensemble PL. However, the PL of the ensemble in principle indeed resembles the PL emission of the individual wires.

To get an insight into the microscopic processes which are involved in amplified stimulated emission in ZnO nanowires we have performed PL measurements at high excitation densities and time resolved measurements of the PL decay. The measurements at high excitation density were performed on the as-grown ensemble (excitation with 5 ns pulses at a wavelength of 363 nm). A large area on the sample was excited and the PL of the whole sample was detected. The PL of the nanowire ensemble was measured as a function

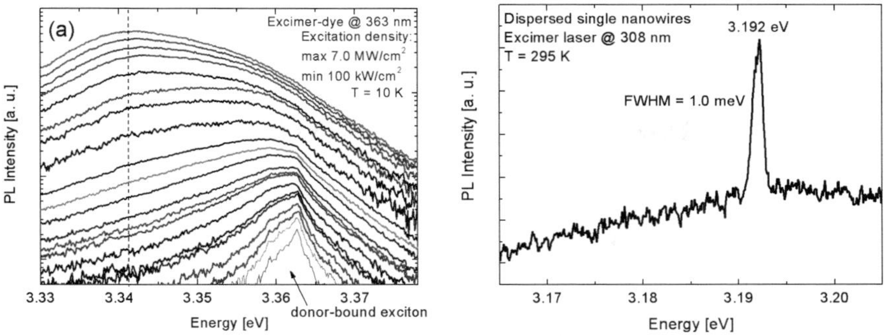

Fig. 3. (a) Excitation-density dependence of the PL intensity of the nanowire ensemble. (b) PL of dispersed single ZnO nanowires at high excitation-density. A sharp line resulting from resonator effects inside the nanowires is observed.

of the excitation density up to 7.0 MW/cm^2 (Fig. 3a). At low excitation densities emission originating from the radiative recombination of donor-bound excitons is detected. With increasing excitation density a second broad peak (dashed line in Fig 3a) most probably resulting from recombination processes in an electron-hole plasma appears and dominates the spectrum at high densities [4, 5]. A super-linear increase of the broad band at excitation densities > 0.7 MW/cm^2 was recorded at the energy position ≈20 meV below the donor-bound exciton. For the emission peak at ≈ 3.363 eV (donor-bound exciton) just a linear increase is obtained even for the highest excitation densities. In Fig. 3b the PL of dispersed nanowires measured at a high excitation-density and room temperature is shown. Due to the higher temperature (T = 293 K) the broad PL band has shifted to lower energies. On the low energy side of the band above a threshold of ≈ 0.5 MW/cm^2 a sharp line with a full width at half maximum of 1.0 meV is observed. This is an indication for amplified spontaneous emission in a single wire together with resonator effects and is due to the fact that a single nanowire acts as an optical cavity.

The time-resolved PL measurements were performed by use of the time-correlated photon-counting technique (TCPC). For excitation the femtosecond laser-pulses of a frequency-doubled Ti-sapphire laser (λ = 362 nm) with a repetition rate of 82 MHz were used. The grey curve in Fig. 4 shows the response of the TCPC system when only the femtosecond excitation pulses of the laser were directly detected. The signal of the nanowire ensemble was measured at the spectral position of the most pronounced donor-bound related PL band (≈3.36 eV). The signal is a convolution of the response function of the system and the real PL decay. An exponential decay of the signal with a decay time of ≈ 250 ps was obtained which is in good agreement to the data in [7] and references there in. Further investigations

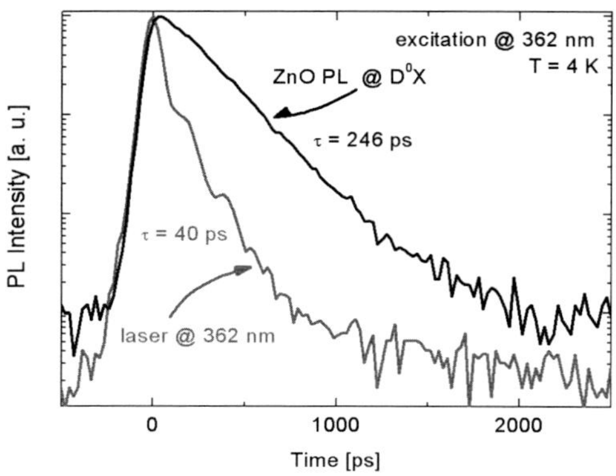

Fig. 4. Time-resolved measurements by use of the time-correlated photon-counting technique. *Black*: Time-resolved donor-bound related PL of the nanowire ensemble. *Grey:* Response function of the system measured with the exciting femtosecond pulses.

of the dynamics as a function of wire morphology and structural properties are in progress.

In summary we presented μ-PL and high-density measurements on ZnO nanowires. Apart from the linewidth the comparison of the ensemble and single nanowire PL shows no detectable difference under the same experimental conditions. Measurements at high excitation density in which a sharp line is oberserved resulting from amplified spontaneous emission together with resonator effects inside a single nanowire were presented. In the end first time-resolved measurements of the PL decay were shown.

The authors thank I. Rückmann and J. Gutowski for fruitful discussions. We also thank S. Börner, W. Schade (Technical University of Clausthal) and S. Müller, C. Ronning (University of Göttingen) for providing the samples.

References

1. Huang, M. H. et al.: 'Room-Temperature Ultraviolet Nanowire Nanolasers', *Science*, **292**, 1897-1899, 2001
2. Yang, P.: 'Controlled Growth of ZnO Nanowires and Their Optical Properties', *Adv. Funct. Mater.*, **12**, 323-331, 2002.
3. Samuelson, L.: 'Self-forming nanoscale devices', *materials today*, **6**, 22-31, 2003
4. Klingshirn, C.: *Semiconductor Optics*, Springer Verlag, 2005.
5. Priller, H. et al.: 'Comparison of linear and nonlinear optical spectra of various ZnO epitaxial layers and of bulk material obtained by different experimental techniques', *phys. stat. sol. (b)*, **241**, 587-590, 2004.
6. Priller, H. et al.: 'Temperature-dependent luminescenc dynamics in ZnO nanorods', *Journal of Luminescence*, **112**, 173-176, 2005
7. Gutowski, J. et al.: 'Optical Non-linearities and Excitation Dynamics in II-VI Bulk and Epitaxial Materials', *Advanced Materials for Optics and Electronics*, **3**, 15-32, 1994

Traditional Hot-Electron MOS Devices for Novel Optoelectronic Applications

T. Dekorsy[1]*, J. Sun[1], W. Skorupa[1], M. Helm[1], L. Rebohle[2] and T. Gebel[2]

1) Forschungszentrum Rossendorf, Institut für Ionenstrahlphysik und Materialforschung, PO Box 510119, 01314 Dresden, Germany
2) nanoparc GmbH, Bautzner Landstraße 45, D-01454 Dresden, Germany
* present address: University Konstanz, Physics Department, Box M700, D-78457 Konstanz, Germany

Summary. We report the realization of highly-efficient light emitting MOS devices which are based on hot-electron excitation of rare-earth ions implanted into SiO_2. The implantation of Gd^+ and Tb^+ ions yields emission wavelengths of 316 nm and 541 nm with external quantum efficiencies up to 1% and 16%, respectively. The observed threshold electric fields for observing electroluminescence are in accordance with the injection of hot electrons via Fowler-Nordheim tunneling into SiO_2 at field strengths in the range of 8-9 MV/cm. The presence of different electroluminescence bands of the Tb-implanted devices allows us to study details of the hot-electron excitation process.

Several approaches toward silicon-based light emitting devices have been pursued in the past with the aim of obtaining the electroluminescence (EL) efficiency necessary for practical applications. The most prominent material systems investigated were porous silicon [1], Si pn diodes [2], silicon nanocrystals in SiO_2 [3], and Er-doped SiO_2 [4]. For optoelectronic applications like microdisplays and on-chip sensors light emission in the visible or UV part of the spectrum are required. In this spectral range Si itself is highly absorbing so light-emission from SiO_2 layers containing the light-emitting species remains the only solution for devices which are fully compatible with standard CMOS semiconductor technology. We demonstrate rare-earth implanted SiO_2 MOS light emitting devices which provide efficient and stable EL from Tb and Gd-doped MOS based EL devices emitting light in the green, and deep UV with relatively high external quantum efficiency above of 16 % (green) [5] and 1 % (UV) [6], respectively.

The MOS structures were fabricated by local oxidation of silicon with a gate oxide and a field oxide of 100 nm and 1 μm thickness, respectively. The gate oxide consists of thermally grown SiO_2. and is implanted with different rare-earth ions at 100 keV, or using double energy of 50 and 110 keV. The implantation was followed by furnace annealing at 800°C-1000°C in flowing N_2 for 1 hour. The top gate electrode is a 100 nm transparent indium-tin-oxide (ITO) layer deposited by RF sputtering, while the bottom electrode is provided through the substrate. The investigated devices have 500 μm diameters.

The rare-earth ions Tb and Gd give strong EL due to their large excitation cross-sections [5,6]. Fig.1 shows the EL spectra of Tb and Gd-implanted devices at an injection current of 10 μA. The electronic transitions from Tb^{3+} yield two groups of peaks from the $^5D_3$ and $^5D_4$ to ^7F_j (j=3-6) levels, respectively, the former in the range below 500 nm, while the latter one is the strongest at a wavelength in the green at 541 nm. The $^6P_{7/2}$ to $^8S_{7/2}$ transition of Gd^{3+} generates a sharp peak in the UV at 316 nm.

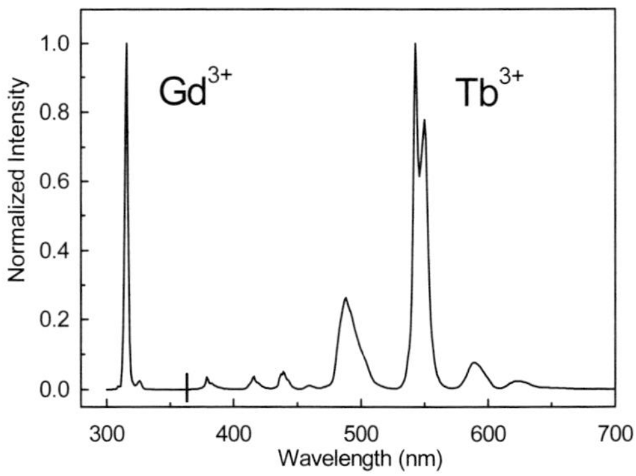

Fig. 1. Normalized electroluminescence intensities of Gd and Tb-implanted MOS based light emitters.

The EL from rare-earth ions embedded in a matrix under high electric field can be excited via two processes: One is the direct impact excitation of the rare-earth ion by energetic hot electrons in the conduction band of the host matrix. The charge states of the rare-earth ions do not change during the excitation process. The other process is the excitation across the host band gap and subsequent energy transfer to the rare-earth ions. In the former process, the relative intensity of the peaks from higher excited levels normally increases stronger with increasing electric field than those at lower energy levels due to an increase of the average energy of the hot electrons [7,8]. In the latter process, since the excitation comes from the energy transfer, the relative intensity of the emission peaks of rare-earth ions is mainly controlled by the thermal equilibrium between different energy levels of the rare earth ions. Therefore, the relative intensity of the different peaks will not change significantly with increasing electric field [9].

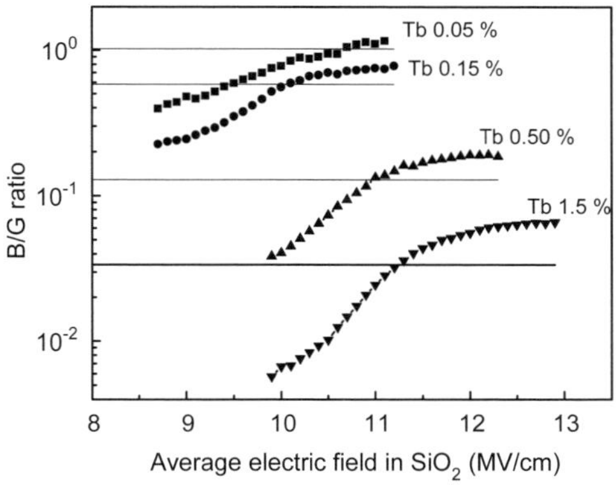

Fig. 2. Ratio of the blue to the green electroluminescence intensity versus the applied electric field in Tb-implanted devices at different Tb concentrations.

In order to check the probable EL excitation process in the Tb-implanted devices, the ratio of the blue EL intensity to the green EL intensity (B/G ratio) is plotted versus the average electric field in the oxide layer for samples with different Tb concentration (Fig.2). An increase of the B/G ratio of the EL spectra (prior to the saturation of excitation) is observed with increasing electric field. This gives strong evidence that the EL excitation is dominated by direct impact excitation from Fowler-Nordheim tunneling injection of hot electrons into the conduction band of SiO_2, since the average energy is increasing at higher electric fields. This leads to the observed increase in the B/G ratio. The relevance of the Fowler-Nordheim tunnel injection into the SiO_2 in these devices has been demonstrated in more detail elsewhere [5]. The strong decrease of the B/G ratio with increasing Tb concentration shows that a cross-relaxation from the higher excitation level $^5D_3$ to $^5D_4$ is also involved in the excitation of the lower lying $^5D_4$ level. Therefore, the peaks of $^5D_4$-^7F_j (j=3-6) transitions were excited by both the impact excitation and the cross-relaxation from

$^5D_3$ states at high Tb concentration. The above excitation mechanism explains an observed change of EL spectra with Tb concentration and annealing temperature in these Tb-doped SiO_2 MOS devices [5]. It also explains the correlation between the onset of the EL and the strong F-N tunneling injection of hot electrons.

If the excitation process is dominated by the direct impact excitation, the change of the B/G ratio of the Tb^{3+} luminescence may reflect an increase of the average energy of hot electrons with increasing the electric field. The hot electrons for EL excitation may have an equivalent average energy similar to the PL excitation photon energy for the EL spectra with the same B/G ratio compared to the PL spectra. The B/G ratio of the PL spectra under optical excitation at 240 nm (5.16 eV) is also marked by the horizontal lines for different Tb concentrations in Fig.2. The intersections indicate that the average hot electrons may have an equivalent average energy around 5.16 eV at an electric field around 10 to 11MV/cm. This value is consistent with the average hot electron energy determined at the same electric field in SiO_2 by different techniques such as vacuum emission, carrier separation and EL [10].

In conclusion highly-efficient Si based MOS light emitters have been demonstrated with emission wavelengths in the UV and green. Their stability and plainness of the fabrication process in conjunction with the compatibility with standard CMOS fabrication technology makes this device concept suitable for future optoelectronic applications.

References

1. Y. Kunemitsu, Phys. Rep. **263**, 1 (1995).
2. J. M. Sun, T. Dekorsy, W. Skorupa, B. Schmidt, and M. Helm, Appl. Phys. Lett. **83**, 3385 (2003).
3. L. Pavesi, L. Dal Negro, C. Mazzoleni, G. Franzò, and F. Priolo, Nature **408,** 440 (2000).
4. F. Iacona, D. Pacifici, A. Irrera, M. Miritello, G. Franzò, F. Priolo, D. Sanfilippo, G. Di Stefano, and P. G. Fallica, Appl. Phys. Lett. **81**, 3242 (2002).
5. J. M. Sun, W. Skorupa, T. Dekorsy, and M. Helm, J. Appl. Phys. **97**, 123513 (2005).
6. J. M. Sun, W. Skorupa, T. Dekorsy, M. Helm, L. Rebohle, and T. Gebel, Appl. Phys. Lett. **85**, 3387 (2004).
7. L. Ma, G. Zhong, and S. Xu, Chinese Lumin. Display **6**, 192 (1985).
8. D.C. Krupka, J. Appl. Phys. **43**, 476 (1972).
9. J. M. Sun, G. Z. Zhong, X. W. Fan, C. W. Zheng, G.O. Mueller, and R. Mueller-Mach, J. Appl. Phys. **83**, 3374 (1998).
10. D. J. DiMaria, E. Cartier, and D. Arnold, J. Appl. Phys. **73**, 3367 (1993).

Investigation of Self- Heating Effects in Individual SOI Devices and Device-Device Interactions

M. Arifuzzaman and D. Vasileska

Arizona State University, Tempe, AZ 85287-5706, USA

Summary. Heating effects in individual devices are investigated using GIGA3D Silvaco simulator. By utilizing the THERMAL3D module of the Silvaco simulation framework, we also investigate heating effects due to device-device interactions. Our simulation results suggest that both factors have to be accounted for if, for example, proper output device characteristics are desired.

1 Introduction

The SOI device exhibits self-heating effects [1], because the device is thermally insulated from the substrate by the buried oxide layer that has very low thermal conductivity. This, in turn, leads to substantial elevation of the lattice temperature near the drain end of the device (hot spot), which consequently modifies the device output characteristics. This means that any state-of-the-art device simulator either way must take into account the heat generation. Thus, thermal and electrical effects need to be coupled via self-consistent calculations.

In this work, we prove the importance of these effects by (a) simulation of an individual SOI device and examining the temperature in the hot spots regions, and (b) by examining the additional heating effects that arise in a circuit, due to the proximity of the devices, that leads to device-device interactions. For isolated devices, the performed non-isothermal and isothermal device simulations lead to very different IV-characteristics which suggests that carrier heating within the device itself plays significant role on its operation. To examine the influence of the thermal effects due to device-device interactions we examine simple CMOS inverter with two different channel lengths. For that purpose, we modeled fully depleted SOI MOSFET with gate length of 50nm and 25nm to compare the results of two technology nodes. Again to observe the device-device interaction when the SOI MOSFET is implemented in the CMOS circuit design for VLSI circuits, we have implemented a simple CMOS inverter with a NMOS as driver and a PMOS as a load. Results of these investigations show more device-device interaction for the 25nm channel CMOS inverter (i.e. elevation of the lattice temperature in the neighboring device) than for the 50 nm.

2 Main Features of the Simulators Used

Giga2D/3D combined with S-Pisces and Blaze device simulators allows simulation of local thermal effects within the Silvaco simulation platform [2]. Models in Giga2D/3D include heat generation, heat flow, lattice heating, heat sinks, and effects of local temperature on physical constants. Thermal and electrical physical effects are coupled through self-consistent calculations. Giga2D/3D is a fully integrated component of the ATLAS device simulation framework. Thermal3D is a general heat-flow simulation module that predicts heat-flow from any power generating devices (not limited to semiconductor devices), typically through a substrate(s) and into the package(s) and/or heat-sink(s) via the bonding medium. Operating temperatures for packaged and heat-sinker devices or systems can be predicted for the design and optimization phase or for general system analysis. The key features of this simulator are as follows (1) it predicts heat flow and temperature rise for many material systems and any number of heat generating sources, (2) models are validated using measured data, (3) there are three models for heat dependent thermal conductivity one can choose from for each of the materials in the system, (4) it utilizes user definable thermal conductivities and coefficients for each material, (5) very fast simulation times allow many combinations to be tried for system design optimization.

3 Simulation Results and Conclusions

On the left panel of Figure 1 we show the temperature contour for 50nm SOI MOSFET device. One can clearly see the existence and the spread of the hot spot region centered at the drain end of the channel. The output characteristics of this device with and without the inclusion of the lattice-heating effects, that are obtained with GIGA3D simulator, are shown on the right panel of Figure 1. The increase in temperature near the drain end of the channel leads to more pronounced phonon scattering which, in turn, degrades the mobility and, therefore, the on current. For bias conditions V_G=1V and V_D=0.5 V there is almost 40 % degradation of current due to lattice heating effects. These effects are expected to be more prominent for smaller device structures where the hot spot region will cover much larger portion of the channel and more drastic mobility degradation is expected.

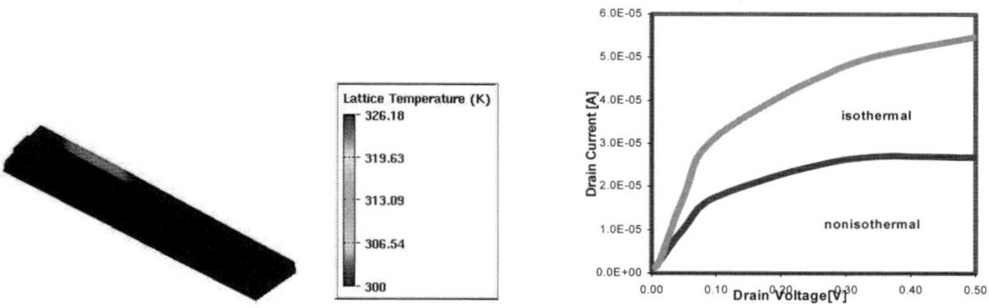

Figure 1. Left Panel-Temperature Contour for 50nm SOI MOSFET, Right Panel-Output characteristics obtained with GIGA3D. Notice drastic changes in drain current when thermal effects are included via GIGA3D Silvaco software.

Having investigated the interplay of heating effects within the device structure itself, we also wanted to examine the influence of heat due to device-device interactions for devices fabricated on the same chip. To accomplish this goal, we utilized the THERMAL3D device simulator, and investigated the temperature distribution in inverters realized in 50nm and 25 nm technology node. As expected, for smaller device technology nodes we observe more lattice heating effects on the stand-by device. This suggests that more clever design will be necessary for smaller technology nodes and possible use of Peltier coolers may be required to extract the excess heat dissipated when the device is in its on-state. The results for the inverter simulations are shown in Figure 2.

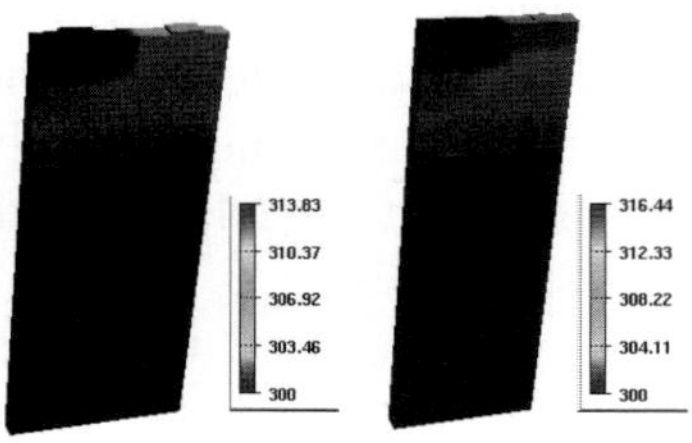

Figure 2. Temperature contour of a CMOS inverter implemented with 50nm (Left Panel) and 25nm (Right Panel) technology node. Notice the additional heat that is coming to the off-device due to strong device-device interactions in a CMOS inverter implemented in 50 nm and 25 nm technology.

4 Future Work

To properly treat heating without any approximations made in the problem at hand, one in principle has to solve the coupled Boltzmann transport equations for the electron and phonon together. More precisely*, one has to solve the coupled electron – optical phonons – acoustic phonons – heat bath problem*, where each sub-process involves different time scales and has to be addressed in a somewhat individual manner and included in the global picture via a self-consistent loop. Both Approach 1 and Approach 2 will use the commercial THERMAL3D simulator in Atlas to obtain the boundary conditions on temperature around the individual device active area including device-device interactions as well. Then:

- Within ***Approach 1***, we will solve self-consistently the BTE for the electrons (thus taking into account hot electron and other non-stationary effects such as velocity overshoot) with the energy balance equations for both the acoustic and optical phonons (which will be derived from the microscopic BTE for phonons).
- The ***Approach 2***, on the other hand, will involve solution of the microscopic BTE for the phonons self-consistently with the microscopic BTE for the electrons and holes. This represents the most general approach to solving the problem, and our group will be the first one to make the attempt of solving it.

In both, Approach 1 and Approach 2, the EMC code for the carrier BTE solution will have to be modified as well. As we have variable lattice temperature in the hot-spot regions, we will have to introduce the concept of regional scattering tables. Using state of the art computers, which have more than 4 GB RAM and more than 2 GHz processor speed, the pre-calculation of these scattering tables will not require much CPU time or memory resources and can be done once in the initialization stages of the simulation for a range of temperatures. Some interpolation scheme will be adopted afterwards for temperatures for which we do not have the appropriate scattering table. In summary, the advantages over the presently used approach and results in Ref. [1] are better transport models for both electrons and phonons.

References

1 E. Pop, K. Banerjee, P. Sverdrup, R. Dutton and K. Goodson, IEDM Tech. Dig., 679 (2001).
2 Silvaco Inc., http://www.silvaco.com.

Measurements of the Electrical Excitation of QH-Devices in the Real Time Domain

G. Vasile[1,2], Ch. Stellmach[1], G. Hein[3], G. Nachtwei[1]

[1]Institut für Angewandte Physik, TU Braunschweig, D-38106 Braunschweig, Germany
[2]National Institute of Research-Development for Cryogenics and Isotopic Technologies, Ro-1000 Rm. Vâlcea, Romania
[3]Physikalisch-Technische Bundesanstalt, D-38116 Braunschweig, Germany

Summary. In these measurements we have investigated the time scale of the excitation of electrons leading to a transition from the quantum Hall state to the dissipative state in the two-dimensional electron system of GaAs/AlGaAs heterostructures with Corbino geometry. The breakdown of the quantum Hall effect occurs after a certain time ($2 \leq \tau_{resp} \leq 20$ ns) that is a function of applied voltage (pulse amplitude), magnetic field and electron mobility.

1 Introduction

In the integer quantum Hall effect (QHE) the longitudinal resistance R_{xx} vanishes while the Hall resistance R_{xy} has plateaus with values $R_{xy} = h/ie^2$, where i is an integer number [1]. If the applied voltage exceeds a critical value, the QHE breaks down (excitation), and if the critical voltage is reduced to subcritical values, the QHE is recovered (relaxation). Experiments with multiterminal QH devices [2-3] have shown that both excitation and relaxation of electrons develop over a certain drifting distance of electrons. Some results concerning these measurements have been explained by an avalanche process of electron heating [4].

However, the applied method provides only an indirect approach to the relaxation times, which are deduced from the stationary profile of the longitudinal resistivity under QH conditions and the assumed drift velocity. Time-resolved (integrating) measurements of the generation and relaxation processes provide a direct access to the corresponding time scales and yield the trend of the generation and relaxation times being approximately equal [5], as predicted by a recent theory of the avalanche heating of electrons [4]. Since from the previous time-integrating measurements of electrical excitation [5] it had not been possible to monitor the individual pulse shape of the sample response, real-time measurements of electrical excitation have been performed in this study.

2 Experimental Setup

The Corbino devices used for these measurements were fabricated from three different wafers (see table 1), with different channel widths w=200, 100 and 50 μm (r_1=100 μm, r_2=300, 200 and 150 μm).

	n_s (cm^{-2})	μ_H (cm^2/Vs)
#8447	2.6×10^{11}	1×10^5
#8789	1.76×10^{11}	8×10^5
#8815	1.7×10^{11}	1.6×10^6

Table 1: Electron density and Hall mobility of fabricated wafers from which the Corbino devices were fabricated (parameters were determined from transport measurements at 1.8 K).

The experimental setup is realized as shown in Fig. 1.

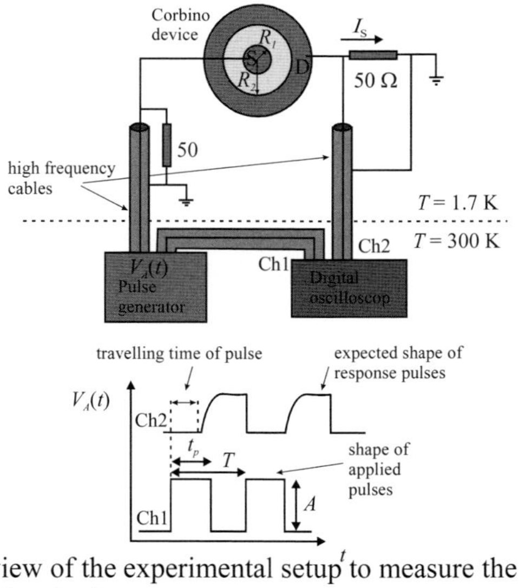

Fig.1. Schematic view of the experimental setup to measure the pulsed-induced breakdown in Corbino devices.

3 Results

By applying electrical pulses on the sample we monitor the response and by fitting with an exponential function (1) we determine the time τ_{resp} (Fig. 2.)

$$y(t) = y_0 + A \cdot \exp\left(-\frac{t}{\tau_{resp}}\right) \tag{1}$$

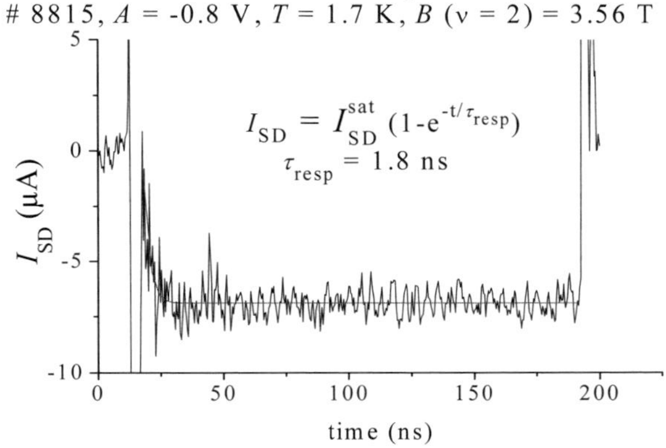

Fig. 2. The measured source-drain current versus time.

By increasing the pulse amplitude the sample response time decreases (Fig. 3.) since the Landau levels bend more and more decreasing the spatial separation between two Landau levels, causing a higher tunneling rate of electrons from initial to final tunneling states.

If the magnetic field is varied from the middle of the second plateau (ν = 2) position to higher or lower values, the QHE breakdown occurs in shorter times (Fig. 4.).

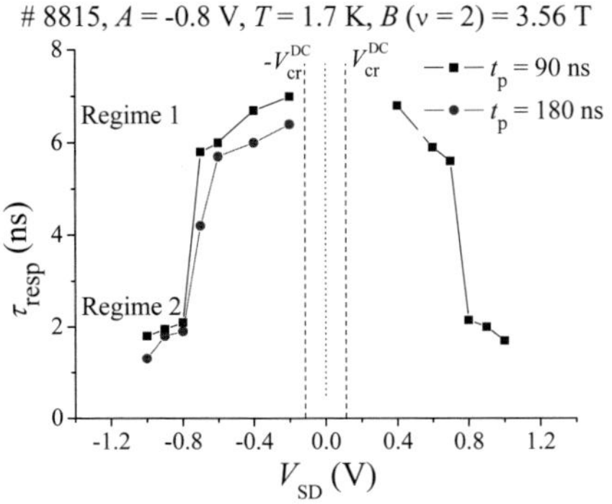

Fig. 3. The response time versus applied voltage (high mobility sample)

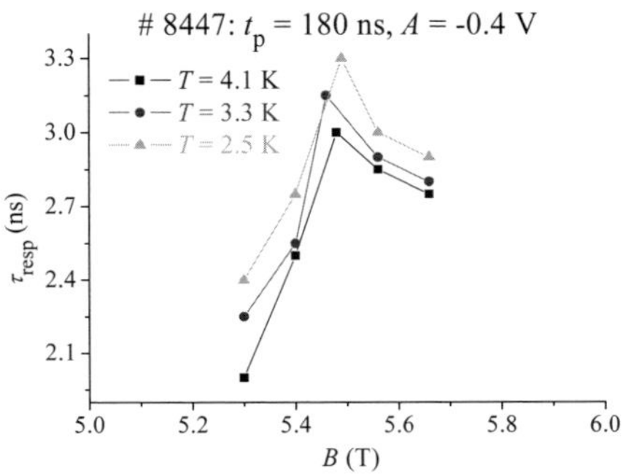

Fig. 4. The magnetic field dependence of the response time.

This can been understood as a result of the decreasing width of incompressible strips with decreasing magnetic field around the integer filling factor where the width reaches its maximum value [6].

4 Conclusions

To summarize, we have investigated the time scale of the excitation of electrons in the two-dimensional electron system of GaAs/AlGaAs heterostructures at the QHE breakdown. The measurements were performed by applying short electric pulses to the QH device with Corbino geometry at filling factor $\nu = 2$, having different electron mobilities. The QHE breakdown starts only after certain times of the order of nanoseconds and is dependent on pulse amplitude and magnetic field, electron mobility. The time scale of excitation (1.8-20 ns) is in agreement with the previous time-integrating electrical excitation measurements [5].

5 Acknowledgments

This work was supported by the Deutsche Forschungsgemeinschaft. The authors thank Prof. Dr. K. von Klitzing, Prof. Dr. R. Gerhardts and Prof. Dr. W. Dietsche for valuable discussions.

References

1. von Klitzing, K., *et al.*: "New Method for High-Accuracy Determination of the Fine-Structure Constant Based on Quantized Hall Resistance", *Phys. Rev. Lett.,* **45**, 494-497, 1980.
2. Kaya, I. I., *et al.*: "Spatial evolution of hot-electron relaxation in quantum Hall conductors", *Phys. Rev. B* **58**, R 7536- R 7539, 1998.
3. Kaya, I. I., *et al.*: "Spatially resolved monitoring of the evolution of the breakdown of the quantum Hall effect: Direct observation of inter-Landau -level tunneling", *Europhys. Lett.*, **46**, 64-67, 1999.
4. Güven, K., *et al.*: "Two-level model for the generation and relaxation of hot electrons near the breakdown of the quantum Hall effect", *Phys. Rev. B* **65**, 155316-1-155316-8, 2002.
5. Sağol, B. E., *et al.*: "Time scale of the excitation of electrons at the breakdown of the quatum Hall effect", *Phys. Rev. B* **66**, 075305-1-075305-7, 2002.
6. Siddiki, A., *et al.*: "Incompressible strips in dissipative Hall bars as origin of quantized Hall plateaus", *Phys. Rev. B* **70**, 195335-1-195335-12, 2004.

Impact Ionization and Avalanche Multiplication in AlGaAs: a Time-Resolved Study

M. Betz,[1] S. Trumm,[1] M. Eckardt,[2] A. Schwanhäußer,[2] F. Sotier,[3] A. Leitenstorfer,[3] M. Hanson,[4] D. Driscoll,[4] A. C. Gossard,[4] S. Malzer,[2] and G. H. Döhler[2]

[1] Physik-Department E 11, Technische Universität München
[2] Max-Planck-Forschungsgruppe für Optik, Information und Photonik, Universität Erlangen
[3] Fachbereich Physik, Universität Konstanz
[4] Materials Department, UCSB Santa Barbara

Summary. Tracing ultrafast modifications of the Franz-Keldysh absorption spectrum of biased $Al_xGa_{1-x}As$ heterostructure diodes, we directly analyze the dynamical build up of a nonequilibrium carrier avalanche due to impact ionization for electric fields F > 350 kV/cm. The timescale of the carrier multiplication is found to be in the order of 10 ps depending on the applied bias. Monte Carlo simulations in a simplified band structure agree well with the experiment.

Band-to-band impact ionization in semiconductors is fundamental in small high-speed devices and is advantageously exploited as charge multiplication, e.g. in avalanche photodiodes. In contrast, it is a detrimental mechanism deteriorating the performance of modern field-effect transistors. Until now, this extreme non-equilibrium process has not been time resolved experimentally and its theoretical picture remains subject of lively discussions [1].

Recently, several studies of femtosecond transport in high electric fields have been published. These experiments rely either on the THz electromagnetic radiation emitted by charge carriers accelerating in an external field [2,3] or on all-optical electroabsorption measurements [4-6]. However, no signature of impact ionization has been extracted from these results.

In this contribution, we directly analyze the dynamical buildup of a non-equilibrium electron-hole avalanche in biased $Al_xGa_{1-x}As$ hetero-structure diodes for electric fields F between 350 kV/cm and 440 kV/cm. The experiment relies on changes of the Franz-Keldysh absorption spectrum due to a partial screening of the applied external field by propagating charge carriers. Most interestingly, we find a surprisingly slow picosecond avalanche dynamics in agreement with simulations in a simplified bandstructure.

The p-i-n-heterostructure employed in this study has been discussed previously [6]. A 350 nm wide intrinsic $Al_xGa_{1-x}As$ ($0 < x < 0.3$) layer is embedded between transparent contacts. A thin GaAs injection region allows for spatially well defined carrier generation via resonant interband absorption. After photoexcitation of the sample, a dipole builds up resulting from the spatial separation of the electron and hole ensembles propagating in the external field. As a result, this field is partially screened modifying the Franz-Keldysh absorption spectrum of the specimen. In particular, the transmission change of the heterostructure is an ideal measure for the number of electron-hole pairs with a spatial separation extending over the intrinsic region of the heterostructure. The experimental implementation takes advantage of a two-color Ti:sapphire laser providing synchronized excitation and probe pulses. A test pulse of 20 fs duration is used to detect the time-dependent optical response of the specimen. It is spectrally dispersed after transmission through the sample and detected with a photodiode.

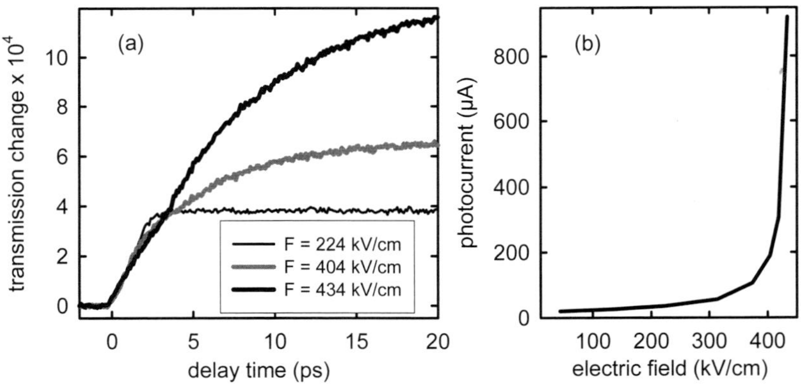

Fig. 1 (a) Transmission changes for a probe photon energy of 1.61 eV observed after excitation with an 80 fs pulse at 1.51 eV for various electric fields F and a temperature of T_L = 4 K. (b) Corresponding photocurrents of the specimen for various electric fields F.

Transmission changes for various electric fields F and a temperature of T_L = 4 K after photoexcitation at $t_D = 0$ are shown in Fig. 1(a). The probe photon energy of 1.61 eV is chosen to be especially sensitive to the transient screening field in the intrinsic zone. For an electric field of F = 224 kV/cm, the transient optical response rises to a saturation value within 2.5 ps. This finding reflects the ultrafast carrier motion from the injection layer across the high-field region and is consistent with previous results [6]. More interestingly, an additional slower signal component is observed for the case of F = 404 kV/cm and F = 434 kV/cm. As evident from Fig. 1(a), also the absolute value of the transmission change is larger as compared to a field strength of F = 224 kV/cm (in contrast, the theory of the Franz-Keldysh effect predicts a reduced amplitude for increasing bias and a constant screening field). Moreover, the photocurrent of the diode increases drastically for electric fields around 400 kV/cm (see Fig. 1(b)). As a consequence, our findings are clearly related to the generation of additional electron-hole pairs due to impact ionization and subsequent avalanche multiplication. Interestingly, establishing the carrier avalanche within the heterostructure requires a relatively long time of 10 ps to 20 ps depending on the external bias.

We now turn to the theoretical analysis of the observed nonequilibrium carrier dynamics. The simulation considers a simplified band structure with parabolic bands including heavy hole, light hole and split-off valence bands as well as conduction bands at the Γ-, L- and X-points of the Brillouin zone. The excitation is modelled according to the femtosecond pump pulse of the experiment. The dominant carrier-phonon scattering mechanisms with material parameters reported in Refs. [7-9] are taken into account.

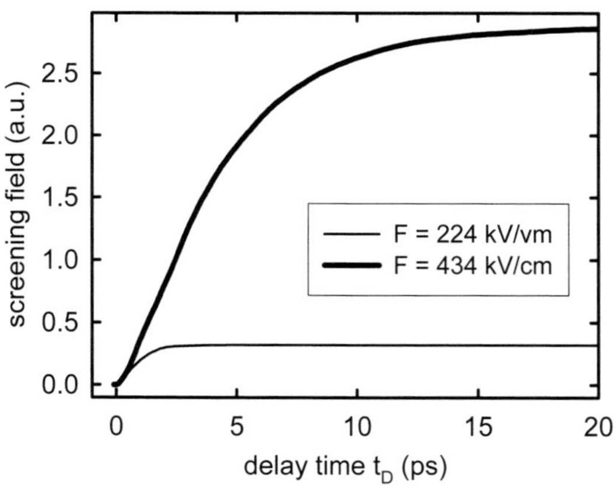

Fig.2 Screening fields within the heterostructure for two different electric fields F extracted from the Monte Carlo simulation of the spatio-temporal carrier dynamics in the sample.

The scattering rate for impact ionization is modelled according to the well-known Keldysh formula [10]: $\Gamma_i = A\,(E - E_{th})^P$. The parameters in this formula are chosen near those reported in [11].

Within this model, we calculate the spatio-temporal carrier dynamics in a Monte Carlo simulation. From these results, it is straightforward to extract the time dependent screening field in the inrinsic region resulting from the spatial separation of the propagating electron and hole ensembles. This quantity is expected to be proportional to the modification of the Franz-Keldysh absorption and, thus, to the transmission change of the experiment. In Fig. 2, typical results of the simulation are shown. For the case of F = 224 kV/cm, a saturation of the screening field at t_D = 2.5 ps is found indicating ultrafast carrier motion and a negligible influence of impact ionization. Similarly to the experiment, an additional slower dynamics on the 10 ps time scale is found for an electric field of F = 434 kV/cm. Considering the simplicity of the model, the agreement between theory and experiment is strikingly well.

In conclusion, we have studied the dynamical buildup of an extreme non-equilibrium carrier avalanche in GaAs for electric fields of F > 350 kV/cm. The typical timescale of the carrier multiplication process is found to be in the order of 10 ps depending on the applied bias and the number of photoinjected carriers. This finding may be an important ingredient for the design of fast avalanche based photodetectors. The dynamics of the charge multiplication is readily understood from a Monte Carlo simulation treating impact ionization on the basis of the Keldysh formula.

References

1. S. Picozzi, et al. Phys. Rev. Lett. **89**, 197601 (2002).
2. J. E. Pedersen, et al., Appl. Phys. Lett. **62**, 1265 (1993).
3. A. Leitenstorfer, et al., Phys. Rev. Lett. **82**, 5140 (1999).
4. H. Heesel, et al., Phys. Rev. B **47**, 16000 (1993).
5. M. Wraback, et al., Appl. Phys. Lett. **79**, 1303 (2001).
6. A. Schwanhäußer, et al., Phys. Rev. B **70**, 085211 (2004).
7. S. Adachi, Physical Properties III-V Semiconductor Compounds, Wiley, New York (1992)
8. K. Hess, Advanced Theory of Semiconductor Devices, Prentice-Hall, Englewood Cliffs, NJ, (1988)
9. C. Jacoboni, L. Reggiani, Review of Modern Physics **55**, 645 (1983)
10. L. V. Keldysh, Soviet. Phys. JETP **34**, 788 (1958)
11. D. Harrison, R. A. Abram, S. Brand, J. of Appl. Phys. **85**, 8178 (1991)

Fermi-Dirac Statistics in Monte Carlo Simulations of InGaAs MOSFETs

K. Kalna, L. Yang and A. Asenov

Device Modelling Group, Department of Electronics & Electrical Engineering, Univeristy of Glasgow, Glasgow, G12 8LT, Scotland, United Kingdom

Summary. The potential performance of 80 and 35nm physical gate length $In_{0.2}Ga_{0.8}As$ MOSFETs are compared with the equivalent Si and strained Si devices using ensemble Monte Carlo simulations. The 80 nm InGaAs MOSFET with a source/drain peak doping of 2×10^{19} cm^{-3} or a very high doping of 5×10^{19} cm^{-3} can outperform the equivalent strained Si MOSFET at both low and high drain biases. However, the 35 nm InGaAs MOSFET gives a performance comparable to the equivalent strained Si MOSFETs even with the source/drain doping of 5×10^{19} cm^{-3} .

1 Introduction

The need to increase the performance of scaled CMOS transistors recreates the interest in a MOSFET based on III-V materials [2] which would profit from high mobility in the channel. An essential step toward the realisation of III-V MOSFET is the development of suitable high-κ gate dielectrics for GaAs [3] exhibiting an 'unpinned' oxide/semiconductor interface with a low density of interface states. Monte Carlo (MC) device simulations [2,4] show that the equivalent III-V MOSFET transistor would outperform the Si and strained Si MOSFETs [2] when scaled down to a metallurgical gate length of 50 nm but the performance will deteriorate for sub-50 nm dimensions [4]. The performance decline is associated with a low density of states in III-V materials when compared to that of Si and Ge [2] which results in lower inversion charge densities for the equivalent gate overdrive.

We study the potential performance of a *n*-type $In_{0.2}Ga_{0.8}As$ channel MOSFET with the 80 nm physical gate length and a high-κ gate dielectric using ensemble MC simulations [5] Our electron transport model is extended by implementing Fermi-Dirac (F-D) statistics via self-consistent calculations of Fermi energy and electron temperature, a roughness scattering at the oxide/semiconductor interface, and by adopting the effective potential (EP) [6] to mimic the quantum confinement in the lowest subband. The simulated performance of the InGaAs MOSFETs is then compared with that of equivalent *n*-type conventional and strained Si MOSFETs [1,7].

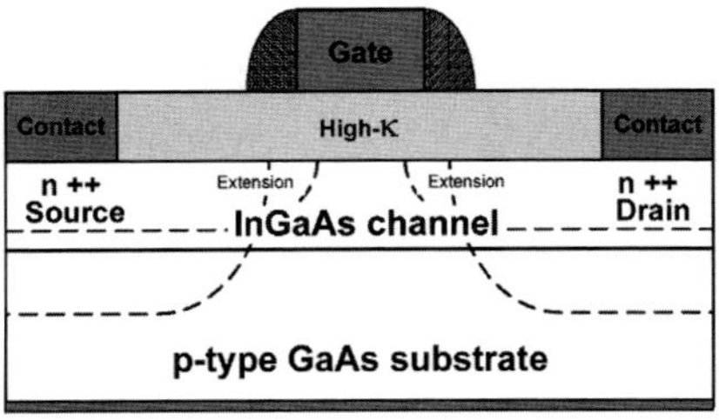

Fig. 1. Cross-section of the 80 nm gate length $In_{0.2}Ga_{0.8}As$ MOSFET considered in MC simulations.

The layout of the $In_{0.2}Ga_{0.8}As$ MOSFETs on a GaAs buffer in Fig. 1 is mimicking the structure of the 67 nm effective channel length conventional and strained Si MOSFETs published by IBM [1] and the 35 nm strained Si MOSFETs by Toshiba [7]. The doping profiles of the InGaAs MOSFETs with 80 nm and 35 nm metallurgical gate lengths are designed to have the same junction depth and subdiffusion [1,7] while the source/drain peak doping is reduced to reflect lower activation in InGaAs. Thus, we have created two doping variants: one with the peak concentration of 5×10^{19} cm^{-3}, representing a maximum technologically possible doping, and another with a peak of 2×10^{19} cm^{-3}, representing a more realistic doping profile.

2 Fermi-Dirac Statistics and Performance Characteristic

The F-D statistics are implemented by self-consistent calculations of Fermi energy E_F and electron temperature T. From the known electron density n and average electron energy $\langle E \rangle$ at each mesh point the following equations are solved in non-parabolic approximation:

$$n(\mathbf{r},t) = \frac{(2mk_BT)^{3/2}}{4\pi^2\hbar^3}\Big[\Gamma(3/2)F_{1/2}(\eta_F) + \frac{5}{2}\alpha\, k_BT \times \Gamma(5/2)F_{3/2}(\eta_F) + (\alpha\, k_BT)^2\,\Gamma(7/2)F_{5/2}(\eta_F)\Big] \quad (1)$$

$$\langle E(\mathbf{r},t)\rangle = \frac{4\pi^2\hbar^3}{n(2m)^{3/2}\sqrt{k_BT}}\Big[\Gamma(5/2)F_{3/2}(\eta_F)+\frac{5}{2}\alpha\, k_BT\,\Gamma(7/2)F_{5/2}(\eta_F) + (\alpha\, k_BT)^2\,\Gamma(9/2)F_{7/2}(\eta_F)\Big] \tag{2}$$

where $\eta_F = [E_F(\mathbf{r},t) - E_C]/k_BT$, E_C is the conduction band energy, and F_α the Fermi function of order α. The obtained E_F and T is then used in F-D distribution function to get an occupation probability of the final state after a scattering and in the static screening model for ionized impurity scattering. The inverse screening length, β, of the static screening model reads $\beta^2 = n(\mathbf{r},t)e^2/(\varepsilon\, k_BT)\times F_{-1/2}(\eta_F)/F_{1/2}(\eta_F)$, where ε is the dielectric constant.

This approach has been first verified in the MC simulations of high electron mobility transistors (HEMTs) [5]. Fig. 2 compares obtained I_D-V_G characteristics with results obtained using Boltzmann statistics calibrated to experimental data [5]. A slight drain current overestimate at $V_G = 0.4$ V for a low drain bias of 0.1 V is caused by the missing remote Coulomb scattering.

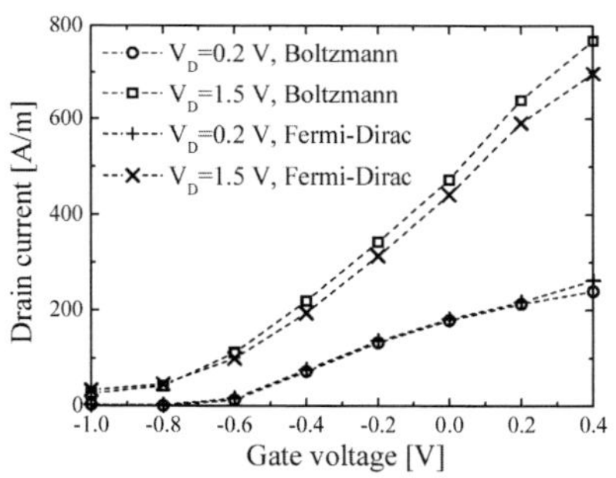

Fig. 2 I_D-V_G characteristics of the 120 nm HEMT with $In_{0.2}Ga_{0.8}As$ channel. The calibrated simulations of Ref. 5 using Boltzmann statistics are also shown for comparison.

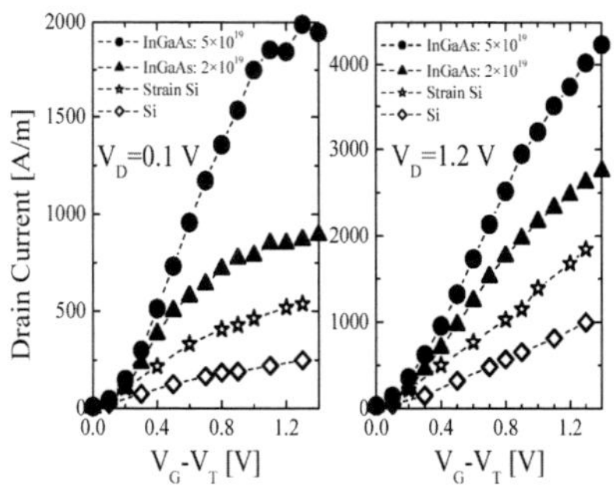

Fig. 3 I_D-(V_G-V_T) characteristics of the 80 nm $In_{0.2}Ga_{0.8}As$ MOSFET compared with the equivalent Si and strained Si MOSFETs. The source/drain peak doping for the $In_{0.2}Ga_{0.8}As$ MOSFET is indicated.

Quantum confinement at the dielectric/semiconductor interface is adopted via the EP approach [6]. A Gaussian parameter *a* of the EP is calibrated [2] against a Poisson-Schrödinger solution in the subthreshold leading to *a*=1.5 *nm* for the 80 *nm* and to *a*=0.8 *nm* for the 35 nm InGaAs MOSFETs.

The interface roughness scattering is based on Ando's model [8] with an exponential auto-correlation function. The corresponding electron scattering rate caused by an interface roughness induced potential reads:

$$\Gamma(E)=\frac{m\Delta^2\lambda^2e^2}{\hbar^2}\frac{F_{trans}^2}{\sqrt{\hbar^2+2mE(1+\alpha E)\lambda^2}}\mathsf{E}\left(\frac{\sqrt{2mE(1+\alpha E)}\lambda}{\sqrt{\hbar^2+2mE(1+\alpha E)\lambda^2}}\right) \quad (3)$$

where E is a complete elliptic integral and F_{trans} is the transversal electric field re-evaluated at each MC step. The RMS height, Δ, of 0.5 nm and the correlation length, λ, of 1.7 nm were considered for the InGaAs MOSFETs.

Fig. 3 shows I_D-(V_G-V_T) characteristics at V_D=0.1 V and 1.2 V. The 80 nm InGaAs MOSFETs with a source/drain peak doping of 2×10^{19} cm^{-3} and 5×10^{19} cm^{-3} would outperform by 65% andmore than 300%, respectively, the equivalent strained Si MOSFET at a low drain voltage of 0.1 V. At a high drain voltage of 1.2 V, the 80 nm InGaAs MOSFETs still outperforms the strained Si MOSFET by more than 200% when the peak doping is 5×10^{19} cm^{-3} and by 65% when this is 2×10^{19} cm^{-3}. The average velocity along a channel of the 80 nm InGaAs MOSFET with the more realistic and very high doping shown in Fig. 4 is also much higher (about 95% and 130% larger, respectively) than that in the Si based MOSFETs.

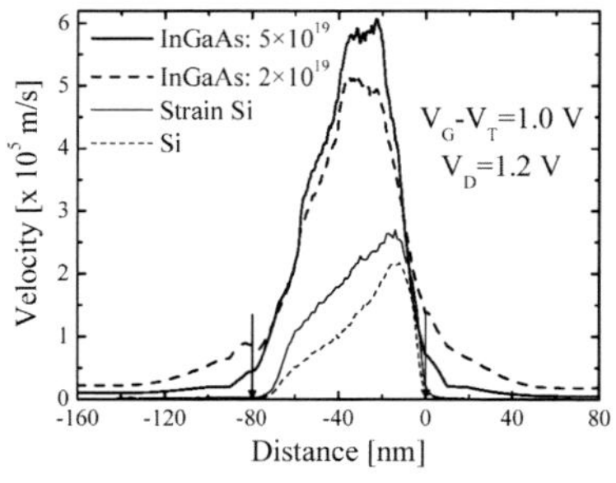

Fig. 4 Average electron velocity along the channel of 80 nm $In_{0.2}Ga_{0.8}As$ MOSFETs compared with that in the equivalent Si MOSFETs (thin lines). The arrows depict the beginning and the end of the gate.

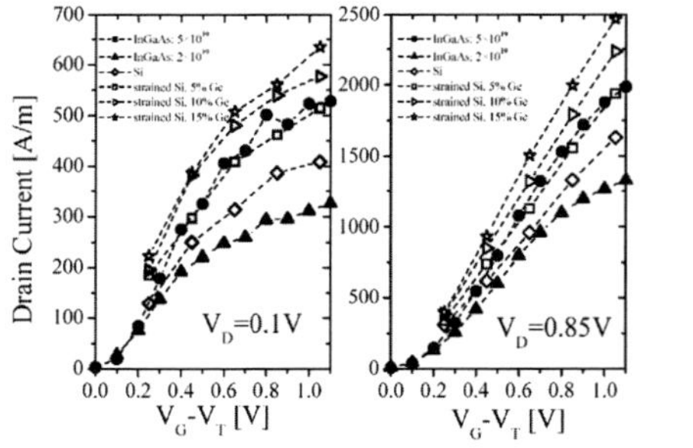

Fig. 5 ID-(VG-VT) characteristics of the 35 nm $In_{0.2}Ga_{0.8}As$ MOSFETs (two variants) compared with the 35 nm Si and strained Si MOSFETs with various Ge content substrates indicated in the legend.

The situation completely changes when InGaAs and Si MOSFETs are scaled down to the 35 nm physical gate length. The 35 nm InGaAs MOSFET cannot outperform the equivalent strained Si MOSFETs with more than a 10% Ge content 'virtual' substrate at both low and high drain voltages as shown in Fig. 5. The 35 nm $In_{0.2}Ga_{0.8}As$ MOSFET with a peak doping of 5×10^{19} cm^{-3} can outperform the conventional 35 nm Si MOSFET by just by 24% while it is outperformed by strained Si MOSFETs with 10 and 15% Ge content 'virtual' substrates. This performance decline is attributed to a relatively low density of states in InGaAs when compared to that in Si or Ge [4] and to the lower possible *n*-type doping in the source/drain.

Using MC device simulations we have demonstrated that $In_{0.2}Ga_{0.8}As$ 80nm MOSFETs may deliver significant improvement in the drive current compared to the present state of art Si based MOSFETs. However, the situation changes for the 35 nm $In_{0.2}Ga_{0.8}As$ MOSFET which cannot outperform the equivalent strained Si MOSFETs with more than a 10% Ge content 'virtual' substrate at any drain biases even with the extremely high source/drain peak doping of 5×10^{19} cm^{-3}. **Acknowledgement**. This work has been supported by EPSRC under Grant No. GR/M93383.

References

1. K. Rim et al., *Symp. VLSI Technology 2001* (Tokyo, Japan), 59, 2001.
2. K. Kalna et al., *Semicond. Sci. Technol.*, **19**, S202, 2004.
3. Z. Yu et al., *Appl. Phys. Lett.* **82**, 2978, 2003.
4. M. V. Fischetti et al., *J.Comput. Electron.* **3**, 287, 2005.
5. K. Kalna et al., *Solid-State Electron.* **46**, 631, 2002.
6. D. K. Ferry, *Superlatt. Microstruct.* **27**, 61, 2000.
7. S. Inaba et al., *IEEE Trans. Electron Devices* **49**, 2263, 2002.
8. D. K. Ferry, *Semiconductor Transport* (Taylor & Francis, N. Y.) p 135, 2000.

Monte Carlo Study of the Suppression of Diffusion Noise

L. Varani*, E. Starikov**, P. Shiktorov**, V. Gruzhinskis**, C. Palermo*, J.-C. Vaissière*, J.-P. Nougier*

* Centre d'Electronique et de Micro-optoélectronique de Montpellier
c.c. 084 – Université Montpellier II
Place Bataillon – 34 095 Montpellier Cedex 5 – France

** Semiconductor Physics Institute
Goshtauto 11
2600 Vilnius – Lithuania

Summary. We investigate the possibility to suppress the diffusion noise in bulk semiconductors by adding to a constant electric field a fluctuating contribution characterized by a gaussian distribution and a characteristic time. To this purpose, Monte Carlo simulations are performed applied to the case of GaAs at room temperature. Numerical results confirm the possibility to suppress the total noise power by an appropriate choice of the externally added noise.

1 Introduction

Since several years an increasing interest is devoted to constructive aspects of noise and fluctuations in different systems. In general, this kind of studies concerns the mutual interplay between a fluctuating random component (the noise) and a deterministic component (the signal). Accordingly, the addition of noise is able to enhance under some specific conditions the response to the signal (e.g. in the case of stochastic resonance). However, very few studies concern the mutual interplay between two fluctuating random signals. In this context, one of the most intriguing results was reported by Vilar and Rubí in 2001 [1]: they showed that adding an external noise to systems with an intrinsic noise may result in a less noisy system. More recent studies have extended the original theory (developed only for the low–frequency spectrum) to the full frequency spectrum of the output signal [2]. An experimental validation of the general theory applied to the case of a microscopic bead held in a noisy optical trap has been proposed in Ref. [3]. The original theory was developed using very general statistical properties of nonlinear systems without involving any microscopic detail of the system under study. Therefore the following question arises: is the general theory applicable to a realistic situation where a bulk semiconductor material is submitted to a fluctuating electric field? It is the purpose of this contribution to clarify these points using Monte Carlo simulations applied to bulk GaAs.

2 Monte Carlo Calculations

Bulk semiconductors such as GaAs can be considered as stochastic systems whose state, characterized by the average electron velocity, depends on some external parameter, that is the applied electric field. In Ref. [1] it is shown that favorable conditions for noise suppression by the addition of a noisy electric field are reached when the spectral density of velocity fluctuations $S(E)$ verifies $\partial^2 S/\partial E^2 < 0$. However, we have shown that reduction of the total noise power is best reached when the variance of velocity fluctuations exhibits a maximum in the considered electric field range [4].

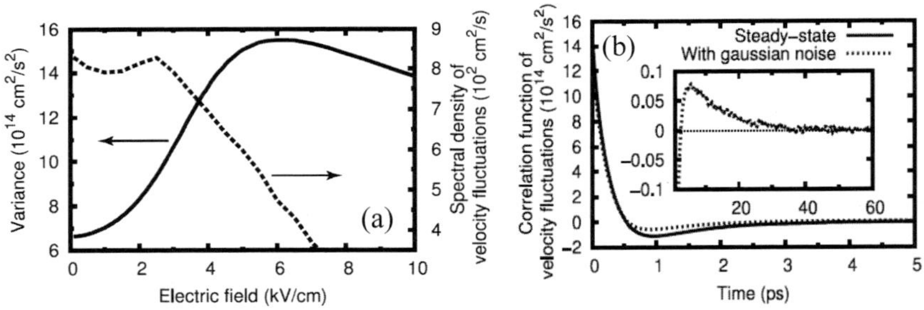

Fig. 1. Variance of velocity fluctuations (continuous lines) and spectral density of velocity fluctuations at zero frequency (dashed lines) as functions of the electric field calculated by the Monte Carlo simulation at temperature T = 300 K for the case of n-GaAs with a donor concentration $N_D = 10^{16}$ cm^{-3} (a). Correlation functions of single particle velocity fluctuations (b). Continuous lines correspond to the calculation with a fixed electric field E=6.5 kV/cm while dashed lines correspond to the calculation with a fluctuating electric field characterized by a standard deviation of 3 kV/cm and a characteristic time of 10 ps. The insert in (b) reports a zoom of the long-time tail of the correlation function.

This condition is verified in bulk GaAs, as shown in Fig. 1(a), and from this result we conclude that the most favorable condition to obtain noise suppression in GaAs, in the sense of a decrease of the total power of the fluctuating signal, should be reached for a static electric field around 6 kV/cm. This expectation is confirmed by the results given for the correlation functions of velocity fluctuations [Fig. 1(b)] and for the frequency dependent spectral density of velocity fluctuations [Fig. 2(a)] obtained for the case of a static electric field equal to 6.5 kV/cm and when a fluctuating electric field (characterized by a standard deviation $\sqrt{\Delta E^2} = 3\,\text{kV/cm}$ and a relaxation time $\tau_F =$ 10 ps) is superimposed to the static field.

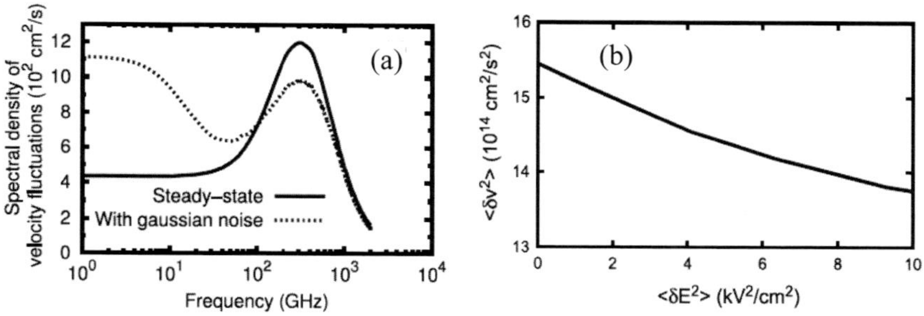

Fig. 2. Frequency dependent spectral density of velocity fluctuations (a). Continuous line corresponds to steady state and dashed line to Gaussian noise addition with parameters $\sqrt{\Delta E^2} = 3\,\text{kV/cm}$ and $\tau_F =$ 10 ps . Variance of velocity fluctuations with Gaussian noise ($\tau_F =$ 100 ps) (b).

We notice that the negative part in the correlation function, which is a typical feature of non-equilibrium conditions [5], is slightly reduced for the case of an externally added noise (we recall that the negative part of the correlation function in the picosecond time scale produces a peak in the spectral density at frequencies in the TeraHertz region). Additionally, we remark the appearance of a long-time positive tail of the correlation function which vanishes for a time corresponding nearly to $3\tau_F$ as one may conclude by treating τ_F as a relaxation time characteristic of field fluctuations.

As a matter of fact, the presence of a fluctuating field introduces an additional correlation in the fluctuations of carrier velocity which gives a positive contribution to the low-frequency spectral density. This is confirmed by the calculation of the Fourier transforms of the correlation functions which are shown in Fig. 2(a). We remark that, while the spectral density is enhanced in the low–frequency region, it is suppressed in the high–

frequency region near the peak associated with intervalley transfers. Due to the logarithmic representation in the frequency domain, this suppression plays a dominant role in the total noise power. This is confirmed by the results shown in Fig. 2(b) which report the integrated spectral density (i.e. $<\delta v^2>$) for a characteristic time τ_F = 100 ps. Indeed, we remark that the variance decreases monotonously by increasing the amplitude of electric field fluctuations thus confirming the possibility of obtaining the suppression of diffusion noise by using an external noise. Here the amplitude of field fluctuations plays the role of an *amplifying* factor for noise suppression. From a microscopic point of view, this suppression arises from the fact that the fluctuating electric field obliges the carriers to visit regions of the momentum space characterized by a smaller variance with respect to the situation in the presence of a constant field.

3 Conclusion

We have investigated the possibility to suppress the diffusion noise in bulk GaAs by using a time dependent electric field fluctuating according to a gaussian distribution with a given characteristic time. The Monte Carlo results show that the suppression of the total noise power is possible and that the most favorable conditions of noise reduction in GaAs are for a fluctuating electric field around 6.5 kV/cm. Additional calculations are needed to extend the model to more complex one-port and two-port devices and eventually to other types of electrical noises.

References

1. J.M.G. Vilar and J.M. Rubí, *Phys. Rev. Lett.*, **86**, 950–953, 2001.
2. D.B. Walton and K. Visscher, *Phys. Rev.*, E **69**, 051110-1–051110-8, 2004.
3. Y. Seol, K. Visscher, D.B. Walton, *Phys. Rev. Lett.*, **93**, 160602-1–160602-4, 2004.
4. L. Varani, C. Palermo, C. De Vasconcelos, J.-F. Millithaler, J.-C. Vaissière, J.-P. Nougier, E. Starikov, P. Shiktorov and V. Gruzhinskis, *Proceedings of UPoN: Unsolved Problems on Noise*, on press.
5. T. Kuhn, L. Reggiani, L. Varani, V. Mitin, *Phys. Rev.*, B **42**, 5702, 1990.

TeraHertz Emission From Nanometric HEMTs Analyzed by Noise Spectra

J.-F. Millithaler[1], L. Varani[1], C. Palermo[1], J. Mateos[2], T. González[2], S. Perez[2], D. Pardo[2], W. Knap[3], J. Lusakowski[3], N. Dyakonova[3], S. Bollaert[4] and A. Cappy[4]

[1]CEM2 - UMR CNRS 5507 - Université Montpellier II - France
[2]Universidad de Salamanca - 37008 Salamanca - Spain
[5]GES – UMR CNRS 5650 – Université Montpellier II – France
[3]IEMN - UMR CNRS 8520 – Avenue Poincaré – 59652 Villeneuve d'Ascq -France
[4]IEP – Warsaw University – Warsaw - Poland

Summary. Recent experiments have shown that High Electron Mobility Transistors can emit electromagnetic radiation in the TeraHertz range. The emission spectra exhibit two peaks: one around 1 THz is sensitive to drain and gate voltages, and another one around 5 THz is fixed. In order to get physical insight into the microscopic mechanism at the basis of the radiation emission we have performed a Monte Carlo (MC) simulation of the measured transistors using the current noise spectra as sensitive probes to detect the presence of electrical instabilities. Numerical results are found to be in good agreement with experiments confirming the presence of an oscillatory dynamics in the TeraHertz range.

1 Introduction

The TeraHertz (THz) domain is a frequency range in the electromagnetic spectrum between microwaves and near-infrared. Its ability to penetrate into several materials as a non-destructive radiation may be usefully employed for imaging applications in security, industry, biology and medicine. However, there is lack of compact and cheap THz sources and standard solid-state devices like transistors could offer great possibilities of integration with other electronic and optoelectronic devices within a single chip [1]. This is the case of InGaAs/InAlAs lattice-matched nanometric High Electron Mobility Transistors (HEMT) where THz emission has been recently observed for the first time [2].

It is evident that a detailed analysis of the physical processes characteristics of such devices requires a microscopic approach due to the nanometric dimensions leading to the appearance of hot carriers phenomena, nonstationary transport, ballistic conditions etc.

As a consequence, in order to investigate the observed emission, we have used MC simulations to obtain the spectrum of the current noise, which is known to be a quantity very sensitive to different carrier instabilities [3]. In this context, we emphasize that, while a standard relaxation dynamics gives rise to a Lorentzian cut-off frequency in the noise spectrum related to the inverse of the relaxation time, any carrier instability responsible for an oscillatory dynamics will be evidenced as a peak.

This work is carried out by means of a semi classical two-dimensional MC model whose validity was already checked for similar devices [4].

2 Experiments

Experiments have been performed on lattice-matched InGaAs/InAlAs High Electron Mobility Transistor (HEMT) on InP substrate with a nominal gate length L_G = 60 nm, a source-drain distance L_A = 1.3 μm and a width W = 50 μm (Fig.1.(a)).

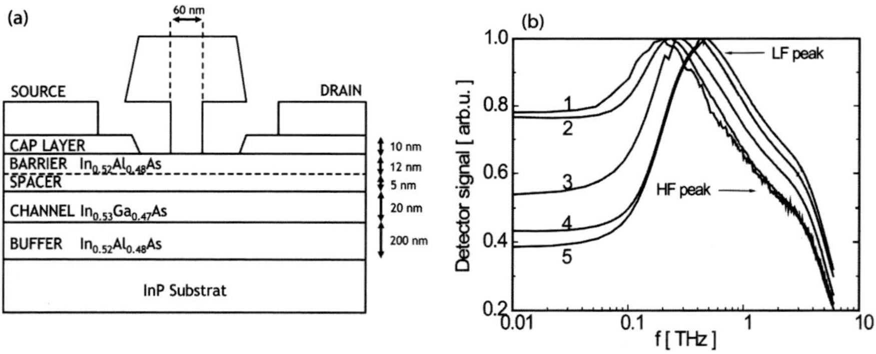

Fig. 1 (a) The schematic structure of the transistor investigated with a T-shaped gate. The dotted line in the barrier region shows the position of the delta doping. (b) HEMT emission spectra for the gate short circuited with the source by a gold wire on the chip. The spectra from 1 to 5 correspond to V_{ds} equal to 0.3, 0.45, 0.6, 0.7 and 0.8 V, respectively.

The spectrum of THz radiation (Fig. 1(b)) observed experimentally by a cyclotron emission spectrometer [2], exhibits two structures: a low frequency (LF) peak (lower than 1 THz) which has been found to be sensitive to gate and drain voltages and a high frequency (HF) peak (around 5 THz) whose frequency is rather fixed.

3 Monte Carlo Results

Figure 2 shows the spatial dependence of the average electron velocity inside the HEMT for different drain voltages V_{ds} and for a constant gate voltage V_{gs}. We notice that, under the gate, the carriers reach a much higher velocity than in the recess and contact parts of the transistor: this can be interpreted as a signature of a ballistic regime. This velocity distribution reflects the observed frequency dependence of the positions of the low frequency peaks in the emission spectra. Figures 3(a) and (b) show the gate current and drain current noise spectra. The most striking result is that the MC calculation of the noise spectra confirms the presence of some oscillatory dynamics in the THz frequency region in good agreement with the experimental results (Fig. 1(b)). We remark in the first case (Fig.3(a)) the presence of two peaks: a low frequency one, around 1-2 THz, which depends on the drain-to-source as well as on the gate-to-source voltages and a high frequency one at a fixed frequency around 5 THz.

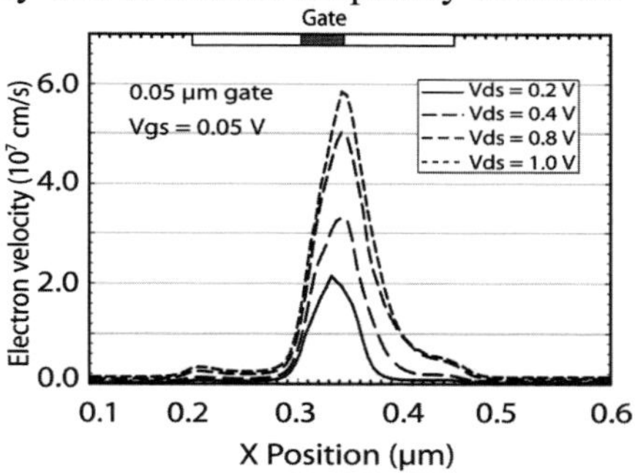

Fig. 2 Spatial dependence of the average electron velocity in the transistor for a constant V_{gs}=0.05 V and for the indicated values of V_{ds}. The lines at the top of the figure show the spatial extensions of the gate and recess regions.

The suppression of the HF peak in the drain current noise spectra in Fig.3(b) seems to indicate that the origin of this resonance should be attributed to plasma oscillations situated in the source-gate part of the transistor [2]. On the other hand, when V_{ds} exceeds a threshold value, simulations show the appearance of Gunn oscillations located in the near-drain part of the channel which could explain the LF peak in the emission spectra. The calculated frequency dependence of the LF peak on the applied voltage as well as the threshold-like behaviour is in good agreement with experiments. The peculiar boundary conditions (short circuiting the gate with the source) which was imposed on the transistor in order to measure the emission spectra [2], could explain the discrepancies between the experiments and the simulations, where such a condition cannot be easily accounted for. Moreover, MC simulations have been performed at room temperature contrary to the experiments made at 4.2K since the numerical model has been validated for devices operating at 300K [4]. However, besides these differences, the measured and simulated spectra exhibit similar features. This indicates that the MC simulation is able to model satisfactorily the behaviour of the transistors.

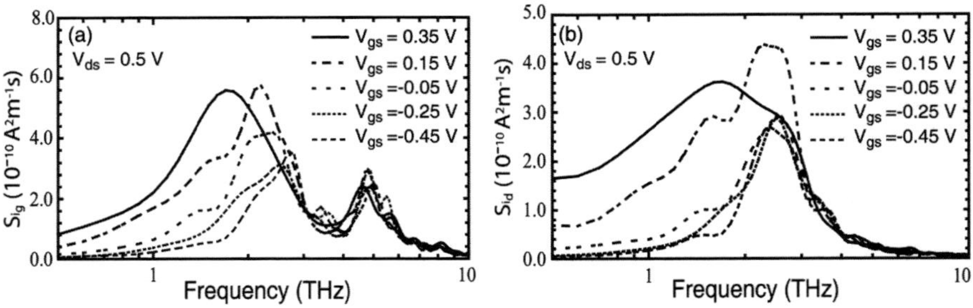

Fig. 3 Gate current noise spectra for the reported V_{gs} and a constant V_{ds} equal to 0.05 V (a) and Drain current noise spectra for the reported V_{gs} and a constant V_{ds} equal to 0.05 V (b).

4 Conclusion

TeraHertz emission from nanometric InGaAs/InAlAs HEMTs has been experimentally measured. The emission spectrum consists of two peaks around 5 THz and 1 THz. The results of Monte Carlo simulations shows that the electron transport under the gate is ballistic and that the gate current noise spectra evidence the presence of peaks corresponding to the observed frequencies. This analysis shows that the frequency behaviour of the emitted radiation can be directly linked to the spectra of current fluctuations. As a consequence, noise spectra are here employed as sensitive indicators of the onset of collective motion of carriers inside the transistor.

References

1. R. E. Miles, P. Harrison, and D. Lippens, Terahertz Sources and Systems, vol. **27**, Kluwer Academic Publishers, 2001.
2. W. Knap, J. Lusakowski, T. Parenty, S. Bollaert, A. Cappy, V. V. Popov, and M. Shur, Appl. Phys. Lett, **84**, 2331 (2004).
3. L. Varani, L. Reggiani, T. Kuhn, T. González, and D. Pardo, IEEE Trans. Electron. Dev., **41**, 1916 (1994).
4. J. Mateos, T. González, D. Pardo, S. Bollaert, T. Parenty, and A. Cappy, IEEE Trans. Electron. Dev., **51**, 521 (2004).

Electron Transport in Novel Sb-based Quantum Cascade Lasers

V. Spagnolo[1], M. S. Vitiello[1], G. Scamarcio[1], D. G. Revin[2], J. W. Cockburn[2]

[1]CNR-INFM Regional Laboratory LIT[3] and Dipartimento Interateneo di Fisica "M. Merlin", Università and Politecnico di Bari

[2]Department of Physics and Astronomy, University of Sheffield, S3 7RH United Kingdom

Summary. We have investigated the transport properties of $In_{0.53}Ga_{0.47}As/AlAs_{0.56}Sb_{0.44}$ quantum cascade lasers operating at $\lambda \sim 4.3$ μm, by means of band-to-band photoluminescence (PL). The evolution of the PL bands with the applied electric field allows a detailed analysis of the injection and emission efficiency. A clear correlation between the device thermal performance and the electron distribution has been found.

There is considerable interest to address the problem of quantum cascade laser (QCL) emission in the wavelength range $\lambda = 3$-5 μm, mainly for free space communication systems and for applications in sensing of many hazardous chemicals, drugs and explosives. The short wavelength operation limit in QCLs results from the low conduction band energy discontinuity of the semiconductor materials typically employed, which imposes a restricted range of inter-subband laser transition energies. Among lattice matched material systems with larger ΔE_C values, the InGaAs/AlAsSb heterostructure is one of the most promising candidate for high performance laser development in the range $\lambda = 3$-5 μm, due to its large ΔE_C value (≥ 1.6 eV) and, particularly, its demonstrated compatibility with well-established InP-based waveguides [1]. The optical characteristics of InGaAs/AlAsSb-based devices depend strongly on the detailed interface structure between the two ternary alloys, which in turn is determined by the growth conditions. Interface degradation is typically reported and it is due to Sb interdiffusion, which leads to the creation of a quaternary alloy $In_xGa_{1-x}As_ySb_{1-y}$ and give rises to radiation active centres.

The formation of this quaternary layers affects the band structure alignment by relaxing the interface strain and changing the interface-dipole contribution to the quantum well profile. The latter effect can significantly change the valence band-offset and eventually convert the type-I band alignment (valence band offset $\Delta E_v = E_v^{InGaAs} - E_v^{AlAsSb} > 0$), as expected in the case of abrupt interfaces [2] and corresponding to vertical band-to-band recombination in InGaAs layers, to the type-II configuration ($\Delta E_v < 0$) characterized by diagonal transitions [3]. Also, the conduction band offset ΔE_c may change, thereby modifying the subband energies and the emission wavelength of the device.

In this work we study the chemical nature of the interfaces in InGaAs/AlAsSb QCL structures by means of band-to-band photoluminescence technique. A detailed analysis of the transport dynamics and the injection/emission efficiency during device operation, explains their differences in thermal performance.

We have investigated two molecular beam epitaxy grown $In_{0.53}Ga_{0.47}As/AlAs_{0.56}Sb_{0.44}$ quantum cascade devices, both operating at 4.3 μm. Sample A essential features are the vertical laser transition with the electrons confined in the upper level by a minigap created by the Bragg reflectors between the active regions. The maximum operating temperature is 270 K and its detailed description is reported in ref. [4]. Sample B active region design is based on a bound to continuum scheme. This device shows a maximum operating temperature of 240K and its detailed description is reported in ref. 1. Our experimental method is based on a microprobe band-to-band photoluminescence (PL) technique [5]. Figure 1a shows PL spectra obtained for sample A at device off as a function of the Kr^+ laser power. At low excitation power the spectra are dominated by a band peaked at ~ 0.95 eV, due to emission from interface-related (TI) transitions. This emission saturates at larger power and the spectra is dominated by band-to-band emission, peaked at 0.975 eV at the highest power. The presence of the TI band demonstrates the formation of quaternary alloy $In_xGa_{1-x}As_ySb_{1-y}$ at the interfaces [2]. Fig. 1b shows the PL spectrum collected at zero-field and at a Kr^+ laser incident power of 350 μW. In Figs. 1c-d are shown the overlap integrals of the envelope functions of the three main band-to-band transitions calculated at zero-field, with the related energies, in the hypothesis of a type II (panel c) or a type I (panel d) band alignment, using the associated conduction and valence band discontinuity [6]. Whatever the band alignment choice, in both cases, the main contributions to the PL spectra are due to transitions between the lower active region conduction subband (j =1) and the levels localized in the adjacent valence energy barriers (k = 1,2,3). Similar results have been obtained also for sample B.

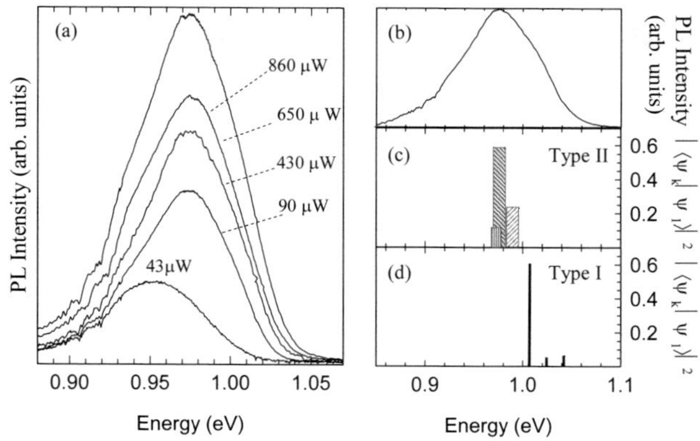

Fig. 1. (a): PL spectra collected at zero-field for sample A as a function of the exciting Kr^+ laser power. (b): PL spectrum taken at zero-field for sample A with an exciting Kr^+ laser power of 0.35 μW. (c), (d): overlap integrals and energy positions of the three main band-to-band transitions associated to the conduction level j =1 and the valence subbands k=1,2,3, as predicted by our Schrödinger–Poisson calculation in the hypothesis of a type II (panel c) and type I (panel d) band alignment, respectively.

The PL lineshape strongly depends from the applied voltage. In all the investigated voltage range the main transition energies as predicted by a type II band structure show a good agreement with the major features of the PL spectra, whereas those predicted by a type I band structure are partially consistent with the spectra, always remaining higher than the experimental PL bands. Thus, the following analysis will be conducted in the hypothesis of a type II band alignment. Figure 2a shows several PL spectra taken for sample A at different applied voltages, starting from the device off up to V=10V. At low voltages (V < 5V) and zero current flowing in the device we observe a progressive reduction of the main PL band intensity, without any change in the energy peak position. Electrons remain localized in the active region ground state j =1 and cannot reach the adjacent injector region due to the effect of an energy difference between the lowest active region subband j =1 and the lowest injector level j=4, which acts as an effective energy barrier. At larger applied voltage (V > 5 V), as current starts to flow in the device, an high energy band emerges in the PL spectra. Indeed, electrons start to populate the injector levels, although the mechanism of depletion of the level j =1 is not yet efficient, due to the effect of the E_4-E_1 energy barrier, and few electrons are injected in the subsequent module.

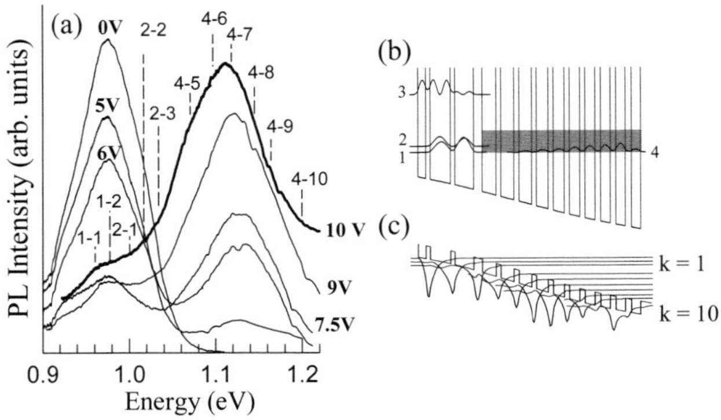

Fig. 2. PL spectra measured for sample A at different applied voltages, from device-off up to V=10V (bold line). Conduction (b) and valence (c) band structure calculated for V=10 V (e). The shaded areas mark the injector miniband. The dashed vertical lines mark the energies of the transitions between levels in the conduction (j) and valence (k) bands calculated at V=10V.

At higher voltages, the energy barrier E_4-E_1 becomes negligible (see Fig. 2b), the population of the level j = 1 decreases, while the population of the state j = 4 further increases, and the PL band, resulting from the convolution of the transitions between the low energy injector state j = 4 and several valence subbands, dominates the spectra. A similar study has been performed for sample B. Figure 3 shows the PL spectra taken for sample B at different applied voltages, up to 9.6 V, in the full conduction regime. The differences in the injector design between the two investigated devices are reflected in the related PL signal.

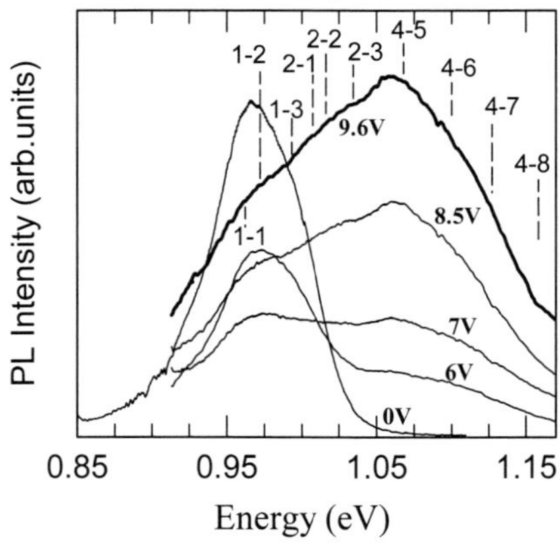

Fig. 3. PL spectra measured for sample B at different applied voltages, from device-off up to V=9.6V (bold line). The dashed vertical lines mark the energies of the transitions between levels in the conduction and valence bands, calculated at an applied voltage of V= 9.6V.

The low energy PL band, due to the allowed transitions between subbands localized in the active region, not only dominates the spectra for low applied voltage (V< 7 V), but remains comparable with that coming from transitions involving the low energy injector state j=4, even at the largest applied voltage, thus demonstrating that electrons still populate the active region lower emission levels, mainly due to their thermal backfilling. This effect, typically due to hot electrons, reduces the population inversion, hence limiting the highest device operating temperature (240 K). In contrast, in sample A the additional pair of well-barrier layers in the injector-Bragg reflector, increases the emission efficiency, hinders the lower emission subbands thermal backfilling and allows a higher temperature operation (270 K).

This work was partly supported by MIUR, project FIRB-RBAU01E8SS and the ANSWER Project STRP 505642-1. VS, MSV and GS acknowledge Q. Yang and J. Wagner for providing mesa devices and for useful discussions.

References

1. Revin, D.G. et al.: 'InGaAs/AlAsSb quantum cascade lasers', *App. Phys. Lett.*, **85**, 3992-3994, 2004.

2. Georgiev, N., Mozume, T.: 'Photoluminescence study of InGaAs/AlAsSb heterostructure', *J. App. Phys*, **89**, 1064-1069, 2001.

3. Nakata, et al.: 'Characterization of GaAsSb/InAlAs quantum-well structures lattice-matched to InP grown by molecular beam epitaxy ', *Journal of Crystal Growth* **99**, 311-314, 1990.

4. Yang, Q. K., et al.: 'Room-temperature intersubband emission from GaInAs–AlAsSb quantum cascade structure', *Electron. Lett*, **40**, 1339-580, 2004.

5. Spagnolo, V., et al.: 'Temperature profile of GaInAs/AlInAs/InP quantum cascade-laser facets measured by microprobe photoluminescence', *App. Phys. Lett.*, **78**, 2095-2097, 2001.

6. We use for $In_{0.53}Ga_{0.47}As$ a band-gap of 0.81 eV, while for $AlAs_{0.56}Sb_{0.44}$ we use band-gap of 2.47 eV . The conduction and valence band discontinuities are respectively ΔE_c = 1.74±0.02 eV and ΔE_v = 0.07±0.02 eV in the hypothesis of a type II alignment and ΔE_c = 1.6 eV and ΔE_v = 0.06 eV for type I band structure.

Quantum Phonon-Limited High-Field Electron Transport in Semiconductors

G. Ferrari, E. Cancellieri, P. Bordone, and C. Jacoboni

National Research Center S3, INFM-CNR
Dipartimento di Fisica, Università di Modena e Reggio Emilia, Via Campi 213/A, I. 41100 Modena, Italy

Summary. A fully quantum theory of phonon-limited electron transport in semiconductors is applied to a homogeneous steady-state situation to investigate the difference between quantum results and the results of a semiclassical theory. The Wigner function is used for the quantum approach, and Monte Carlo simulations are performed in both semiclassical and quantum theories. In the considered case, hot-electron transport in a simple silicon model at 77 K, very little difference has been found since collisional broadening changes the possible final states of the electronic transitions without altering in a significant way the total scattering rate and therefore the momentum relaxation efficiency of phonon scattering.

1 Introduction

Quantum effects are nowadays very common in electron transport in semiconductors. Besides effects due to the space scale used in nanostructures, where the dimensions of the systems are comparable with the coherence distance of the electron dynamics or even with the electron wavelengths, also effects due to the time scales can be expected. In fact, the scattering rate at high fields are of the order of 10^{13} s^{-1} and the corresponding collisional broadening, of the order of 10^2 K, is expected to play a significant role. In the present paper we investigate this problem by considering electron transport in a homogeneous hot-electron simulation and comparing quantum results with the results of the more standard quasi-classical Boltzmann equation (BE).

The Wigner function (WF) is a theoretical tool that proved to be very effective for quantum transport calculations [1] and it allows a Monte Carlo (MC) simulation very similar to the traditional MC approach to solve the BE. In this paper the WF approach has been applied in a version suitable to study a homogeneous system in steady-state conditions.

2 Theory

The electron transport equation for the WF has the form

$$\mathcal{L} f_W = \mathcal{S} f_W = \mathcal{A} f_W + \mathcal{B} f_W + \mathcal{C} f_W + \mathcal{D} f_W$$

where f_w is the WF generalized to include phonon variables as described in [1]; $\mathcal{L}$ is the standard Liouville operator that appears also in the l.h.s. of the BE. $\mathcal{S}$ is the integral operator that contains the electron-phonon interaction. The operator in the r.h.s. of the BE is known as the scattering integral and contains in-scattering and out-scattering terms. In the Wigner equation it contains four terms with emission and absorption phonon operators acting on the left and on the right of the density matrix, or equivalently, on the left and right arguments of the WF. These four terms are indicated above as $\mathcal{A}$, $\mathcal{B}$, $\mathcal{C}$, $\mathcal{D}$. Here we show one of them when an emission occurs on the left argument of the WF:

$$\mathcal{A} = \sum_q F(q) e^{i(qr-\omega_q t)} \sqrt{n_q+1} f_W\left(r, p - \frac{\hbar q}{2}, \{n_1, \dots, n_q+1, \dots\}, \{n'_q\}, t\right)$$

and $F(q)$ is the coupling constant between electrons and phonons. Through the standard procedure that makes use of the path variables the above equation can be transformed into an integral equation that can be formally integrated and then expanded in its Neumann expansion [1]:

$$\begin{aligned} f_W = f_W^{(0)}(t) &+ \int \mathcal{A}(t') f_W^{(0)}(t') dt' + \int \mathcal{B}(t') f_W^{(0)}(t') dt' + \int \mathcal{C}(t') f_W^{(0)}(t') dt' \\ &+ \int \mathcal{D}(t') f_W^{(0)}(t') dt' + \int \mathcal{A}(t') \int \mathcal{A}(t'') f_W^{(0)}(t'') dt' dt'' + \dots \qquad (1) \end{aligned}$$

where $f_w^{(0)}(t)$ represents the ballistic translation of the WF at time t; each $\mathcal{X}(t)$ corresponds to a vertex of interaction at time t (with $\mathcal{X}=\mathcal{A}$, $\mathcal{B}$, $\mathcal{C}$ or $\mathcal{D}$). In order to evaluate diagonal terms in the phonon population starting from a diagonal form, each phonon process must contain two vertices. Four possible combinations of vertices are to be considered that correspond to virtual and real emission and absorption of a phonon mode q.

Virtual scattering contains the information that will lead to the self energy Σ and therefore to the lifetime of the electron states (scattering out in the BE). This means that all the terms containing virtual processes between any two vertices will contribute to the "dressing" of the propagation of the WF with a factor like

$$e^{-\Sigma \Delta t} \qquad (2)$$

The series in Eq.(1) leads to a series expansion of such exponential. For $\Delta t > \Sigma^{-1}$ the series expansion may require many terms and since they are of alternating signs the single terms may become much larger than the final sum. This implies that they must be calculated with great precision. This is the reason why for large times, larger than the momentum relaxation time, the Neumann expansion (as any type of path-integral algorithm) requires very large statistical sampling (beside the fact that each integral contains oscillating functions). The basic idea of self scattering is to approximate Σ in the exponent of Eq. (2) with an approximate value Γ:

$$e^{-\Sigma \Delta t} = e^{-\Gamma \Delta t} e^{(\Gamma - \Sigma)\Delta t}$$

and then evaluate only the second exponential by sampling its power expansion. If Γ is close to Σ the series will be rapidly converging and if $\Gamma > \Sigma$ all terms will be positive, even though single samples may always be negative. The above idea is realized, both in the BE and in the transport equation for the Wigner function, by defining $\tilde{f} = f e^{\Gamma t}$. The equation for $\tilde{f}$ is then transformed into its integral form with the usual procedure. Returning to the original function, the integral equation takes the form:

$$f_W(t) = f_W^{(0)}(t) e^{-\Gamma t} + \int_0^t \Gamma e^{-\Gamma(t-t')} f_W(t') dt' + \int_0^t e^{-\Gamma(t-t')} S f_W(t') dt' \quad (3)$$

The Neumann expansion can now be performed and the Wigner-path interpretation [1] restored, observing that single real scattering contributions (of second order) must be sampled together with single self-scattering events.

2 Results

A Monte Carlo sampling has been performed for a simple semiconductor model based on Si ($m^*=0.32m_0$; optical-phonons equivalent temperature = 450 K; e-ph coupling constant=8×10^{10}eV/m; T = 77 K). Results are shown in Fig. 1 and 2. Fig. 1 shows the drift velocity and mean energy as functions of applied field. The simulation time has been taken t=300 fs, and the initial condition has been taken equal to the steady-state semiclassical distribution at the same field strength. Longer times may be necessary to reach steady state with quantum interaction. However, after 300 fs the difference between quantum and semiclassical results is very small. The analysis has shown that this is due to the fact that collisional broadening changes the possible final states of the electronic transitions without altering in a significant way the total scattering rate and therefore the momentum relaxation efficiency of phonon scattering.

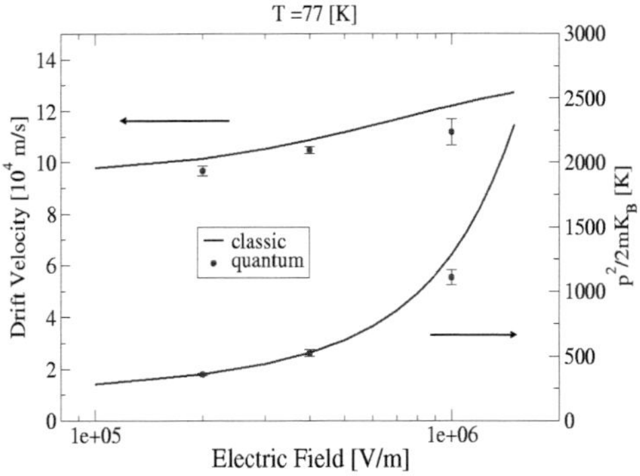

Fig. 1. Drift velocity and mean $p^2/2mK_B$ as a function of applied field. Semiclassical results (continuous line) are compared with quantum results (dots) 300 fs after the interaction has been switched from semiclassical to quantum (see text). The anomalous behavior at lower fields is due to the absence of acoustic phonon scattering.

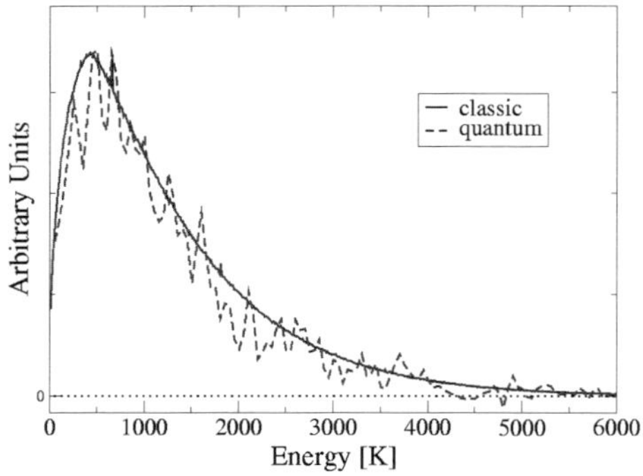

Fig. 2. Electron energy distribution at a field strength of 1 MV/m at T=77 K. Semiclassical results (continuous line) are compared with quantum results 300 fs after the interaction has been switched from semiclassical to quantum.

Fig. 2. Shows the electron distribution as a function of energy where, again, the difference between semiclassical and quantum results is very small. The analysis has also shown that the intracollisional field effect is very small at the considered field strengths.

This work has been partially supported by the U.S.Office of Naval Research (contract No. N00014-98-1-0777 / N00014-03-1-0289).

References

1. Jacoboni C. and Bordone P.: ‘The Wigner-function approach to non-equilibrium electron transport’, Rep. Prog. Phys. **67**, 1033-1071 (2004).
2. Nedjalkov M., Kosina H., Selberherr S., Ringhofer C. and Ferry D.: ‘Unified particle approach to Wigner-Boltzmann transport in small semiconductor devices’, Phys. Rev. B **70**, 115319-115335 (2004).
3. Wigner E.: ‘On the Quantum Correction For Thermodynamic Equilibrium’, Phys. Rev. **40**, 749-759 (1932).
4. Frensley W.: ‘Boundary conditions for open quantum systems driven far from equilibrium’ Rev. Mod. Phys. **62**, 745–791 (1990).

Transit Time and Velocity Distribution Functions in Decananometer Gate-Length SOI MOSFETs

M. J. Martín and R. Rengel

Departamento de Física Aplicada, Universidad de Salamanca. Spain.
e-mail: mjmm@usal.es

Summary. A Monte Carlo investigation of transport in sub-100 nm gate length SOI MOSFETs is presented. Each super-particle within the channel is followed individually: scattering mechanisms undergone, distance travelled, transit time, etc. are accounted for the calculation of average statistical quantities and their distribution functions at different channel positions, together with their dependence on the gate length. In this way, the influence of non-stationary phenomena and quasiballistic transport is evaluated.

1 Introduction

As the dimensions of MOSFET devices approach the decananometer range, Silicon-on-Insulator (SOI) MOSFET technology has become the most feasible solution to avoid the problems of conventional bulk transistors [1]. To fully exploit the possibilities of SOI devices, it is mandatory to properly link the main macroscopic quantities of interest to the internal physics governed by microscopic events. In this work, we have performed a Monte Carlo (MC) device simulation [2] of decananometer gate length Fully-Depleted SOI MOSFETs, developing the algorithms and procedures necessary to carry out a statistical investigation of charge transport following the methodology proposed by other authors in [3].

2 Results

FDSOI MOSFETs have been simulated with an active layer thickness t_{si} of 15 nm. The active layer p-doping is 10^{18} cm^{-3}, and the source and drain wells are $2\cdot10^{19}$ cm^{-3} N^{+} doped. The gate oxide thickness (t_{ox}) is 2 nm, and the buried oxide thickness is 200 nm. The metal gate workfunction is chosen to be 4.05 eV. Since our main purpose is to evaluate the effect of scaling of L_g, scaling of t_{ox} is not considered to avoid undesired gate current leakage effects. Several values of L_g have been considered: 90 nm, 65 nm, 45 nm, 30 nm and 20 nm. Fig. 1(a) shows the average transit time of carriers traversing the channel of the 90 nm transistor, together with the transit time (t_t) distribution function (DF) for V_{GS} = 1.0 V and several V_{DS} from 0.2 to 1.0 V [Fig. 1(b)]. At low V_{DS} (triode regime, for which the inversion layer extends through the entire channel) a noticeable reduction in the transit time is observed when increasing the drain potential for all the values of V_{GS} (with an identical exponential decay). Under these conditions, the homogeneous increase in the average longitudinal electric field along the channel reduces notably t_t since the device behaves practically like a linear resistor and transport is highly diffusive, with the DF showing a wide spectrum. However, when the device enters the saturation regime of operation, the decrease of the transit time with V_{DS} is less pronounced. The existence of the pinch-off region controls the time it takes to electrons to cross the channel: when increasing V_{DS}, the additional drain potential is not translated in a faster transit time of carriers through the pinch-off region since this area already shows a non-stationary and nearly ballistic behaviour, with a significant velocity overshoot. The DF shows a well-defined peak, and as V_{DS} is raised its profile becomes narrower.

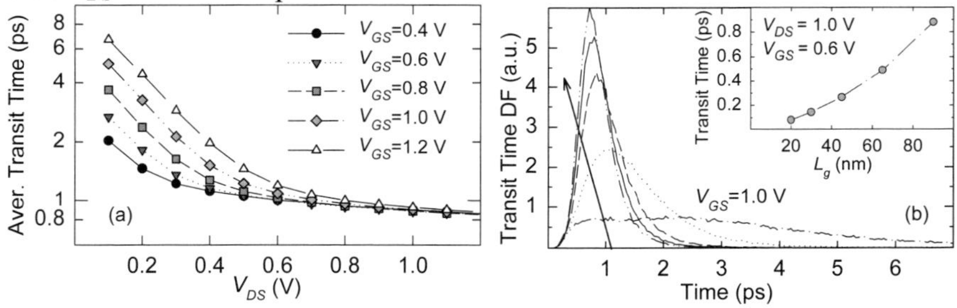

Fig 1. (a) Average transit time of carriers crossing the channel (L_g = 90 nm) as function of V_{DS} and for several V_{GS} conditions. (b) Transit time distribution function. Transit time versus L_g in saturation conditions is displayed on the inset.

Let's check now the evolution of the transit time of carriers when decreasing the gate length [inset of Fig. 1 (b)], in saturation conditions. When the device length is scaled down, the transit time of carriers is obviously reduced; however, as it can be observed, the decrease is not linear, what indicates that carrier transport conditions are being modified for the smallest devices. To analyze this behaviour, we have examined the profile of the average longitudinal velocity along the channel in saturation [Fig. 2 (a)]. The value of velocity saturation in bulk material is also included as a reference. For the longest device a velocity overshoot appears in the second half of the channel (near the drain); this is the signature of non-stationary phenomena. As L_g is reduced (indicated by an arrow in the graph), the percentage of the channel where carriers operate under non-stationary conditions increases (more than 80% for the shortest devices). The total average path R travelled by carriers [Fig. 2(b)] has been also computed. As L_g is reduced, R is closer to the corresponding gate length, which indicates a more effective transport of carriers in the longitudinal dimension.

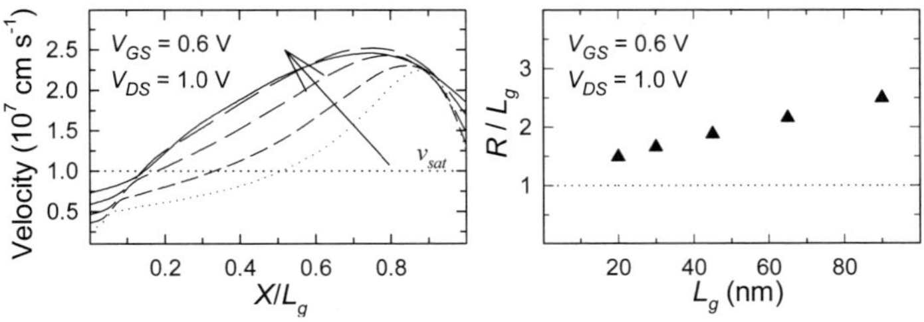

Fig 2. (a), Velocity vs. normalized X position and (b) Total traveled distance vs. L_g

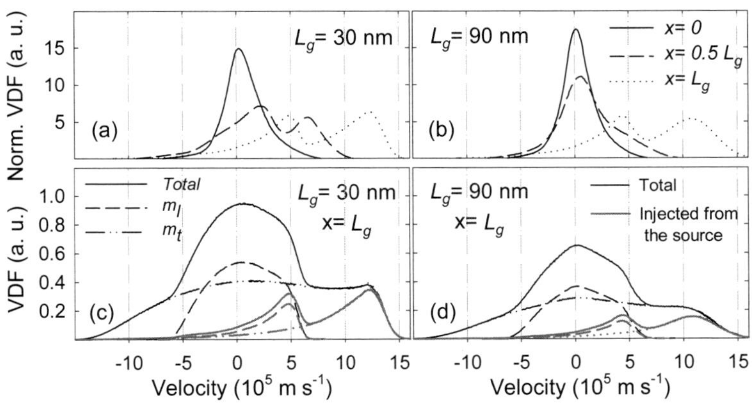

Fig.3. Normalized VDF for **(a)** L_g=30nm and **(b)** L_g=90 nm and tree different relative positions in the channel. **(c)** and **(d)** show differences between VDF of total particles and the ones injected from the source.

In Fig. 3 (a) and (b), the normalized values of the longitudinal velocity distribution function (VDF) of source-injected carriers crossing the channel are shown for the 90 nm and 30 nm devices. Near the source-substrate junction (solid-line), the distribution function shows a quasi-gaussian profile slightly shifted towards positive velocities. At the drain end of the channel (dotted line) two peaks can be clearly identified, the slowest and the fastest associated to carriers in longitudinal and transversal X valleys, respectively. However, the differences at the middle of the channel (dashed line) are noticeable. While for the longest device the VDF only shows one defined peak (and a significant tail for positive velocities), for the 30 nm transistor the two peaks are already evidenced, in good agreement with the appearance of non-stationary phenomena already discussed (Fig. 2 (a)). Considering the total number of electrons in each cell in the calculation of the VDF (Figs 3 (c) and (d)) it is possible to obtain valuable information about backscattered electrons from the drain (mainly concentrated near the substrate-drain junction) that increase the negative velocity values. The elevated electric field near the drain and surface scattering mechanisms rapidly re-orient electrons towards the positive *X* direction.

L_g (nm)	τ (fs)	λ (nm)
20	20.7	7.9
30	19.4	7.1
45	17.9	5.9
65	16.5	5.1
90	14.5	3.9

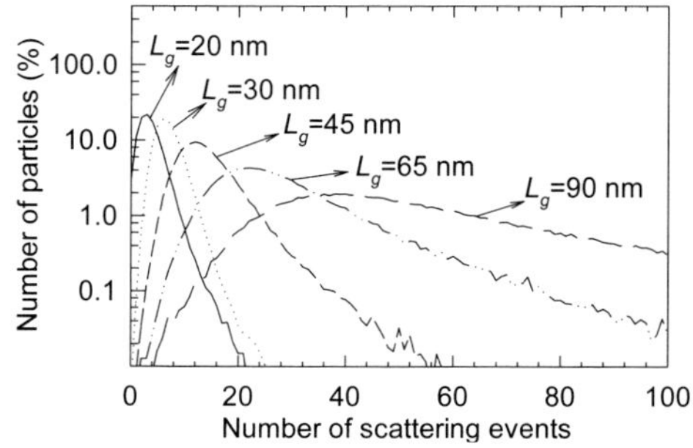

Table I. Estimation of τ and λ for different L_g values.
Fig.4. Number of particles versus number of scattering mechanisms suffered.

As the gate length is reduced, the quasiballistic character of transport is strengthened, as it is shown in Fig. 4. In fact, for L_g = 20 nm electrons only suffer as an average 3.8 scattering mechanisms in their flight across the channel. Accordingly, the mean free path (λ) and the time between scattering events (τ) increases (Table I), thus enhancing the non-stationary character of transport.

Acknowledgements

This work has been funded by the Research Project SA008B05 from the Consejería de Educación de la Junta de Castilla y León.

References

1. Colinge J-P *Silicon-on-Insulator Technology: Materials to VLSI* 2nd edn, Norwell: Kluwer, 1997.
2. Rengel R. *et al.*: 'A physically based investigation of the small-signal behaviour of bulk and fully-depleted silicon-on-insulator MOSFETs for microwave applications'. Semicond. Sci. Technol. 19 (2004) 634–643
3. Saint Martin J, *et al.*: 'On the ballistic Transport in Nanometer-Scaled DG MOSFETs', IEEE Trans. On Electron Dev. 51 (2004) 1148-1155.

Collision of Fano Resonances in a Molecular Ring

E. R. Hedin[1], A. M. Satanin[1,2] and Y. S. Joe[1]

[1]Center for Computational Nanoscience, Department of Physics and Astronomy, Ball State University, Muncie, IN 47306, USA
[2]Institute for Physics of Microstructures, RAS, GSP-105, Nizhny Novgorod, 603950, Russia

Summary. We examine new effects arising from the collision of Fano dipole resonances as affecting the transmission through an Aharonov-Bohm (AB) ring. The behavior of the zeros of the transmission amplitude as a function of magnetic field is investigated for various coupling integrals between quantum dots in the ring. It is shown that the collision of two Fano-dipoles forms a new quasi-particle, which behaves as a coupled object called a Fano-quadrupole.

1 Model of the System and Approach

The interference of localized and propagating waves gives asymmetric Fano resonances [1] in the transmission of quasi-one-dimensional systems. The scattering amplitude near the Fano zero-pole in the complex energy plane behaves like a dipole, where the pole plays the role of the particle and the zero plays the role of the hole [2]. Under conditions giving collision (collapse) of the hole and the particle, the transmission of the nanostructure may change dramatically [3]. Manipulating the operation of these resonances in the transmission gives new methods for controlling the conductance by means of quantum interference.

In the present work, we consider a model (see Fig. 1) which may be created by modern semiconductor nanotechnology. The sites in the ring are considered as quantum dots (or molecules). We treat the electron transfer between the dots in the framework of the tight-binding approximation by introducing matrix elements $V_{n,m}$ ("overlap integrals"), taking into account only nearest neighbor sites. The Schrödinger equation in the standard tight-binding approximation is

$$-\sum V_{n,m}\psi_m + \varepsilon_n\psi_n = E\psi_n . \quad (1)$$

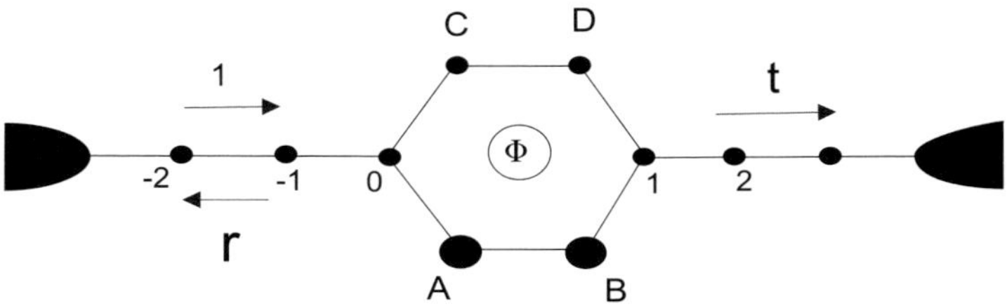

Fig. 1. Geometry of the ring with Ohmic electrodes and magnetic flux. (All the coupling integrals between sites are V_0, except as stated in the text).

The solution of Eq. (1) gives the transmission amplitude (see [4] for details),

$$t(E,\Phi) = \frac{iV_0 \sin\theta e^{i\theta} N(E,\Phi)}{D(E,\Phi)}. \quad (2)$$

Here, $\theta = ka$, where k is the wave vector, connected with the energy by $E = -2V_0 \cos ka$, and a is the distance between the dots. Φ indicates the magnetic flux through the ring, which shifts the phase of the electron wave function through the well-known AB effect. The quantum dots A and B have energies levels ε_A and ε_B. The peaks and zeros in the transmission are defined by equations $D(E,\Phi) = 0$ and $N(E,\Phi) = 0$, respectively. The numerator is in general a complex function dependent upon the coupling integrals and site energies. The equation for the transmission zeros in our model is solved analytically.

2 Results and Discussion

We present here numerical results for the particular case with $V_u = 0.3$ (the value of the coupling integral between sites in the upper arm of the ring), $\varepsilon_A = 0.8$, and $\varepsilon_A = -\varepsilon_B$. We shall use the coupling integral V_0 as a unit of energy, and we will use the dimensionless matrix element V_{AB} as a tuning parameter. In Fig. 2, the calculated transmission through the ring is depicted for the weak, critical, and strong coupling regimes. For the weak-coupling regime, $V_{AB} = 1$, we see two nearly independent Fano resonances for which there are two zeros on the real energy axis and two poles in the complex energy plane. Overlapping of the Fano dipoles occurs at a critical value of V_{AB}, which is seen in Fig. 2 for $V_{AB} = V_{AB}^c = -3.13$.

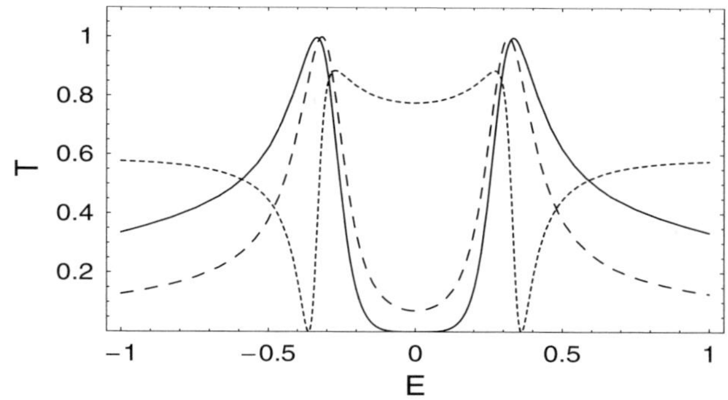

Fig. 2. Fano resonances of the transmission as a function of electron energy for different values of the interaction parameter: $V_{AB} = 1.0$ (dotted line), $V_{AB} = -3.13$ (solid line), and $V_{AB} = -6.0$ (dashed line).

The critical value of V_{AB} is calculated by requiring the energy for the transmission zero to be single-valued and real. When V_{AB} becomes more negative than the critical value, both the zeros leave the real energy axis and move away from the axis in opposite directions in the complex plane. In Fig. 2, we also display the transmission for the strong coupling regime ($V_{AB} = -6$), where the zeros have moved into the complex plane and the Fano resonances are merged into two coupled dipoles (called here a Fano quadrupole).

The variation of V_{AB}^{c} with V_u is shown in Fig. 3 for various values of the combined site energies $\varepsilon_A + \varepsilon_B$. The dependence of V_{AB}^{c} with V_u is not particularly sensitive to the energy product, $\varepsilon_A \varepsilon_B$.

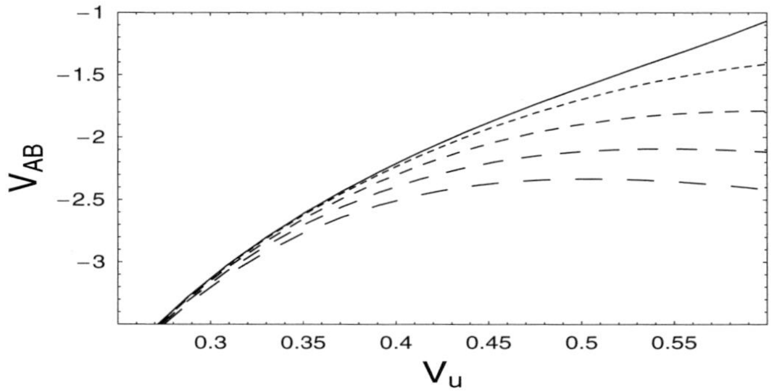

Fig. 3. The variation of the critical value of V_{AB} as a function of V_u for $\varepsilon_A + \varepsilon_B$=0, 2.5, 5.0, 7.5, and 10.0 (upper to lower curves). $\varepsilon_A \varepsilon_B = -0.64$.

The coupled Fano resonance in the transmission can be tuned by the magnetic flux threading the AB ring. We have calculated the trajectories of the zeros, shown in Fig. 4, by using the explicit formula for the transmission amplitude, Eq. (2). The qualitatively different behavior of the zeros as a function of flux for the weak and strong coupling regimes is clearly seen. For weak coupling, the zeros move in separate "orbits" near their associated poles, and for strong coupling, the zeros move in a common orbit around the two coupled poles, indicating the behavior of a Fano quadrupole. For coupling near the critical value, the zeros traverse a figure-8 type pattern in the complex energy plane, as seen in Fig. 4.

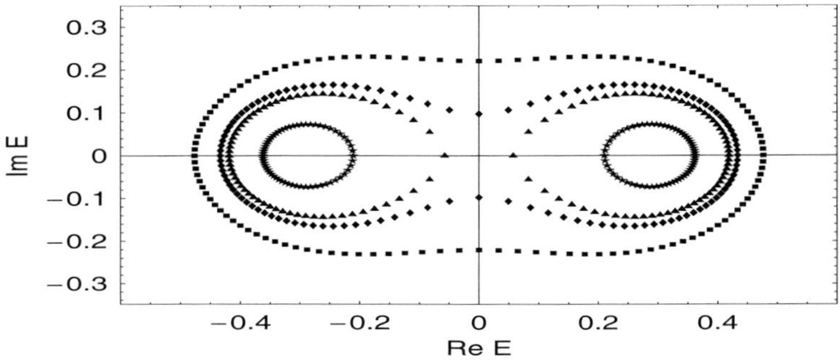

Fig. 4. Trajectories of the zeros in the complex energy plane as a function of magnetic field (flux) for $V_{AB} = -0.5, -3.0, -3.5, -5.0$ (from inner to outer curves).

In conclusion, we have shown how Fano dipoles arising from the interference of states in coupled quantum dots and in the reference arm of an AB ring merge at critical values of tunable system parameters and form a coupled object termed a Fano quadrupole.

This work is supported by the Indiana 21st Century Research and Technology Fund. The work of AMS was supported in part by the Russian Basic Research Foundation Grant No. 05-02-16762.

References

1. U. Fano, Phys. Rev. **124**, 1866 (1961).
2. Z. Shao, W. Porod, and C. S. Lent, Phys. Rev. B **49**, 7453 (1994).
3. C. S. Kim, A.M. Satanin, Y. S. Joe, and R. M. Cosby, Phys. Rev. B **60**, 10962 (1999).
4. A.M. Satanin, E.R. Hedin, and Y. S. Joe, Phys. Lett. A, in press (2005).

Simulation of Domain Formation in p-Si/SiGe Quantum Cascade Structures

Z. Ikonic, P. Harrison and R. W. Kelsall

School of Electronic and Electrical Engineering, University of Leeds, Leeds LS2 9JT, United Kingdom

Summary. Domain formation in p-doped Si/SiGe quantum cascades is considered using a carrier scattering transport framework. The hole flow along the cascade is described via scattering between quantized states belonging to neighbouring periods, caused by phonons, alloy disorder, and carrier-carrier interactions. The generation of either periodic or of nonperiodic domains is studied in uniformly or modulationally doped cascades, and criteria for their appearance are found. The domains in modulationally doped cascades have a relatively smaller effect on the energy structure than in uniformly doped ones.

1 Introduction

Biased semiconductor quantum well cascade structures, interesting for intersubband photodetectors, quantum cascade lasers, and Bloch oscillators are prone to the formation of electric field domains. The homogeneous field is broken, due to charge redistribution over individual wells. The effect is related to negative differential resistivity (NDR), and has been addressed in a number of papers [1-6]. The NDR is usually ascribed to resonant tunneling between quantized states of adjacent wells. However, current thought puts scattering, rather than resonant tunneling, as the dominant transport mechanism in quantum cascade lasers [7]. We here consider the formation of stationary domains in p-type Si/SiGe cascades.

2 Theoretical Considerations

Hole transport is described via ``upstream" and ``downstream" scattering between quantized subbands in neighbouring wells, which are calculated using the fully anisotropic 6X6 k.p method. The processes taken into account are deformation potential (acoustic and optical phonons), alloy disorder, and carrier-carrier scattering. The scattering rates between all pairs of states within a well, or between adjacent wells, Fig.1, are calculated as a function of the electric field. Carrier heating is accounted for via energy balance approach [8]. In order to explore domain formation the system of rate equations [8] is modified to allow for inhomogeneous electric field.

To solve the rate equations, appropriate boundary conditions at the contacts have to be specified. There have been a few different approaches, from simply assuming the contacts to be equivalent to the adjacent real wells [4], or imposing ``Ohmic" boundary conditions [1], to having a more complex structure which required a self-consistent calculation [5]. The type of contact boundary conditions affects the precise criteria for domain formation, but this is most prominent in short structures.

The conventional picture of domains is that a cascade is split into a low-field and a high-field region, one expanding at the expense of the other with varying field. Some more recent there detailed studies [6] report that a homogeneous carrier and field distribution in an n-type multiple quantum well photodetector evolves into a period-doubled configuration, to form periodic space-charge / field domains that include two structural periods of the cascade. In effect, one can then impose periodic boundary conditions on subband populations and field.

3 Results and Discussion

Results are here given for a simple cascade having 4.41 nm $Ge_{0.3}Si_{0.7}$ wells and 2.15 nm wide Si barriers, grown on a $Ge_{0.2}Si_{0.8}$ virtual substrate. It has two low-lying (and populated) subbands per period, the ground HH1 and the first excited, LH1 subband, spaced by 28 meV. At a field of 42 kV/cm the HH1 subband from the preceding (higher) well and the LH1 subband of the next (lower) well are aligned at the zone center, but off it the alignment field changes due to different dispersions of HH1 and LH1 subbands. The calculated current-voltage dependence, assuming a uniform field, is shown in Fig.1 for different values of doping. The NDR occurs between 60 and 70 kV/cm, away from where the simple picture of zone-center alignment and resonant tunnelling would put it. Being mostly brought about by carrier-carrier scattering, NDR is more prominent at higher doping.

In calculations using contact boundary conditions we have modelled the contacts as wells structurally identical to those inside the cascade. Concerning the carrier distribution over subbands of the ``contact wells", these could be taken as equivalent to the distribution in the adjacent well, or that only the ground subband of the ``contact well" was populated. In either case, the bias-dependent scattering rates used inside the cascade were also employed for calculating the scattering into or out of the contacts.

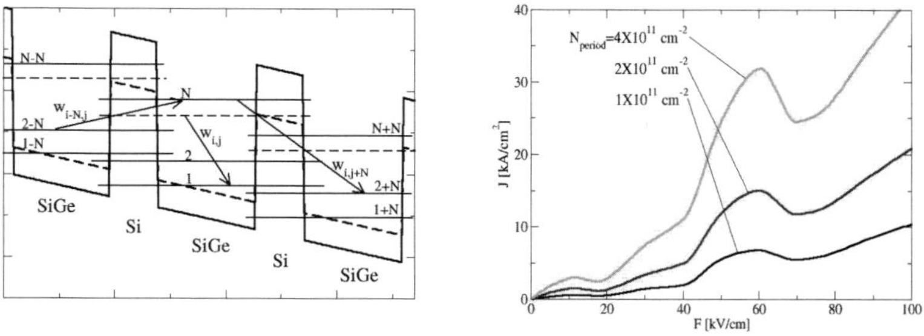

Fig. 1. The scattering transport model of the quantum cascade structure (using the upside down energy scale for holes, left), and the current density J vs. bias field F calculated for the homogeneous cascade described in the text, for different values of doping density per period (right).

Nonperiodic domains were found at the doping level of approximately P=2.4X10^{11} cm^{-2}, slightly depending on the type of contact boundary conditions used. This compares rather well with the approximate analytic criterion for threshold [1], which gives 2X10^{11} cm^{-2}. In Fig.2 we plot the population of the two subbands around the domain boundary, which separates the low- and high-field regions. The boundary is not very sharp, but actually extends over a few structural periods, and becomes narrower as the doping increases from its threshold value. The population of the upper (LH1) subband is considerably smaller in the high-field region.

As for the periodic domains, in a homogeneously doped cascade only domains with a period of 2 unit cells were found, at about the same threshold density as for the contact-related domains, i.e. at 2.4X10^{11} cm^{-2} (Fig.2). It is interesting to note that the variation of subband population for the two types of domains is quite different: for periodic domains the lower subband population is more strongly modulated than that of the upper.

The modulation doping (the doping period including two or more structural periods, with a modulated – i.e. unevenly distributed over individual periods - doping density) strongly influences the domain structure. Whilst increasing the modulation depth reduces slightly the range of parameters for which domain formation occurs, such domains also become ``milder'': the difference between electric fields in the initial and period-doubled configurations is much smaller than for homogeneous doping. The same applies when increasing the modulation doping ``super-period''.

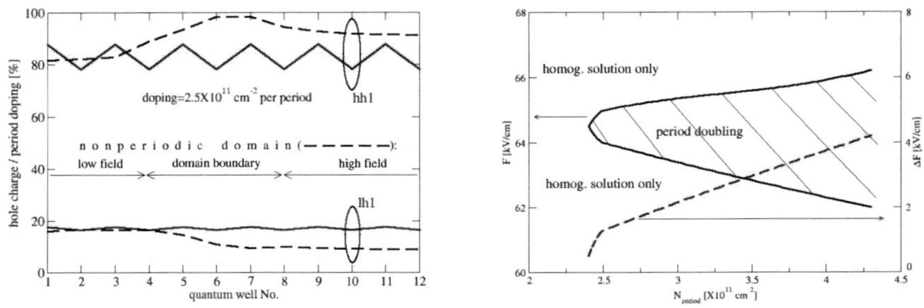

Fig. 2. The charge density in the HH1 and LH1 subbands of quantum wells around the domain boundary in the nonperiodic case (dashed), and in the period-doubled cascade (solid line), with the doping density of $2.5X10^{11}$ cm^{-2} (left), and the range of cascade parameters where period-doubling domain formation occurs (cross-hatched area), and the size of the electric field perturbation (dashed line) (right).

This work was supported by DARPA/USAF contract No. F19628-99-C-0074, and EPSRC (UK) Grant GR/S27528/01.

References

1. A. Wacker, 'Vertical transport and domain formation', in *Theory of transport properties of semiconductor nanostructures*, Ed. E. Scholl, Chapman and Hall, London, 1998.
2. A. Wacker, 'Semiconductor superlattices: A model system for nonlinear transport', *Phys. Reports,* **357**, 1-111, 2002.
3. L. L. Bonilla, H. T. Grahn, ``Non-linear dynamics of semiconductor superlattices", *Rep. Prog. Phys*., **68**, 577-684, 2005.
4. F. Prengel, A. Wacker, E. Scholl, 'Simple model for multistability and domain formation in semiconductor superlattices', *Phys. Rev. B,* **50**, 1705-1712, 1994.
5. R. Aguado, G. Platero, M. Moscoso, L. L. Bonilla, 'Microscopic model for sequential tunneling in semiconductor multiple quantum wells', *Phys. Rev. B*, **55**, 16053-16056, 1997.
6. V. Ryzhii, I. Khmyrova, M. Ryzhii, R. Suris, C. Hamaguchi, 'Phenomenological theory of electric-field domains introduced by infrared radiation in multiple quantum well structures', *Phys. Rev. B*, **62**, 7268-7274, 2000.
7. R. C. Iotti, F. Rossi, 'Nature of charge transport in quantum-cascade lasers', *Phys. Rev. Lett*., **87**, 146603 (1-4), 2001.
8. Z. Ikonic, P. Harrison, R. W. Kelsall, 'Self-consistent energy balance simulations of hole dynamics in SiGe/Si THz quantum cascade structures', *J. Appl. Phys*., **96**, 6803-6811, 2004.

Calculation of Optical Gain and Electron Relaxation Rates in Single- and Double-Phonon Resonant Quantum Cascade Lasers in a Magnetic Field

J. Radovanović,[1] A. Mirčetić,[2] V. Milanović,[2] Z. Ikonić,[3] D. Indjin,[3] P. Harrison,[3] and R. W. Kelsall[3]

[1] Institute of Physics, University of Belgrade, Serbia and Montenegro,

[2] Faculty of Electrical Engineering, University of Belgrade, Serbia and Montenegro,

[3] School of Electronic and Electrical Engineering, University of Leeds, UK

Summary. We have explored the possibility of modulating the optical gain in the active region of mid-infrared QCLs by means of external magnetic field, which strongly influences the relaxation processes, in particular the LO phonon assisted intersubband transitions. The additional carrier confinement, induced by the field, leads to an increase in the upper laser level carrier lifetime, which results in pronounced oscillations of the optical gain. The described model was applied to two structures designed for λ~9μm emission.

1 Introduction

Quantum cascade lasers (QCL) represent a new group of semiconductor light sources whose operation is based upon intersubband optical transitions between size-quantized states formed within periodic multiple quantum well type structures [1]-[3]. The excited state lifetimes in such systems are very short (~1ps) because they are dominated by the electron-longitudinal optical (LO) phonon scattering process, so typical threshold currents in QCLs are much larger than is common for conventional interband lasers. However, it has been pointed out that increased excited state carrier lifetimes could be achieved by reducing the dimensionality of the system, e.g. by applying a strong magnetic field perpendicular to the layers. The continuous two-dimensional (2D) subbands are then split into series of discrete Landau levels (LL) [3]-[5], whose energies depend on the field. Since the scattering rates between states sensitively depend on their energy spacing, different relaxation channels in a multilevel system can be selectively enhanced or inhibited by varying the magnetic field, allowing for field-induced modulation of the population inversion and optical gain.

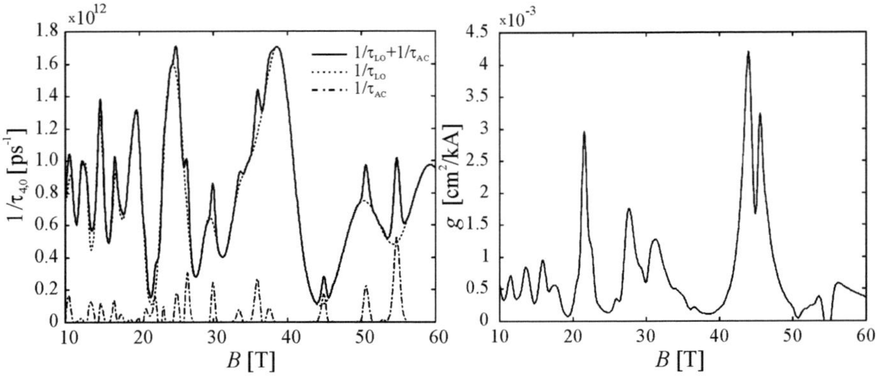

Fig. 1. The total electron relaxation rate due to emission of optical and acoustic phonons as a function of magnetic field, for transitions from state (4,0) into LLs belonging to the three lower states (left) and the optical gain (per unit of injection current) at T=77K (right), for the structure described in Ref. [6].

In this work we find the relaxation rates between Landau levels in the QCL active region in a strong magnetic field, due to both optical and acoustic phonons. Since Landau levels are formally discrete, and both the optical and acoustic phonons are taken non-dispersive, we introduce the level broadening to be able to calculate the scattering rates, as described in more detail in Ref. [7]. Zero dispersion is a cruder approximation for acoustic phonons, but still acceptable in view of their relatively minor role, and gives a peak-like structure of relaxation rates, which would otherwise be more diffuse. The electron distribution over the states, and optical gain, are then found by solving the set of rate equations. Since the injector regions consist of a multitude of states with small energy separations, the electron relaxation in these regions is not expected to be sensitive to this additional quantization, so it has not been considered in these calculations.

2 Theoretical Considerations

When a four quantum well active region of the GaAs/$Al_{0.45}Ga_{0.55}As$ QCL, described in Ref. [6], is subjected to a strong magnetic field perpendicular to its layers, each of the continuous subbands $E_n(k_{||})$ is split into a series of discrete Landau levels $E_{n,l} \approx E_n + (l+1/2)\hbar\omega_c$, where $\omega_c = eB/m^*$ stands for the cyclotron frequency and l is the Landau level index. Depending on the value of the magnetic field, the configuration of these states changes significantly, facilitating the suppression of main non-radiative energy relaxation processes (emission of LO and acoustic phonons).

Since the optical gain depends on the population inversion $N_{S4} - N_{S3}$, one has to find the electron distribution over all the states in the active region. This is obtained by solving the system of rate equations. We assume that electrons arrive in the active region by a constant current, and are injected into only a limited number of LLs of the excited laser state, i.e. into levels (4,0);...;(4, l_{max4}), where the value l_{max4} is determined so that all the levels above (4, l_{max4}) may be considered as almost empty at a given temperature, if the carrier distribution over LLs was equilibrium-like.

3 Numerical Results

The first analyzed structure, described in Ref. [6], is a result of the optimization of layer widths so to achieve maximal gain. Its active region is designed to have a double-LO (2LO) phonon depopulation mechanism, which is predicted to yield higher gain and lower threshold currents.

The total electron relaxation rate in the active region due to the emission of optical and acoustic phonons as a function of magnetic field, for transitions from the state (4,0) into LLs belonging to the three lower subbands is shown in Fig. 1 (left). The discrete states are magnetically tunable and resonance conditions for the emission of LO phonons from the upper laser state can be created by an appropriate choice of the magnetic field value. However, if there are no levels situated approximately $\hbar\omega_{LO}$ below the state (4,0) this type of scattering is suppressed, an therefore the upper laser level lifetime is increased as well as the optical gain (Fig. 1 (right)).

The second analyzed QCL structure has the triple quantum well (TQW) active region, emitting at $\lambda \sim 9\mu m$ [2]. The dependence of the optical gain on the applied magnetic field for this structure is given in Fig. 2, while the inset shows its influence on non-radiative relaxation mechanisms.

4 Conclusion

We have set up a rate equations model and analyzed the optical gain in the active region of a quantum cascade laser in a magnetic field perpendicular to the structure layers. The model shows that the effect of magnetic field induced modulation is less prominent in the 2LO phonon structure than it is in the classical TQW active region, which is a consequence of the increased number of states in the active region.

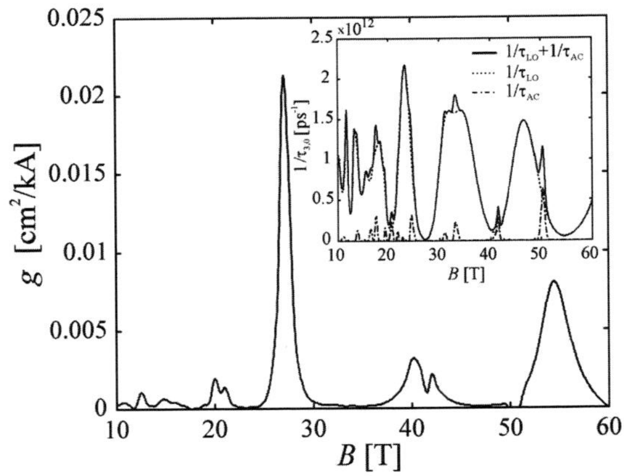

Fig. 2. The optical gain (per unit of injection current) at T=77 K and the total electron relaxation rate due to emission of optical and acoustic phonons, as a function of magnetic field, for transitions from state (3,0) into LLs belonging to the two lower states (inset), for the structure described in Ref. [2].

References

1. C. Sirtori, P. Kruck, S. Barbieri, P. Collot, J. Nagle, M. Beck, J. Faist, U. Oesterle, *Appl. Phys. Lett.*, **73**, 3486-3488, 1998.
2. H. Page, C. Becker, A. Robertson, G. Glastre, V. Ortiz, C. Sirtori, *Appl. Phys. Lett*, **78**, 3529-3531, 2001.
3. T. Chakraborty and V. Apalkov, *Advances in Physics*, **52**, 455-521, 2003.
4. D. Smirnov, C. Becker, O. Drachenko, V. V. Rylkov, H. Page, J. Leotin, and C. Sirtori, *Phys. Rev. B*, **66**, 121305(R), 2002.
5. C. Becker, C. Sirtori, O. Drachenko, V. Rylkov, D. Smirnov, J. Leotin, *Appl. Phys. Lett.,* **81**, 2941-2943, 2002.
6. A. Mirčetić, D. Indjin, Z. Ikonić, P. Harrison, V. Milanović, R. W. Kelsall, *J. Appl. Phys.*, **97**, Art. No. 084506, 2005.
7. J. Radovanović, V. Milanović, Z. Ikonić, D. Indjin, P. Harrison, *J. Appl. Phys.*, **97**, Art. No. 103109, 2005.

Curvature-Dependent Conductance Resonances in Quantum Cavities

G. J. Meyer, R. H. Blick and I. Knezevic

Department of Electrical and Computer Engineering
University of Wisconsin-Madison

Summary. Conductance in planar quantum cavities is known to exhibit resonances which depend on the dimensions of the cavity and the applied electric and magnetic fields. We demonstrate that these resonant features are also highly sensitive to the curvature of the cavity, and suggest that these curved quantum cavities, fabricated on flexible substrates, have potential applications as MEMS components or highly-responsive sensors.

Quantum-mechanical properties of non-planar two-dimensional (NP2D) systems have been discussed for a number of years, and multiple approaches have been developed [1,2]. This topic has received renewed interest, however, since the development and refinement of fabrication methods for NP2D devices in the quantum regime [3,4]. There has been much recent theoretical work describing their novel transport properties [5], and experimental results are only beginning to come in [6].

In this paper, we investigate ballistic transport in a flexible 2D electron system, and show that the conductance is highly sensitive to the structure's local curvature. In planar quantum cavities, it is well established [7] that low-field ballistic conductance will exhibit resonant features. Resonant maxima and antiresonant minima in conductance are attributed to quantum-mechanical states with very high and very low transmission (respectively) between contacts. In flexible quantum cavities, we show that the existence and location of such resonances can be tuned through the manipulation of the cavity's curvature.

We consider a simple quantum cavity that can be rolled into a partial cylinder, and study the effect of curvature on its conductance (Fig. 1). The dimensions of this tethered cylindrical cavity are 300x200 nm, with 60x60 nm contacts which "pin down" the center of the cavity so that it remains flat (Fig.1). The 2DEG is assumed to exist in a thin layer of GaAs. All conducting electrons are assumed to be at the Fermi energy, which is 0.5t (2.85 meV) , where t = $\hbar^2 / 2M^* a^2$, the hopping energy of the mesh used in the numerical calculations. Such a device could readily be fabricated using known techniques [3] where a thin-film heterostructure is grown atop a sacrificial substrate. As the sacrificial layer is etched away underneath, the forces at the heterojunction cause the film to bend toward the layer under tensile strain.

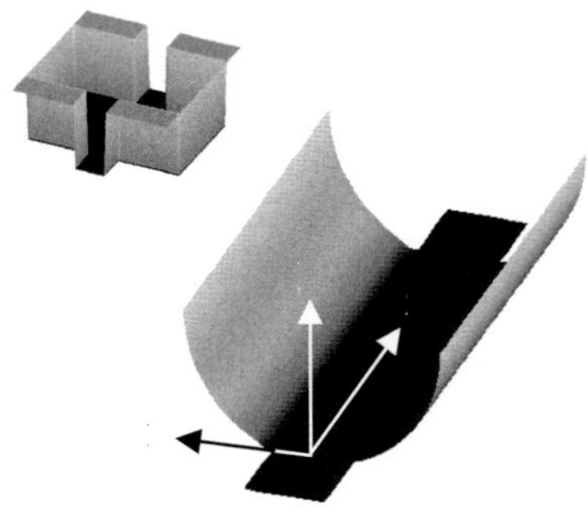

Fig.1 A cylindrical quantum cavity with its potential V_{edge} (top left).

The confinement method for quantum-mechanical modeling of curved two-dimensional electron gases (2DEGs) involves the introduction of an infinitely high, infinitely narrow one-dimensional (1D) potential well into the three-dimensional (3D) Schrödinger equation, effectively trapping the electrons onto the 2D surface of interest. This surface is parameterized using coordinates q_1 and q_2, while q_3 is defined along the normal to the surface at all points. Using separation of variables, $\Psi(q_1,q_2,q_3,t)$ is rewritten as $\psi(q_1,q_2,t)\Psi_N(q_3,t)$ where Ψ_N is assumed to be the ground state. In the case of a cylindrical surface, like the one we are interested in (Fig. 1), ψ obeys the time-independent Schrödinger equation

$$\frac{\hat{p}^2}{2M^*}\psi + V_{edge}\psi + V_g\psi = E\psi \ .$$

The Schrödinger equation for a cylindrical cavity[1] is identical to the equation for a flat cavity [confinement potential V_{edge} (Fig. 1) defines the lateral edges of the cavity], with the addition of an attractive potential $V_g = -\hbar^2/8M^*R_C^2$ of purely geometric origin, where R_C is the local radius of curvature.[2] The presence of a magnetic field can be accounted for through the use of Peierls' phase factor [8], $\psi = \psi_{B=0}\exp\{-i\Phi_\perp/\Phi_0\}$. The resultant 2D Schrödinger equation is discretized, and Usuki's transfer matrix method [9] is employed to calculate conductance and probability density.Baseline calculations were made for a flat cavity in a magnetic field, and the first resonance feature was observed at 0.08 Tesla. Upon

[1] In general, as elaborated by daCosta [1], the kinetic energy operator bears explicit dependence on the components of the metric tensor inherited from the 3D metric, but reduces to the common form given above for a cylindrical surface.

[2] For our tethered cylinder, there is formally a singularity in the equation of motion due to the infinite radius of curvature at the junction where the curved and flat portions meet. However, in a real device this will not be the case, so as an approximation this singularity is ignored.

curving the cavity, both the shape of the valley and the magnetic field required to generate it were altered, as shown in Fig. 2.

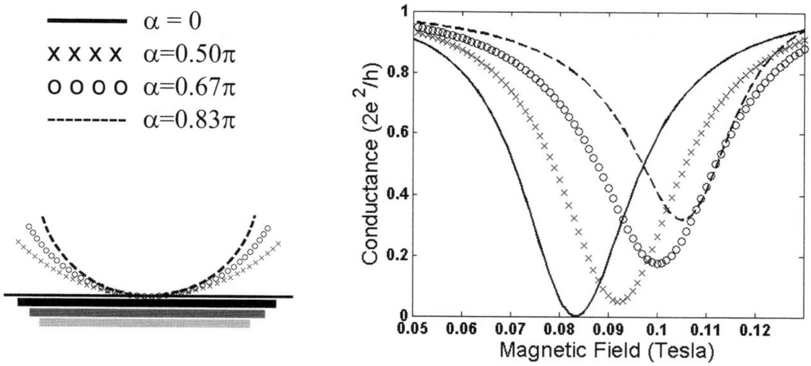

Fig.2 Conductance valley for four different curvatures (right), partially explained by the shrinking footprint (left) and therefore reduced flux through the cavity.

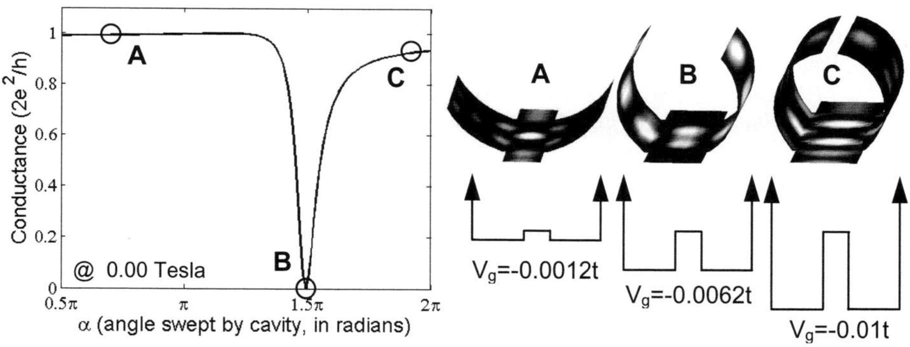

Fig.3 Curvature induces strong resonance even in the absence of a magnetic field. This effect is attributed to the geometric potential.

At low curvature, the shift of this antiresonance is largely attributable to reduced magnetic flux through the cavity. At higher curvatures, however, this correspondence breaks down as other effects become more prominent. Moreover, there is an increase in the minimum of the valley that is not explained by a simple reduction in total flux. With increasing curvature, $B_\perp$ becomes increasingly nonuniform, remaining strong in the center of the cavity while attenuating at the lateral edges, which become more and more parallel to the magnetic field. This causes flattening of the valley. Second, a geometric potential is induced in the wings of the cavity, altering the wavefunction and its ability to couple to the contacts. This is the reason for the leftward shift.

The most exciting result of this study is the introduction of resonant states by curvature alone, in the absence of a magnetic field (Fig.3). The geometric potential, even though relatively small at these dimensions (approximately E_F /80), was sufficient to produce a well-defined antiresonance. The conductance antiresonance (left panel of Fig. 3) corresponds to a minimum of transmission between leads: in panel B on the right-hand-side, the probability density in part of the cavity connecting to the leads is zero, and this particular wavefunction is an eigenfunction for the closed cavity problem (i.e., leads removed). This effective decoupling from the leads is known [7] to correspond to conductance minima in flat cavities.

The cumulative effects of curvature (the geometric potential, the non-uniform $B_\perp$, and the decrease in total flux) result in a quantum cavity whose conductance can be extremely sensitive to its shape. This suggests possible application as a minute force-detecting MEMS component, or as a sensor for chemical agents that can affect the curvature-inducing strain.

[1] da Costa R C T 1981 *Phys.Rev.*A **23** #4 1982-7.

[2] Jensen H and Koppe H 1971 *Ann.Phys.* **63** 586; Matsutani S 1992 *J.Phys.Soc.Jpn.* **61** #143109 55-63; Kaplan L, Maitra N T and Heller E J 1997 *Phys.Rev.*A **56** #4 2592-9.

[3] Prinz V Ya, Seleznev V A, Gutakovsky A K, Chehovskiy A V, Preobrazhenskii V V, Putyato M A and Gavrilova T A 2000 *Physica* E **6** 828-31; Vorob'ev A B and Prinz V Ya 2002 *Semicond.Sci.Technol.* **17** 614-6.

[4] Grundmann M 2003 *Applied Physics Letters* **83** #12 2444-6; Schumacher O, Mendach S, Welsch H, Schramm A, Heyn Ch and Hansen W 2005 *Applied Physics Letters* **86** #143109 1-3.

[5] Chaplik A V, Romanov D A and Magarill L I 1998 *Superlattices and Microstructures* **23** #6 1227-30; Chryssomalakos C, Franco A and Reyes-Coronado A 2004 *Eur.J.Phys.* **25** 489-502; Marchi A, Reggiani S, Rudan M, and Bertoni A 2005 *Phys.Rev.*B **72** #035403 1-10.

[6] Vorob'ev A B, Prinz V Ya , Yukecheva Yu S and Toropov A I 2004 *Physica* E **23** 171-6; Mendach S, Schumacher O, Heyn Ch, Schnull S, Welsch H and Hansen W 2004 *Physica* E **23** 274-9.

[7] Bird J P, Akis R, Ferry D K, de Moura A P S, Lai Y-C and Indlekofer K M 2003 *Rep.Prog.Phys* **66** 583-632; Akis R, Ferry D K and Bird J P 1996 *Phys.Rev.*B **54** #24 705-15; Sheng W-D and Xia J-B 1996 *J.Phys.Cond.Matter* **8** 3635-45.

[8] Peierls R E 1933 *Z.Phys.* **80** 763; Nazareno H N and de Brito P E 2001 *Phys.Rev.*B **64** #045112 1-6.

[9] Usuki T, Saito M, Takatsu M, Kiehl R A, and Yokoyama N, 1995 *Phys.Rev.*B **52** #11 8244-58.

Mid-Infrared Optical Absorption in Germanium Under Intense Laser Fields

H. Furuse, Y. Nakata, H. Kubo and N. Mori

Department of Electronic Engineering, Osaka University, Suita City, Osaka 565-0871, Japan

Summary. We have measured optical transmittance through germanium in the mid-infrared region, $\lambda = 6\ \mu m - 12\ \mu m$, using a free electron laser (FEL) at Osaka University. We find that the transmission is strongly suppressed under high-intensity FEL excitation. We also find that the FEL wavelength dependence of the transmission spectrum is very weak.

1 Introduction

In the nonlinear and non-equilibrium conditions, electrons in atoms and solids show quite different behavior from those in the linear and equilibrium conditions. For example, atoms under high-intensity laser excitation exhibit above-threshold ionization [1], high-order harmonic generation [2], and the laser-assisted photoelectric effect [3]. In crystalline solids, Chin et al. have clearly demonstrated unusually large induced absorption in GaAs strongly driven by intense ultrashort mid-infrared (MIR) laser fields [4]. This ultrafast electroabsorption is closely related to the high-frequency Franz-Keldysh effect [5]. Here we report the observation of MIR optical absorption in germanium (Ge) under intense laser fields generated by a free electron laser (FEL). In spite of the fact that Ge is transparent in the MIR region and is widely used for optical components, such as lens and a beam splitter, we have observed strong suppression of optical transmission under high-intensity FEL excitation.

2 Experimental Method

The experiments were carried out using the FEL facility of the Institute of Free Electron Laser (iFEL), Osaka University at Hirakata, Japan. The FEL is a coherent optical source using a relativistic electron beam in a magnetic field as a gain medium. The FEL system consists of a relativistic electron accelerator, an undulator in which the electrons emit the synchrotron radiation, and an optical resonator. By virtue of its simple gain medium, the FEL has unique advantages, the broad wavelength tunability and high-intensity output power. Figure 1 shows a typical FEL output and a schematic diagram of the measurement setup. The MIR FEL beam consists of trains of 5 ps-pulses (micro-pulses) with 44.8 ns-separation. The train continues for about 20 μs (macro-pulse) with a repetition rate of 10 Hz. We have measured optical transmittance through a 3 mm-thick bulk Ge sample in the MIR region, λ = 6 μm – 12 μm. The FEL beam is focused onto the sample using a ZnSe lens. The polarizer was located between the ZnSe lens and the sample for adjusting the irradiation power. After passing through the Ge sample, the transmitted power was measured by a power meter.

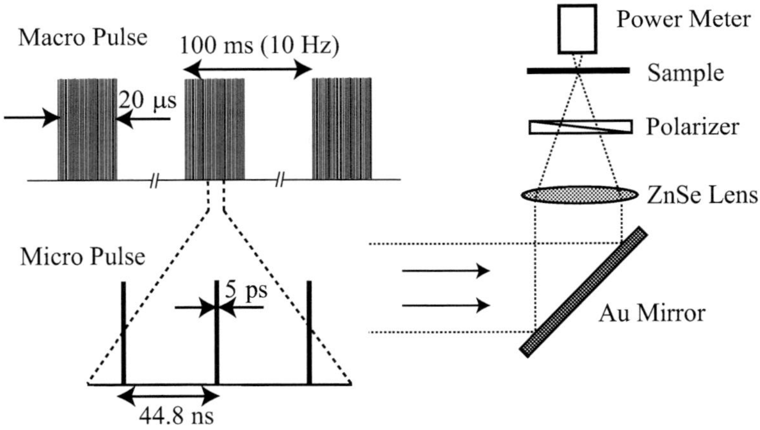

Fig. 1. A typical FEL pulse structure and a schematic diagram of the measurement setup.

3 Results and Discussion

Figure 2 shows the transmittance of Ge as a function of the FEL wavelength, λ, and the FEL average power, P_{ave}, at room temperature. Solid circles show the actual data points and the surface is generated by an interpolation between the data points. In Fig. 2, we also plot the transmission spectrum measured by FTIR absorption spectroscopy. The values of P_{ave} for the measurements were taken to be $P_{ave} = 2, 3, 5, 8$, and 11 mW. As can be seen in Fig. 2, the transmission of the MIR light is strongly suppressed as P_{ave} increases. The electric-field strength is estimated to be $E \sim 1.7$ MV/cm for $P_{ave} = 11$ mW when the beam spot size is 0.06 mm. Although the transmission strongly depends on P_{ave}, the λ-dependence is found to be rather weak. In order to check whether this decrease in the transmittance is due to optical absorption, we have measured transmission and reflection power simultaneously. Figure 3 shows a correlation between T + R and T (T is transmittance and R is reflectivity) for various values of P_{ave} and λ (Note that the detector for measuring R could not be placed close to the sample which results in rather scattered data points of R). We find that T + R decreases as T decreases. Hence, we conclude that the observed decrease in the transmittance is due to optical absorption in Ge. Such absorption may be possible when one take into account of the oscillatory motion of the Bloch electron driven by the intense FEL field when modeling the band-to-band absorption process [6]. Theoretical study of this scheme is under investigation and will be reported in the future.

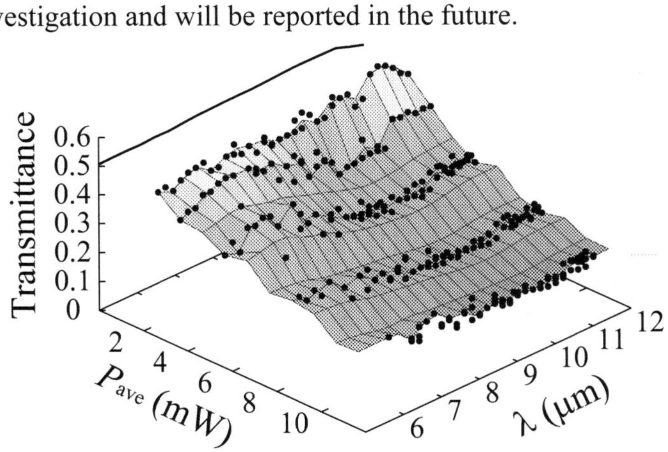

Fig. 2. Transmittance of a 3mm-thick Ge sample as a function of the FEL wavelength, λ, and the FEL average power, P_{ave}, at room temperature. Solid curve shows the transmission spectrum measured by FTIR absorption spectroscopy.

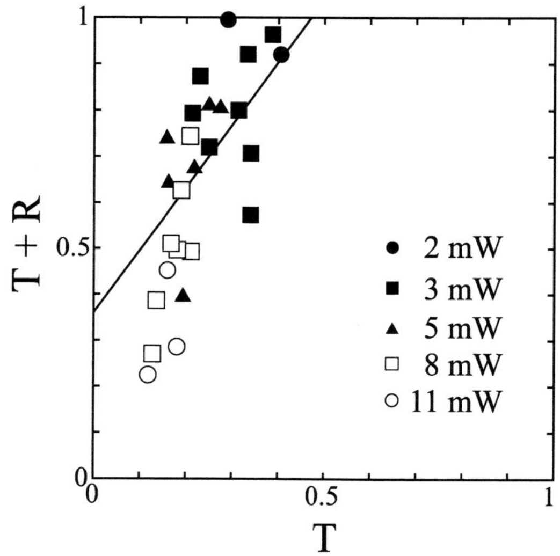

Fig. 3. Correlation between T + R and T for various values of P_{ave} and λ. Solid line shows a theoretical result. Note that the theory assumes a normal incident beam while the FEL beam is slightly tilted in the transmittance and reflectivity measurements.

4 Summary

In the present study, we investigated the optical transmission of Ge under intense laser fields in the mid-infrared region using a FEL. We observed a dramatic decrease in the transmission and reflection below the band edge of Ge. We also find that the FEL wavelength dependence of the transmission spectrum is very weak.

References

1. P. H. Bucksbaum, J. Opt. Soc. Am. B **4**, 760 (1987).
2. A. Rundquist et al., Science **280**, 1412 (1998).
3. T. E. Glover et al., Phys. Rev. Lett. **76**, 2468 (1996).
4. A. H. Chin et al., Phys. Rev. Lett. **85**, 3293 (2000).
5. Y. Yacoby, Phys. Rev. **169**, 610 (1968).
6. L. V. Keldysh, Sov. Phys. JETP **20**, 1307 (1965).

Interface Related Radiative Recombination on a Type-II Broken-Gap Single GaInAsSb/InAs Heterojunction

K. A. Korolev[1, 3], K. D. Moiseev[2], V. A. Berezovets[2], M. P. Mikhailova[2], Yu. P. Yakovlev[2], R.V. Parfeniev[2], C. J. Meinning[4], B. D. McCombe[4]

[1]Institute of Radio-engineering and Electronics of RAS, Moscow, Russia
[2]A. F. Ioffe Physico-Technical Institute of RAS, St. Petersburg, Russia
[3]Dpartment of ECE, Tufts University, Medford MA, 02155 USA
[4]Department of Physics, SUNY at Buffalo, Buffalo NY, 14260 USA

Summary. Magneto-photoluminescence studies have been performed on the type-II broken-gap GaInAsSb/InAs single heterostructures with 2D-electon channel at the interface containing two occupied energy subbnads. Photoluminescence spectra manifest a set of pronounced emission bands in the spectral region of 0.3-0.5 eV and it has been investigated in magnetic fields up to 10 T.

1 Introduction

The type-II broken-gap single p(n)-GaInAsSb/p-InAs heterojunction is a heterostructure with the two-dimensional electron gas (2DEG) formed in the potential well on the p-InAs side of the heterointerface [1]. The flexibility in the structure parameters, such as epilayer composition, doping concentration of contacting materials, applying an external bias, etc allows us to obtain a 2DEG with widely varying properties [2]. This paper presents observation and study of infrared magneto-photoluminescence (PL), provided by 2D-electrons localized at the type-II broken-gap single heterointerface for the first time.

2 Samples and Experimental Setup

The epitaxial layers of the wide-gap $Ga_{0.94}In_{0.06}As_{0.13}Sb_{0.87}$ solid solutions (E_G=0.724 eV at T=7 K) with a mirror-like surface were obtained lattice-matched to the InAs substrate by liquid phase epitaxy. The quaternary solid solution doped with Te was grown on p-InAs substrate heavily compensated with Mn. 2D-electron channel with two energy subbands (E_1 and E_2) with a sheet concentration of $n_1=9.2\times10^{11}cm^{-2}$ and $n_2=4.77\times10^{11}cm^{-2}$,

respectively, computed from Shubnikov-de Haas oscillations in magnetic fields up to 14 T at 1.5 K, was found at the heterointerface.
PL measurements were performed with a BOMEM DA3.01 Fourier Transform Spectrometer equipped with a CaF_2 beam splitter and a liquid-nitrogen-cooled InSb photovoltaic detector. A fiber-pigtailed laser diode with λ_{ex}=840 nm (hν=1.476 eV) and output power up to 1 W was used for photo-excitation of the samples. Magneto PL measurements were made in a superconducting split-coil optical magnet system capable of the field up to 10 T. The measurements were carried out in the Faraday configuration, i.e. the magnetic field axis and the direction of the light were parallel to each other and both were parallel to the growth direction of the samples.

3 Results and Discussion

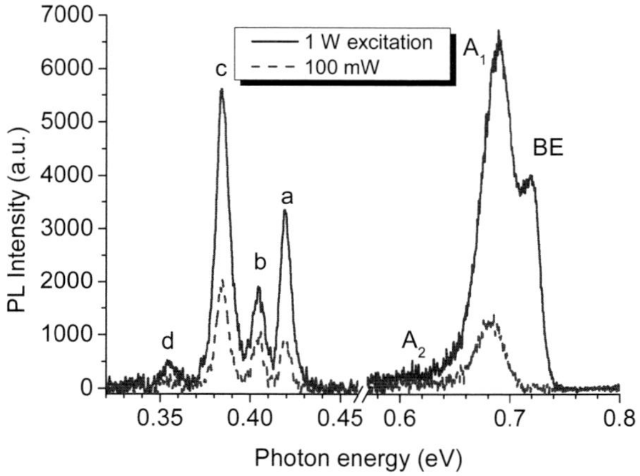

Fig. 1. PL spectra for the type II broken-gap n-$GaIn_{0.06}As_{0.13}Sb$/p-InAs single heterostructure at low and high excitations in zero magnetic field at 6.8 K.

The samples under study exhibited intense photoluminescence at 4.2 K in two spectral ranges, high-energy (0.55-0.80 eV) and low-energy (0.30-0.45 eV), associated with two different spatial regions of radiative recombination (Fig. 1). The high-energy part of the PL spectrum contains a pronounced emission band with maximum at the photon energy of about $h\nu_{A1}$=0.684 eV and full width at half maximum (FWHM) of 28 meV. The small diamagnetic shift for A_1 line towards high energies with the increasing of a magnetic field was observed. The intensity of the high-energy bands and their shape are almost independent of magnetic field. High-energy emission bands manifest the features of a typical bulk-related magneto-PL. Due to lack of the space we skip detailed high-energy PL analysis and discuss low-energy part of PL spectra only.

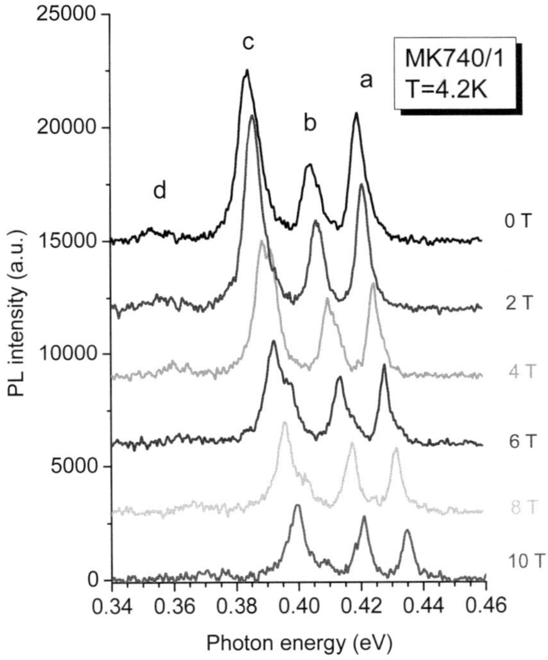

Fig. 2. Low-energy part of PL spectra for the type II broken-gap n-$GaIn_{0.06}As_{0.13}Sb$/p-InAs single heterostructure at different magnetic fields at 4.2 K.

Low-energy range PL spectra exhibited three sharp pronounced emission bands labeled “a”, “b”, “c”, and “d”, respectively. These peaks ($h\nu_a$=0.419 eV, $h\nu_b$=0.404 eV and $h\nu_c$=0.384 eV, $h\nu_d$=0.355 eV) reveal relatively high intensity, and are extremely sharp with FWHM of 7 meV. The narrow emission band seems to be due to strong localization of the carriers involved in the recombination process. The low-energy part of the PL spectra indicates that the radiative recombination occurs near the type-II broken-gap n-GaInAsSb/p-InAs heterointerface. An important feature of all low-energy lines is that the spectral position of the emission band maximum remains the same with the varying of pump intensity. No blue shift of the emission bands with increasing excitation level was found. It means that the electron phase space is not filling up to the Fermi energy and the radiative transitions are due to confined states at the heteroboundary.

By applying the magnetic field (see Fig. 2), low-energy lines “a”, “b” and “c” (with Lorentzian-like shape) become narrower and their FWHM decrease to 4 meV, on average. The result of the deconvolution of these lines by Lorentzian distribution reveals very good matching with the experimental curves allows us to conclude that in a magnetic filed the radiative transitions occurs through localized states [3].
In low magnetic fields up to 2 T the rise of PL intensity was observed. This rising is accompanied by an extremely small shift of PL peaks by 1 meV towards higher energies with the increasing of the magnetic field and the broadening of the lines width by two times. In higher fields (B>4 T) the intensity of “a” and “c” peaks decreases with the increasing magnetic field. The transitions associated with Landau levels N=0 and N=1 becomes resolved starting with 4 T and their FWHM comes back with the value of about 4 meV. It explains the rising of PL intensity in low-field range due to a presence of two close situated emission bands connected with different Landau levels. Lines “a”, “b” and “c” manifest a fundamental, interface-related property of the type- II GaInAsSb/InAs heteroboundary and for these emission bands we deal with the radiative recombination involving the electrons localized in the quantum well at the GaInAsSb/InAs interface. The n-$GaIn_{0.06}As_{0.13}Sb$/p-InAs heterojunction creates a similar situation of band bending at the heterointerface to that previously reported for InAs-$Al_{0.1}Ga_{0.9}Sb$ single quantum well structure [4]. In contrary to the latter the type-II broken-gap GaInAsSb/p-InAs heterostructure can realize the coexistence of electron-like and hole-like Fermi surfaces at the single interface. Such samples should demonstrate unusual properties due to mixed character of the InAs conduction band and the GaInAsSb valence band.

References

1. Moiseev, K. D., et al.: 'Quantum magnetotransport at a type II broken-gap single heterointerface', *Surface Science*, **482-485**, 1083-1089, 2001.
2. Mikhailova, M. P., et al.: 'Interface-induced optical and transport phenomena in type II broken-gap single heterojunctions', *Semicond. Sci. Technol.*, **19**, R109-R128, 2004.
3. Christen, J., et al.: 'Line shapes of intersubband and excitonic recombination in quantum wells: Influence of final-state interaction, statistical broadening, and momentum conservation', *Phys. Rev. B* **42**, 7213-7219, 1990.
4. Kono, J., et al.: 'Far-infrared magneto-optical study of two-dimensional electrons and holes in InAs/AlGaSb quantum wells', *Phys. Rev. B* **55**, 1617-1636, 1997.

Drift and Diffusion in Superlattices Within the Wannier-Stark Approach

M. Rosini and L. Reggiani

National Nanotechnology Laboratory of INFM, Dipartimento di Ingegneria dell'Innovazione, Università di Lecce, via Arnesano, 73100 Lecce, Italy

Summary. Drift and diffusion properties of electrons in superlattices in the negative differential mobility region and beyond are investigated by Monte Carlo simulations based on the Wannier-Stark approach. At the highest applied fields we found that the diffusion coefficient departs from Einstein law to achieve values controlled by a characteristic length comparable with the lattice period. This size effect is accompanied by quantum effects associated with intra- and inter-band transitions assisted by phonon interactions typical of the hopping transport regime.

1 Introduction

Superlattices (SLs) have received a relevant scientific and technological interest, owing to their nonlinear electrical and optical properties. In particular, they exhibit a strong negative differential conductivity (NDC) regime [1]. In this regime, charge transport can be described in terms of hopping between Wannier-Stark (WS) states thus SLs represent a good model for studying hopping conduction in semiconductor structures.

The aim of this paper is to investigate the drift velocity and the longitudinal diffusion coefficient through the spatial spread of a carrier ensemble. To this purpose, we have performed a set of Monte Carlo simulations. Under thermal equilibrium the diffusion coefficient follows the Einstein relation $D_z = 2\mu\langle\varepsilon\rangle/(3e)$ where e is the electron charge, μ is the carrier mobility, and $\langle\varepsilon\rangle$ their mean energy which for thermal electrons is given by $\langle\varepsilon\rangle = (3/2)K_B T$ with K_B the Boltzmann constant and T the temperature. Far from equilibrium conditions the Einstein relation is in general no longer valid and in the case of WS-hopping transport another formulation for the diffusion coefficient has been proposed in the literature

[2]. In this paper we will investigate to which extent the diffusion coefficient in a SL compares with the Einstein and WS-hopping models.

2 Theoretical Model and Results

The physical system considered here is a Si/SiO_2 SL with lattice parameter L=3,1 nm, where Si is grown along the (100) direction. The unperturbed Hamiltonian for the biased SL in the effective-mass approximation writes:

$$\mathrm{H} = \frac{\hbar^2}{2}\nabla_i \left(\frac{1}{m(z)}\right)_{ij} \nabla_j + W_{SL}(z) - eEz \tag{1}$$

where $m^{-1}(z)|_{ij}$ is the effective mass tensor of the band minima of the bulk, depending on the coordinate z, W_{SL} is the Kroenig-Penney potential for the SL and E is the applied electric field. The eigenfunction of this Hamiltonian are the Wannier-Stark functions with eigenvalues:

$$\varepsilon_n^\nu = \varepsilon_n - \nu eEL + \frac{\hbar^2}{2m_x}k_x^2 + \frac{\hbar^2}{2m_y}k_y^2 \tag{2}$$

where νL is the distance of the ν-th well from the origin, and n is the energy level in each well.

The interaction with phonon is treated within the Fermi golden rule. the scattering mechanisms are with optical phonons, in particular, we have considered [3] the deformation potential optical mode of Si, and the optical modes of SiO_2 , both polar and deformation potential. Moreover, we have applied the approximation of confined optical phonons: Si phonons are confined in the silicon layer, SiO_2 phonons are confined in the SiO_2 layer. To simulate the transport of an ensemble of uncorrelated electrons in a SL, with many minibands, we implemented a Monte Carlo code. In the finite difference scheme, the drift velocity and the longitudinal diffusion coefficient are found as

$$v_z(t) = \frac{1}{N\delta t}\sum_{i=1}^{N} \delta z_i(t) \tag{3}$$

$$D_z(t) = \frac{1}{2}\frac{\delta\left\langle (\Delta z_i)^2 \right\rangle}{\delta t} \tag{4}$$

with N the number of simulated carriers, δt the time step, and $\Delta z = z - \langle z \rangle$ the instantaneous fluctuation of the carrier position in the field direction.

The result for the drift velocity is reported in Fig. 1. The NDC regime, that appears at intermediate fields, is fully explained by the simple Esaki-

Tsu model [1] or by more complex theories [4]. The sudden increase of the drift velocity at high fields is the consequence of two kinds of phenomena: interband transitions, that are responsible for the positive increase, intraband resonances with phonons, that are responsible for the localized peaks (separate contributions are given in Fig. 1). In particular, the main resonance peak (indicated by the arrow) appears at $eEL = \hbar\omega_{Si}$, where $\hbar\omega_{Si} = 62\ meV$ is the energy of the most important optical phonon in Si.

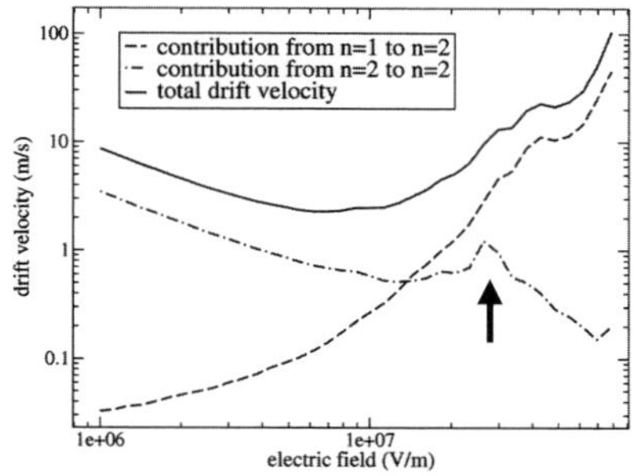

Fig. 1. Electron drift velocity in the SL. The dashed and dot dashed curves report the intrasubband and intersubband contributions, respectively. The arrow indicates the main resonance peak.

The result for the diffusion coefficient is reported in Fig. 2. Here the curve obtained directly from the simulation is compared with those obtained using the Einstein relation and the following formula

$$D_z = \frac{v_z L}{2} \coth \frac{L}{L_E} \tag{5}$$

obtained from a WS calculation [2], where $L_E = 2K_BT/(eE)$. Both Einstein and WS expressions, are in good agreement with the simulations, in the NDC region. However, both strongly underestimate the diffusion in the highest field region of the plot. Indeed the WS formula is obtained exactly for a system with only nearest neighbor (NN) coupling and for an equilibrium distribution function. By contrast, the present model considers coupling beyond the NN and a distribution function which in general is a far from equilibrium one. To account for these differences, we have introduced a fitting parameter L^*, replacing L in Eq. (5), with the meaning of a mean hopping distance per jump. In our simulation $L^*=3\ L$. Interestingly enough, for $L^* < L_E$ the associated diffusion coefficient D_z^* reproduces the Einstein law and the SL exhibits a classical behavior. For $L^* > L_E$, diffusion is controlled by this size parameter and the SL exhibits the quantum behaviour of the WS diffusion.

3 Conclusions

By investigating drift and diffusion in SLs under strong electric fields we found the possibility to identify two different transport regimes. The first regime, occurring together with negative differential mobility, exhibits semiclassical results, where the Einstein law is well reproduced. The second regime, occurring at the highest fields, is dominated by WS hopping diffusion and exhibits peculiar quantum effects. Both regimes are controlled by a characteristic

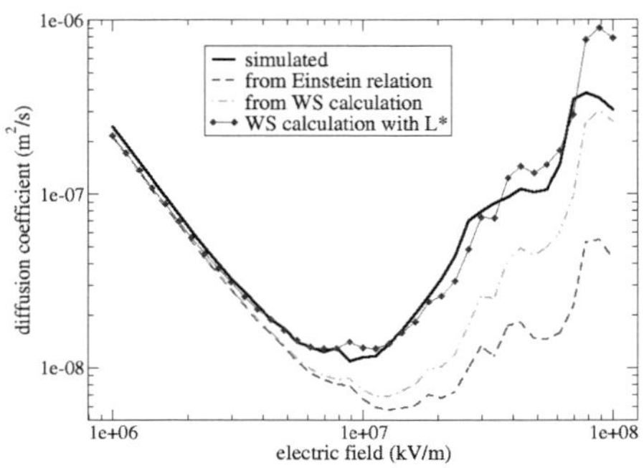

Fig. 2. Diffusion coefficient in the SL. The simulated curve (continuous line) is compared with those obtained from the Einstein relation (dash-dotted line), those from the relation of Ref.[2] (dashed line) and those with $L^*=3$ (diamonds).

length parameter L^*, containing the geometrical and physical properties of the SL, and whose value when compared with the period of the SL L is found to determine the crossover between different transport regimes.

This work is supported by MIUR under the project "Noise models and measurements in nanostructures".

References

1. Esaki, L., Tsu, R.: 'Superlattices and negative differential conductivity in semiconductors', *IBM J. Res. Dev.*, **14**, 61-65, 1970
2. Bryksin, V. V.,Kleinert, P.,: 'Theory of quantum diffusion in biased semiconductors', *J. Phys.: Condens. Matter*, **15**, 1415-1425, 2003.
3. Rosini, M., Jacoboni, C., Ossicini, S.: 'Monte Carlo simulation of electron transport in Si/SiO_2 superlattices: vertical transport enhanced by parallel field', *Phys. Rev. B*, **66,** 155332, 2002.
4. Wacker, A., Jauho, A. P.: 'Quantum transport: the link between standard approaches in superlattices', *Phys. Rev. Lett.*, **80** , 369-372, 1998.

Ballistic Transport in Arbitrary Oriented Nanowire MOSFETs

M. Bescond*, N. Cavassilas**, L. Raymond**, A. Asenov*

* Dept of Electronics and Electrical Engineering, University of Glasgow G12 8LT, Glasgow, Scotland, United Kingdom
** L2MP, UMR CNRS 6137, 49 rue Joliot-Curie, BP 146, 13384 Marseille Cedex 13, France

Summary. A general method is applied to the calculation of effective masses in arbitrary oriented semiconductor nanowires. The results give the effective mass expressions for each valley and several channel orientations. We then study the ballistic transport in Si and Ge <100>-oriented nanowire MOSFETs by using a self-consistent 3D approach based on the non-equilibrium Green's function formalism.

1 Introduction

Near the end of the present edition of the ITRS in 2018 [1] MOSFETs will reach the sub-10 nm dimensions. There is, however, a consensus that new device architectures and materials are needed to continue the downscaling after the 45 nm technology node. In that context, Ge is a promising material for nanoCMOS applications due to higher carrier mobility [2-4], but no study has been reported in the case of nanowire architectures operating in the ballistic regime. This work applies a general method [5,6] to calculate the effective mass tensor in nanowires with arbitrary crystallographic orientations. We then study, using the non-equilibrium Green's function formalism [7,8], the ballistic transport for two channel materials (Si, Ge) in nanowire MOSFETs fabricated on a <100>-wafer.

2 Theoretical Aspects

We use a general method, based on the work of Stern *et al.* [6], which calculates the effective mass tensor in nanowire MOSFETs. In a 3D description, the full time-independent Schrödinger equation is given by:

$$H_{3D}\Psi(x,y,z) = [T_{3D} + V(x,y,z)]\Psi(x,y,z) = E\,\Psi(x,y,z), \tag{1}$$

where $\Psi(x,y,z)$ is the 3D wave function, $V(x,y,z)$ is the potential energy, and T_{3D} is the kinetic-energy operator. Assuming an ellipsoidal parabolic energy band approximation, the effective mass tensor is no longer diagonal for an arbitrary wire orientation and the expression of the kinetic-energy operator becomes:

$$T_{3D} = \frac{-\hbar^2}{2} \times \left(\omega_{XX} \frac{\partial^2}{\partial x^2} + \omega_{YY} \frac{\partial^2}{\partial y^2} + \omega_{ZZ} \frac{\partial^2}{\partial z^2} + 2\omega_{XY} \frac{\partial^2}{\partial x \partial y} + 2\omega_{YZ} \frac{\partial^2}{\partial y \partial z} + 2\omega_{XZ} \frac{\partial^2}{\partial x \partial z} \right) \quad (2)$$

where ω_{ij} is the reciprocal effective-mass tensor in the coordinate system (x,y,z). We first consider a constant potential energy $V(y,z)$ along the transport direction x and write the 3D wave function as follows:

$$\Psi(x, y, z) = S(y, z) e^{ik_x x} . \quad (3)$$

In other words, we consider an infinitely long nanowire with a constant transverse confinement in the y-z plane. the transverse part of the wave function $S(y,z)$ can be also rewrite as:

$$S(y, z) = \sigma(y, z) e^{ik_x (\alpha y + \beta z)} , \quad (4)$$

where α and β are two adjustable parameters which cancel the coupled derivatives associated with y and z in the 3D Hamiltonian. Substituting Eq.(3) and (4) in Eq.(1), we obtain a new expression of the Schrödinger equation:

$$-\frac{\hbar^2}{2} \left(\omega_{YY} \frac{\partial^2 \sigma(y,z)}{\partial y^2} + \omega_{ZZ} \frac{\partial^2 \sigma(y,z)}{\partial z^2} + 2\omega_{YZ} \frac{\partial^2 \sigma(y,z)}{\partial y \partial z} \right) + [V(y,z) - E'] \sigma(y,z) = 0 , \quad (5)$$

where E' is given by:

$$E = E' + \frac{\hbar^2 k_x^2}{2} \frac{\omega_t^2 \omega_l}{\omega_{YY} \omega_{ZZ} - \omega_{YZ}^2} = E' + \frac{\hbar^2 k_x^2}{2 m_{trans}} , \quad (6)$$

where ω_t, ω_l are respectively the opposite of the transverse and longitudinal effective masses of the conduction band minima ($\omega_t = 1/m_t$, $\omega_l = 1/m_l$), and m_{trans} is the new effective mass along the transport direction:

$$m_{trans} = \frac{\omega_{YY} \omega_{ZZ} - \omega_{YZ}^2}{\omega_t^2 \omega_l} . \quad (7)$$

The energy associated with the quantum confinement directions is clearly decoupled from that associated with the channel axis x. As a result we can treat the transport using a self-consistent 3D Green's function simulator expressed in the mode-space approach [7-10] which consists of separating the 2D confinement plane from the transport direction.

3 Results and Discussion

We consider a square cross-section gate-all-around nanowire MOSFET with a channel length L=9nm, a semiconductor cross-section T×T=4×4nm, an oxide thickness T_{OX}=1nm, and a source-drain reservoir doping $N_{S/D}=10^{20}$ cm^{-3}.

Table 1. Effective mass tensor as well as subband degeneracy for two important semiconductor nanowire orientations on a <100> wafer.

Wire orientation	Minimum type	ω_{YY}	ω_{ZZ}	ω_{YZ}	m_{trans}	*Deg.*
$<100>$	$\Delta 1$	$\frac{1}{m_t}$	$\frac{1}{m_t}$	0	m_l	2
$<100>$	$\Delta 2$	$\frac{1}{m_l}$	$\frac{1}{m_t}$	0	m_t	2
$<100>$	$\Delta 3$	$\frac{1}{m_t}$	$\frac{1}{m_l}$	0	m_t	2
$<100>$	$\Lambda 1$	$\frac{m_t+2m_l}{3m_l m_t}$	$\frac{m_t+2m_l}{3m_l m_t}$	$\frac{m_t-m_l}{3m_l m_t}$	$\frac{2m_t+m_l}{3}$	2
$<100>$	$\Lambda 2$	$\frac{m_t+2m_l}{3m_l m_t}$	$\frac{m_t+2m_l}{3m_l m_t}$	$\frac{m_l-m_t}{3m_l m_t}$	$\frac{2m_t+m_l}{3}$	2
$<110>$	$\Delta 1$	$\frac{1}{m_l}$	$\frac{1}{m_t}$	0	m_t	2
$<110>$	$\Delta 2$	$\frac{m_t+m_l}{2m_l m_t}$	$\frac{1}{m_t}$	0	$\frac{m_t+m_l}{2}$	2
$<110>$	$\Delta 3$	$\frac{1}{m_t}$	$\frac{m_t+m_l}{2m_l m_t}$	0	$\frac{m_t+m_l}{2}$	2
$<110>$	$\Lambda 1$	$\frac{m_t+2m_l}{3m_l m_t}$	$\frac{2m_t+m_l}{3m_l m_t}$	$\frac{\sqrt{2}}{3}\frac{m_t-m_l}{m_l m_t}$	m_t	1
$<110>$	$\Lambda 2$	$\frac{m_t+2m_l}{3m_l m_t}$	$\frac{2m_t+m_l}{3m_l m_t}$	$\frac{\sqrt{2}}{3}\frac{m_l-m_t}{m_l m_t}$	m_t	1
$<110>$	$\Lambda 3$	$\frac{m_t+2m_l}{3m_l m_t}$	$\frac{1}{m_t}$	0	$\frac{m_t+2m_l}{3}$	2

Table 1 shows the effective mass tensor resulting from the previous approach for two wire orientations and all the electron valleys: the four-fold degenerated Λ-valleys and the six-fold degenerated Δ-valleys (the Γ-valley being defined by a spherical constant energy minimum).

Figure 1 illustrates the square modulus of the transverse eigenstates of the Λ- and Δ-valleys for a <100>-oriented Ge nanowire. The term ω_{YZ} of the reciprocal effective mass tensor in the Λ-valleys couples the transverse directions and induces the alignment of the eigenstates along the diagonal. Figure 2-left compares the total current-voltage characteristics corresponding to the Si and Ge <100>-nanowire MOSFETs. We find that the Ge roughly provides the same current as the Si. Figure 2-right sheds light on this behavior by plotting the electron subbands for each valley of the Ge. For the considered cross-section, the transverse effective masses of the Λ-valleys (m_{YY}=m_{ZZ}=0.117×m_0, where m_0 is the free electron mass) produce a higher energy splitting than those of the Δ2-valleys (m_{YY}=0.2×m_0, m_{ZZ}=0.95×m_0) whose subbands become energetically lower and participate more efficiently in electron transport. Moreover, the Δ2-valleys of the Ge have an effective mass along the transport direction very close to the one of the Δ2-valleys in Si (0.2×m_0 against 0.19×m_0 respectively).

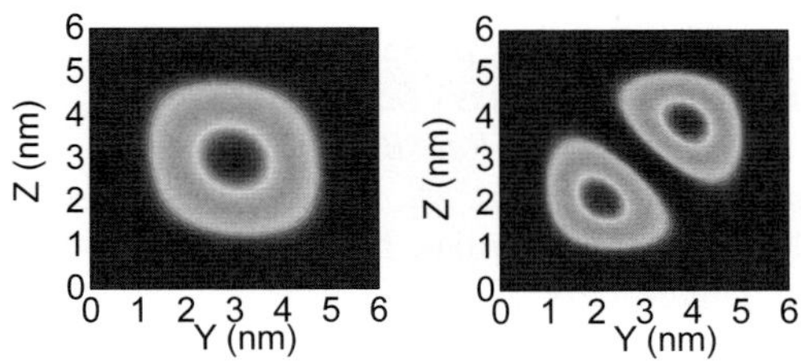

Fig. 1: Square modulus of the first two eigenstates of the Λ-valleys of a Ge nanowire oriented along the <100> direction.

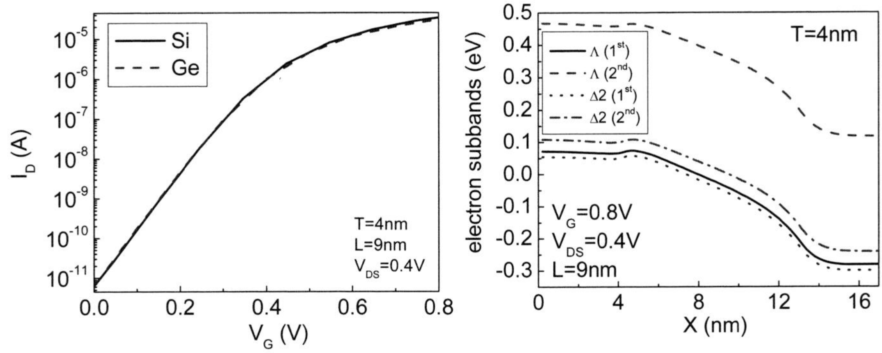

Fig. 2-Left: Comparison of the I_D-V_G characteristics calculated for a Si and Ge Gate-all-around MOSFET <100>-oriented. L=9 nm, T=4 nm, V_{DS}=0.4 V.
-Right: Electron subband positions in a 4 nm cross-section Ge nanowire.

4 Conclusion

We have applied a general approach to calculate the effective mass tensor in semiconductor nanowires for all valleys and arbitrary crystallographic orientations. Effective mass tensor has been detailed for two technologically important wire orientations. Within this approach, the influence of the coupling term of the effective mass tensor on the transverse eigenstate of the wire has been illustrated. We also have calculated the current-voltage characteristics in gate-all-around Si and Ge nanowire MOSFET in the case of an <100>-oriented wire. We have shown that the strong transverse confinement makes the Δ-valleys in Ge energetically lower, which now starts to have a significant influence in the transport.

References

1. http://public.itrs.net
2. Rahman, A, Ghosh, A., Lundstrom, M., *IEDM Tech. Digest*, p.471, 2003.
3. Beysserie, S., Aboud, S., Goodnick, S., Thornton, T., Saraniti, M., *Phys. Status Solidi B* **241**, 2297 2004.
4. Laux, S. E., *IEDM Tech. Digest*, p.135, 2004.
5. Pikus, G.E., Bir, G.L., *Soviet Phys.- Solid State* **1**, 1502, 1960.
6. Stern, F., Howard, W. E., *Phys. Rev. B* **163**, 816, 1967.
7. Wang, J., Polizzi, E., Lundstrom, M., *IEDM Tech. Digest*, p.695, 2003.
8. Bescond, M., Nehari, K., Cavassilas, N., Autran, J.L., Munteanu, D., Lannoo, M., *IEDM Tech. Digest*, p.617, 2004.
9. Wang, J., Polizzi, E., Lundstrom, M., *J. Appl. Phys.* **96**, 2192, 2004.
10. Bescond, M., Cavassilas, N., Kalna, K., Nehari, K., Raymond, L., Autran, J.L., Lannoo, M., Asenov, A., *IEDM Tech. Digest*, to be published, December 2005.

Scanning Tunneling Microscopy of Ultrathin Silicon-on-Insulator

[1]P. P. Zhang, [1]E. Tevaarwerk, [1]B. N. Park, [1]D. E. Savage, [2]G. Celler, [1]I. Knezevic, [1]P. G. Evans, [1]M. A. Eriksson, and [1,a]M. G. Lagally

[1]University of Wisconsin-Madison, Madison, WI, 53706

[2]Soitec USA, 2 Centennial Dr, Peabody, MA 01960

[a]Correspondence should be addressed to: lagally@engr.wisc.edu

Summary. We present near-atomic-resolution scanning tunneling microscopy (STM) images of the surface of a 10 nm thick Si template layer in silicon-on-insulator (SOI), demonstrating that ultra-thin SOI, which is typically described as fully depleted of charge carriers, can indeed be imaged. We attribute the ability to image to our cleaning process, which results in a Si (001) free of oxide and defects. Electronic conduction in this type of very thin Si film is enabled by the interaction of Si (001) surface bands caused by the Si (001) 2x1 reconstruction with the bulk Si bands.

Silicon-on-insulator (SOI) has emerged as an important substrate for Si device technology, offering enhanced speed and reduced power consumption for field-effect transistors in comparison to conventional bulk-Si MOSFETs [1]. As the Si layer becomes very thin, interfaces begin to determine its electrical transport properties.

Usually the silicon template layer in SOI will be bounded by two SiO_2 layers, as the silicon template layer is covered either by native oxide or by gate oxide. The Si/SiO_2 interfaces play an important role in changing the film electronic properties. At a Si-SiO_2 interface, there exist charge trapping sites with energy levels within the bandgap (1.1eV). The interface trap density of states (D_{it}) typically exhibits a U shape across the Si band gap, with D_{it} of order $10^{11}cm^{-2}eV^{-1}$. Interface traps in the upper half of the band gap behave as acceptors, while those in the lower half behave as donors [2]. These interface charge traps arise from structural or oxidation–induced defects and are mainly in the form of Si "dangling bonds".

As the Si layer gets very thin, the interfaces will eventually deplete this layer of free carriers and the conductivity vanishes. To demonstrate this effect, we analyzed the sheet resistance of Si template layers with thicknesses ranging from 15nm to 200nm, sandwiched between a native oxide and the buried oxide, using the van der Pauw technique. We used Unibond™ SOI wafers from SOITEC, comprised of a silicon handle wafer, a 3μm buried oxide layer (BOX), and a 200 nm silicon top or "template" layer. Both the template and handle wafer are crystalline silicon with a (001) orientation, boron doped with a resistivity between 14-22 Ω-cm, corresponding to a carrier density of $\sim 10^{15}/cm^3$. We deliberately choose 3μm BOX to reduce the oxide leakage current. In order to obtain Si templates with different thicknesses, we thinned the top Si layer by dry thermal oxidation at 1050°C, followed by wet chemical etching in HF. As shown in Figure 1, the measured sheet resistances are all higher than the resistance predicted for bulk Si for this nominal doping level. For Si template layer thicknesses below 30 nm, the sheet resistance ranged between $10^{10}\Omega$/square and $10^{11}\Omega$/square. Consistent with our prediction, holes in the film are getting trapped at the interfaces and the Fermi level is therefore pulled towards midgap.

Recent work using scanning tunneling microscopy (STM) concluded that SOI (001) with template layer thicknesses below ~35 nm is fully depleted, because the ability to create an image in STM ceased [3,4]. In contrast to these conclusions, we have been able to demonstrate STM imaging of the clean surface of SOI (001), with nominal doping level of $10^{15}/cm^3$ and Si template layers as thin as 10nm. The surface quality, rather than the Si thickness alone, dictates whether carrier density is high enough to enable STM imaging.

To investigate the clean Si (001) surface of SOI, the top oxide must be removed in ultrahigh vacuum (UHV). Surface preparation consists of *ex-situ* and *in-situ* cleaning. *Ex-situ*, the SOI is triple IMEC cleaned. A final piranha clean terminated the surface with a thin (1-2nm) protective surface oxide. The sample is then introduced into an ultrahigh-vacuum STM, which has a base pressure below 1×10^{-10} torr. Traditional *in-situ* preparation of bulk Si surfaces at 1500 K is not possible for ultra-thin SOI, because the Si template layer will dewet and agglomerate into 3D Si islands [5]. Instead, we deposit several monolayers of Si or Ge at 700°C, which reduce the surface SiO_2 to the volatile SiO or GeO. Finally we flash the sample to 800°C (still below the critical temperature for dewetting) for two minutes, quench it to and anneal it at 600°C for 30 minutes, radiation cool it, and then image it.

Figure 2 shows STM images of the clean SOI (001) surface. The 2x1 reconstruction via the formation of dimers occurs even for 10 nm thick Si template layers. The intrinsic anisotropic stress (tensile along the dimer bonds; compressive along the dimer rows) leads to the alternate orthogonal (2 x 1) and (1 x 2) terraces on the surface, terminated respectively by S_B and S_A steps. This dimerized surface leads to two-dimensional bands, a filled π band and an empty π^* band, separated by a ~0.5 eV gap [6]. The density of states in the two-dimensional bands is around 10^{15} $cm^{-2}eV^{-1}$. Our low-thermal-budget surface cleaning method allows us to remove the protective oxide without breaking up the fragile Si template layer and produces very-low-defect-density surfaces. The fact that we can image very thin SOI in STM implies that the depletion of the thin Si layer is not important. Admittedly only one surface is now covered with oxide, but the trap states at this one interface are more than sufficient to maintain the full bulk depletion of the Si layer. It is the presence of the surface bands due to the dimer reconstruction that enables the conduction necessary to image the surface in STM. It is not, however, the conduction through the surface bands directly that makes the film conduct, as one might at first imagine. It is the position in energy of the surface bands relative to the bulk bands, and the high density of states in the bulk valance band and the surface conduction band that enables the conduction [7]. The bulk doping density and the density of interface states on the back Si/SiO_2 interface are largely irrelevant for electronic transport in ultrathin template layers in SOI with a clean reconstructed surface.

In summary, we show that STM imaging of the clean surface of SOI (001) with nominal doping levels of $10^{15}/cm^3$ and silicon template layers as thin as 10 nm is entirely possible, even though such films are depleted of charge carriers and should behave like intrinsic Si. The surface bands caused by reconstruction enable electronic transport. We predict that only disruption of the surface bands can hinder imaging of SOI (001), no matter how thin the Si film is or how low the bulk doping level is. Such disruption can come via surface disorder or chemisorption that in Si (001) breaks the pi-bonded dimer chains that produce the surface bands. In systems that do not have surface bands, we predict that making the layer thin enough to cause bulk depletion will inhibit imaging. Moreover, we believe our surface cleaning method with low thermal budget should be applicable to actual device fabrication, because the same deposition technique and source material can be used both to clean and to grow epitaxial layers for raised source and drain [8], as well as to image the surface with STM and perform other electron-transport-dependent measurements.

This research has been supported by DOE, NSF-MRSEC, and AFOSR..

Fig. 1. Sheet resistance measured from van der Pauw method versus film thickness (circles with error bars). The solid curve is the predicted sheet resistance calculated from the nominal bulk doping density ($10^{15}/cm^3$).

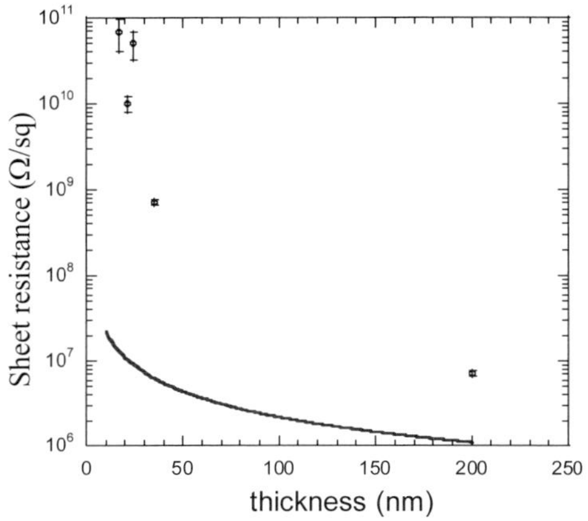

Fig. 2. Filled-state STM image of the surface of a 10nm thick Si template layer of bonded SOI (001) with native oxide removed by 3 ML of Ge. V_{sample} = -2V, 100 nm x 100 nm view.

References

1. Celler, G. K. and Cristoloveanu, S.: '*Frontiers of silicon-on-insulator', Journal Of Applied Physics.*, **93** , 4955-4978, 2003.
2. Schroder, D. K.: *Semiconductor Material and Device Characterization. Chapter 6.*, John Wiley & Sons, Inc. New York. 1998.
3. Lin, K. C., et al.: '*Surface characterization of silicon on insulator material*', *Applied Physics Letters.*, **72**, 2313-2315, 1998.
4. Sutter, P., Ernst, W. and Sutter, E.: '*Scanning tunneling microscopy on ultrathin silicon on insulator (100)*'. *Applied Physics Letters.*, **85**, 3148-3150, 2004.
5. Nuryadi, R., Ishikawa, Y. and Tabe, M.: '*Formation and ordering of self-assembled Si islands by ultrahigh vacuum annealing of ultrathin bonded silicon-on-insulator structure*'. *Applied Surface Science.*, **159**, 121-126, 2000.
6. Northrup, J. E.: '*Electronic-Structure of Si(100)C(4x2) Calculated within the Gw Approximation*'. *Physical Review B.*, **47**, 10032-10035,1993.
7. Zhang, P. P., Tevaarwerk, E., Park, B. N., Savage, D. E., Celler, G., Knezevic, I., Evans, P. G., Eriksson, M. A., and Lagally, M. G.: '*Electronic Transport in Nanometer Silicon-on-insulator*'. Submitted to Nature.
8. Choi, Y. K., et al.: '*Nanoscale ultrathin body PMOSFETs with raised selective germanium source/drain*'. *IEEE Electron Device Letters.*, **22**, 447-448, 2001.

Effect of Regular and Irregular Potential Perturbations in Mesoscopic Cavities

P. Marconcini and M. Macucci

Dipartimento di Ingegneria dell'Informazione, Università di Pisa,
Via Caruso 16 – I-56122 Pisa

Summary. We present a numerical investigation of the shot noise suppression in a mesoscopic cavity containing hard-wall obstacles or potential fluctuations. We show that, while in the absence of obstacles the suppression factor is 1/4, in the presence of a random distribution of strong scatterers it becomes 1/3. If instead the scatterers are regularly distributed, the value of the Fano factor depends on the relative position of the obstacles with respect to the apertures.

1 Introduction

As was shown by Schottky, when charge carriers move independently (giving rise to a Poissonian process), the shot noise associated with a current I is equal to $S_I = 2e|I|$ (where e is the elementary charge). Phenomena of shot noise suppression, with respect to such a value, are often found in mesoscopic structures, due to the correlations between charge carriers resulting from Fermi exclusion or from Coulomb interaction. In particular, great effort has been devoted to the theoretical and experimental study of diffusive conductors [1,2], for which the suppression factor (Fano factor) has been proved to be 1/3 if $l << L << Nl$ (with L being the length of the conductor, l the elastic scattering length and N the number of propagating modes), and of mesoscopic cavities [3,4], in which the Fano factor is equal to 1/4.

In particular, mesoscopic cavities are regions a few microns wide delimited by constrictions; a suppression factor 1/4 is found if the constrictions are symmetric and much narrower than the cavity width. Such a result is true also for classically non chaotic shapes of the cavity and is mainly due to discontinuities of the potential at the constrictions and to the resulting diffraction, which determines a quantum chaotic dynamics of the charge carriers [5,6].

We investigate the case in which the cavity contains, in addition, regular or irregular arrays of hard-wall obstacles or potential fluctuations associated with the discrete nature and random position of donors, with the aim of understanding how this type of scattering influences the shot noise behavior.

2 Numerical Method and Model

For our numerical simulations we have applied the recursive Green's function formalism [7]. The structure is subdivided into transverse slices, in each of which the potential is constant along the longitudinal direction. We have computed the Green's function for each separate slice and then we have recursively composed the Green's functions of the slices using the Dyson equation. From the overall Green's function matrix, it is straightforward to obtain the transmission matrix t of the structure and thus the conductance G using the Landauer-Büttiker formula, the shot noise power spectral density S_I following Büttiker [8] and the Fano factor by computing the ratio of S_I to $2e|I|=2e|V|G$ (where V is the externally applied voltage).

In our simulations we have considered a 5 μm long and 8 μm wide rectangular cavity defined by hard-wall boundaries, for various values of the constriction width. The structure has been discretized with a mesh of about 200×800 points.

We have uniformly averaged over 61 values of the Fermi energy in the range between 9.03 meV and 9.24 meV. Averaging has been performed separately for the numerator and the denominator of the Fano factor expression, as it would be done in the actual measurement process.

3 Numerical Results

For an empty rectangular cavity the Fano factor equals 1/4 also in the case of a perfectly regular rectangular hard-wall geometry, because, as already stated, the main source of diffraction is at the interface between the wide and the narrow regions.

We have then added a random distribution of hard-wall obstacles inside the cavity (in the case of Fig.1(a), 240 square obstacles with edge size equal to 200 nm). The coordinates of each obstacle are generated as a pair of uniformly distributed random numbers.

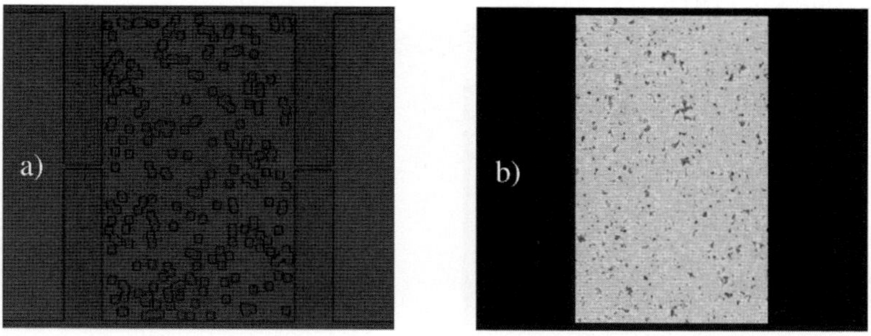

Fig. 1. Mesoscopic cavity with a random distribution of 240 hard-wall scatterers (a); or with potential fluctuations due to the presence of discrete dopants (b).

If the obstacles are sufficiently opaque and not too large, we obtain a Fano factor equal to 1/3, thus recovering the result for diffusive conductors. We point out that, although the presence of random scatterers leads to a distribution of transmission eigenvalues analogous to that of a cavity (i.e. a binomial distribution), the Fano factor raises to 1/3 if the conditions for diffusive transport are satisfied.

A more realistic source of random scattering in a cavity is represented by the potential fluctuations caused by the presence of ionized dopants in the delta-doping layer (Fig. 1(b)). Their effect on the potential has been modeled with a semi-analytical technique, with the inclusion of the screening (for the point-like charges corresponding to the ionized dopants) due to the 2DEG, adapting the theory of Stern and Howard [9] to the case of gallium arsenide. The resulting Fano factor is intermediate between 1/4 and 1/3 (0.29), because the fluctuations of the potential are of the order of a few millielectronvolts and therefore do not represent opaque enough obstacles for electrons at the Fermi energy.

Then we have considered the effect of a regular array of hard-wall obstacles inside the cavity (Fig. 2). In particular, we have focused on square obstacles of various sizes, with separation equal to the obstacle size in both directions.

As long as the obstacles are large compared with the constrictions, we observe a reduction of the Fano factor below 1/4 if the central “corridor” between obstacles is aligned with the constrictions (Fig. 2(a)) and an enhancement if a row of obstacles lies exactly between the constrictions (Fig.

2(b)). This can be easily understood in terms of an increase or decrease of direct (noiseless) transmission between the constrictions.

This result is preserved if the width of the vertical channels separating the obstacles is reduced with respect to the obstacle width and that of the horizontal channels.

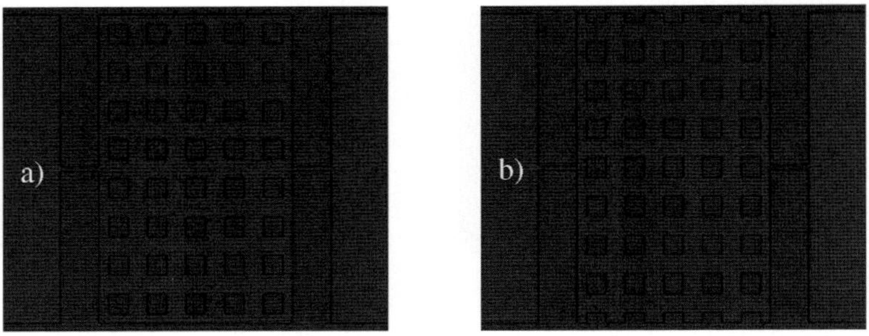

Fig. 2. Mesoscopic cavity with a regular array of hard-wall obstacles, with a channel between obstacles (a) or a row of obstacles (b) aligned with the apertures defining the cavity.

We acknowledge financial support from the Italian Ministry of Education, University and Research (MIUR) through the PRIN project "Excess Noise in Nanoscale Devices" and the FIRB project "Nanotechnologies and Nanodevices for the Information Society"

References

1. Beenakker C.W.J., Büttiker M., *Phys. Rev. B* **46**, 1889 (1992).
2. Liefrink F., Dijkhuis J.I., de Jong M.J.M., Molenkamp L.W., van Houten H., *Phys. Rev. B* **49**, 14066 (1994).
3. Jalabert R.A., Pichard J.-L., Beenakker C.W.J., *Europhys. Lett.* **27**, 255 (1994).
4. Oberholzer S., Sukhorukov E.V., Strunk C., Schönenberger C., Heinzel T., Holland M., *Phys. Rev. Lett.* **86**, 2114 (2001).
5. Marconcini P., Macucci M., Iannaccone G., Pellegrini B., Marola G., *cond-mat*/0411691 (2004).
6. Aigner F., Rotter S., Burgdörfer J., *Phys. Rev. Lett.* **94**, 216801 (2005).
7. Macucci M., Galick A., Ravaioli U., *Phys. Rev. B* **52**, 5210 (1995).
8. Büttiker M., *Phys. Rev. Lett.* **65**, 2901 (1990).
9. Stern F., Howard W.E., *Phys. Rev.* **163**, 816 (1967).

Simulation of Electronic/Ionic Mixed Conduction in Solid Ionic Memories

Hyuck In Kwon and Umberto Ravaioli

Beckman Institute, University of Illinois at Urbana-Champaign

Jong Duk Lee

Seoul National University

Summary. The electronic/ionic mixed conduction phenomena in solid ionic memory devices are investigated. The steady state distributions of electronic carriers and metal chemical potentials within mixed conductors are obtained by numerically solving the Nernst-Plank-Poisson equations. The results are compared with analytical solutions derived under the fixed ion concentration approximation.

1 Introduction

Solid ionic memory devices have been proposed as promising candidates for next-generation non-volatile memory devices due to their high scalabilities, ease of fabrication and low operation voltages [1]. Unlike conventional silicon non-volatile memories, solid ionic memory devices are based on the electrochemical control of nanoscale quantities of metal ions and electronic carriers in thin films of mixed conductors. In these devices, information is stored via metal filament formation between two electrodes. The device is based on a metal base covered by a mixed conductor film where metal ions and electronic carriers can move, particularly under an applied electric bias. Ionic and electronic transports are necessary to form the filament where electronic current can be induced. We study here the combined electronic/ionic mixed conduction phenomena in solid ionic memory devices. By solving the Nernst-Plank-Poisson equations numerically, we obtained the steady state distributions of electronic carriers and metal chemical potentials within mixed conductors.

2 Results and Discussion

The mixed conductor will, for simplicity of notation, be presented as XY. It is assumed to conduct one kind of ions X^+ and quasi-free electrons e^-. The Y sublattice is assumed to be rigid. The ions interact with the electrons to form neutral atoms:

$$X^+ + e^- \leftrightarrow X \tag{1}$$

The concentration of halls will be assumed negligibly small. In solid ionic memory devices, the ionic current vanishes in the stationary state, and it follows that

$$\nabla\eta_i = \nabla\mu_i + q\nabla\varphi = 0 \tag{2}$$

where η denotes the electrochemical potential, μ is the chemical potential, q is the absolute electronic charge, and φ is the electrical potential.
Equation (1) and (2) relates η_i, η_e, and μ_X:

$$\nabla\eta_i + \nabla\eta_e = \nabla\eta_e = \nabla\mu_e - q\nabla\varphi = \nabla\mu_X \tag{3}$$

The applied voltage V is related to $\nabla\eta_e$ by

$$-qV = \eta_e(E(L)) - \eta_e(E(0)) \tag{4}$$

where E(L) and E(0) are ion blocking electrode and reference (reversible and permeable) electrode, respectively.
Equation (2), (3), and (4) yield

$$-qV = \nabla\mu_X = \nabla\mu_e - q\nabla\varphi = \nabla\mu_e + \nabla\mu_i \tag{5}$$

Based on Boltzmann statistics, equation (5) can be expressed as

$$-qV = kT\ln(n(L)/n(0)) + kT(N(L)/N(0)) \tag{6}$$

where k is the Boltzmann constant, T is the temperature, n(L), N(L) are the electron and metal ion concentration at the boundary with ion blocking electrode, and n(0), N(0) are the electron and metal ion concentration at the boundary with reference electrode.
Based on the charge neutrality approximation, the concentration of excess ions (δN) formed by the applied voltage can be assumed to have the same value with that of excess electrons.

$$N(L) = N(0) + \delta N,\ n(L) = n(0) + \delta N \tag{7}$$

Inserting equation (7) into equation (2) and (6), we obtain

$$\nabla\varphi = -\nabla\mu_i/q = -(kT/q)\ln((N(0) + \delta N)/N(0)) \tag{8}$$

and

$$\delta N = [-(n(0) + N(0)) + \{(n(0)+N(0))^2 - 4(1 - e^{-qV/kT})\, n(0)N(0)\}^{1/2}]/2 \quad (8)$$

Using equation (7) and (8), the boundary conditions for the Nernst-Plank-Poisson equations can be determined by Dirichlet-type as a function of temperatures, applied biases, and the chemical compositions of mixed conductors under the reference electrode.

Fig. 1 plots the electrical potential difference between two electrodes as a function of the ratio of metal ion concentration to electron concentration under the reference electrode. It shows that the electrical potential difference between two electrodes gradually deceases as the ratio increases.

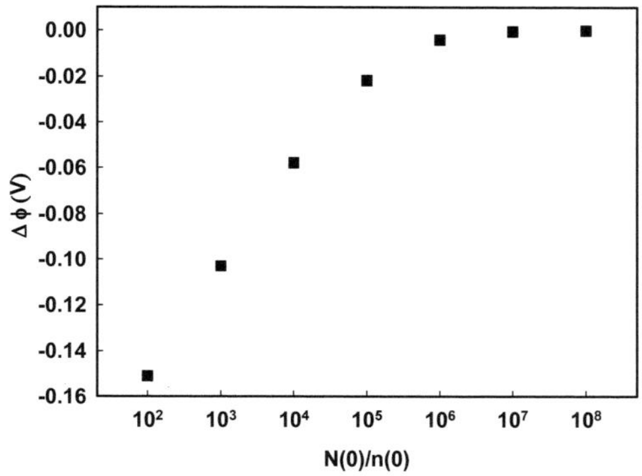

Fig. 1. The difference of electrical potentials between two electrodes in solid ionic memory devices. The values are obtained from equation (7) and (8) under the conditions: V = -0.5 (V) at the ion blocking electrode, T = 500 (K).

Fig. 2 plots the electron concentrations and metal chemical potential differences as functions of metal ion concentration under the reference electrode. Simulations have been performed by adapting the computational platform PROPHET [2] with following parameters: $D_{X+} = 3.0 \times 10^{-5}$ (cm^2/s), $D_{e-} = 7.0$ (cm^2/s), $\varepsilon = 8.7$, L = 100 (μm), $n(0) = 10^{14}$ (cm^{-3}), T = 500 (K), V = -0.5 (V). The results show that the distribution of electron concentrations and metal chemical potential differences within solid ionic memory devices strongly depend on the chemical composition of the reference electrode. It also shows that the numerical results fit the analytical solutions well when the ratio of metal ion concentration to electron concentration under the reference electrode is high. The analytical solutions are obtained based on the assumption that the chemical potential of metal ion has the constant value within the mixed conductor [3].

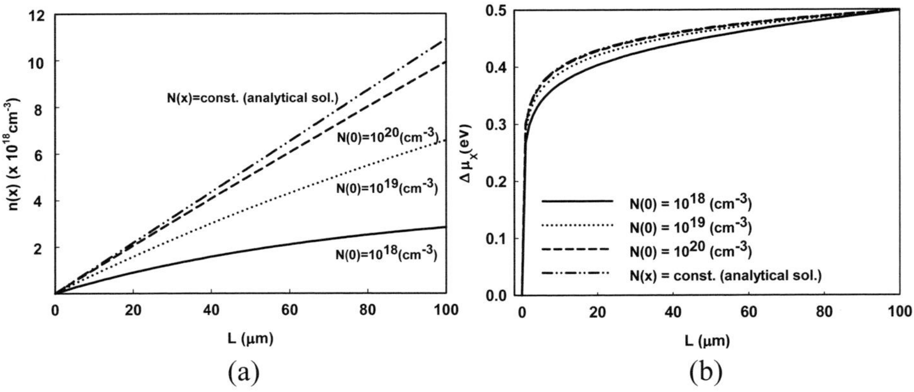

Fig. 2. The steady state distribution of (a) electron concentrations and (b) metal chemical potential differences as functions of metal ion concentration under the reference electrode.

3 Conclusions

A mixed electron/ion Nernst-Plank-Poisson simulation has been performed for solid ionic memory structures using the program PROPHET. The simulation results show that the distribution of electron concentrations and metal chemical potential differences within solid ionic memory devices strongly depend on the chemical composition of the reference electrode. Work is in process to incorporate quantitative models for the actual formation of metallic filaments and the corresponding electronic conduction through the filament. This work was supported by the Network for Computational Nanotechnology NSF grant EEC-0228390 and the Korean Research Foundation Grant funded by the Korean Government (KRF-2005-214-D00288).

References

1. Sakamoto, T., Sunamura, H., Kawaura, H., Hasegawa, T., Nakayama, T., and Aono, M.: 'Nanometer-scale switches using copper sulfide', *App. Phys. Lett.*, **82**, 3032-3034, 2003.
2. https://www.nanohub.org/simulation_tools/prophet_tool_information
3. Janek, J., and Korte, C.: 'Electrochemical blackening of yttria-stabilized zirconia – morphological instability of the moving reaction front', *Soild State Ionics*, **116**, 181-195, 1999.

Full-Band Modeling of Magnetic Semiconductors

S. Beysserie[+], I. Remond[+], S. Goodnick*, and M. Saraniti[+].

[+]Department of Electrical and Computer Engineering, Illinois Institute of Technology, Chicago, IL.

[*]Department of Electrical Engineering and Center for Solid State Electronics Research, Arizona State University, Tempe, AZ.

Summary. In this work we extended our Cellular Monte Carlo (CMC) device modeling code to include the computation of the energy states in dilute magnetic semiconductors using an ab-initio code based on the Korringa, Kohn and Rostoker coherent potential approximation method (KKR-CPA). The CMC simulator was modified to explicitly model spin transport in bulk materials using a full-band representation of the electronic structure. The last part of this work was devoted to the implementation of the major scattering mechanisms responsible for spin flip in III-V semiconductors for which results are presented.

1 Introduction

Dilute magnetic semiconductors (DMS) [1] have been of great interest as potential spin polarized injectors in spin devices. Due to the novelty of the interest in magnetic semiconductors, very few modeling tools are available to guide experimental research. A suitable code for the simulation of spin injection and transport requires not only a robust computation of the electronic structure and energy states over a wide variety of DMS, it also needs to account for the dynamics of spin polarization and spin relaxation, and the consequent implementation of the scattering mechanisms responsible for spin flip in bulk semiconductors such as GaAs. In the first part of this work, we present the techniques used to extract the band structure from DMS materials using an existing density-functional code. We then discuss the implementation of major spin flip mechanisms of conduction electrons in GaAs within the full-band framework of our Cellular Monte Carlo (CMC) particle-based simulator [2, 3], and present simulation results obtained in bulk GaAs at various temperatures.

2 Extraction of Fermi surfaces in DMS

In order to extend the modeling possibilities of our CMC code to DMS materials, we coupled it with an existing *ab-initio* code [4, 5] based on the Korringa, Kohn and Rostoker coherent potential approximation method (KKR-CPA). This robust and flexible method is able to virtually model any alloy or semiconductor material using only a single-site atomic potential obtained by a limited set of atomic input parameters. While this code is both efficient and easy to use, it does not provide the energy states as eigenvalues of the electronic Hamiltonian, but it outputs instead the Bloch spectral function (BSF), representing the density of states as a function of energy for each momentum value. The BSF computed by the KKR-CPA needs therefore to be post-processed in order to obtain a representation of the energy states suitable for particle-based simulation tools. In a perfectly periodic structure, the BSF can be expressed as a sum of delta functions centered about the eigenvalues corresponding to each band of the material. In order to identify degenerate eigenvalues, both the peak value and its integral must be evaluated and compared to the expected values of a non-degenerate band. In the more complex case of DMS materials, the Bloch spectral function follows a more complicated and irregular behavior than in perfect crystals as it can be seen on the three dimensional representation of the BSF shown in Figure 1 for the case of $Ga_{0.95}Mn_{0.05}N$ where atoms of Mn randomly substitute atoms of Ga with a probability of 5%.

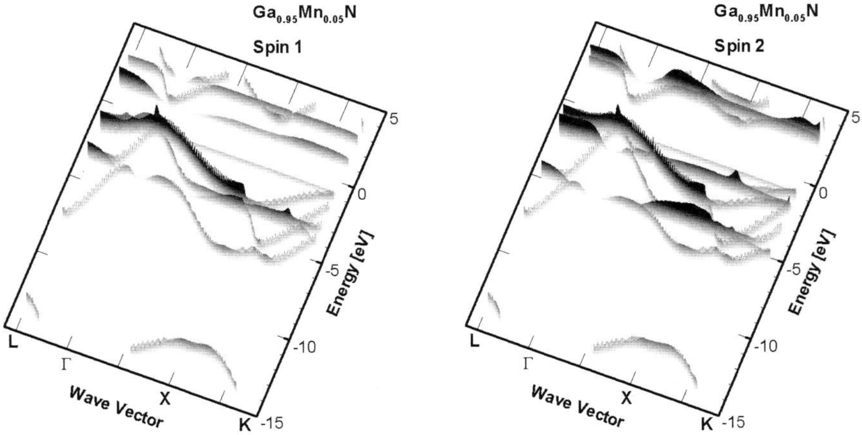

Fig. 1. Bloch Spectral Function for both spin polarizations of $Ga_{0.95}Mn_{0.05}N$ as a function of energy and momentum. The irregular behavior of the BSF is particularly visible around the Γ point.

The order of degeneracy is necessary to compute the velocity and density of states (DOS) within the traditional full-band framework. While the degeneracy can no longer be computed in DMS, as a first approximation we can still extract the energy values and use partial derivatives to compute the velocity. The resulting DOS will then be compared to that obtained directly with AkaiKKR.

3 Simulation of Spin Polarized Transport

The accurate representation of spin polarized transport is another crucial aspect for the simulation of spin devices. In order to discriminate spin states, both the band structure and the tables of phonon scattering are tabulated for each spin state. Once the carrier dynamics can be modeled in a spin-dependent fashion, the implementation of the major mechanisms responsible for carrier spin flip allows the study of spin dynamics. Three spin relaxation mechanism are of particular interest for the study of magnetic properties of III-V semiconductors [6, 7], namely the Elliott-Yafet (EY) [8, 9], the D'Yakonov-Perel (DP) [10], and the Bir-Aronov-Pikus (BAP) mechanism. The EY mechanism originates from the ordinary momentum scattering if the lattice ions induce spin-orbit coupling in the electron wave function, while the DY occurs in materials lacking inversion symmetry in which spin-up and spin-down electrons are not degenerate. The BAP mechanism only exists in the presence of holes and is therefore neglected in the present work, which focuses on *n*-type materials.

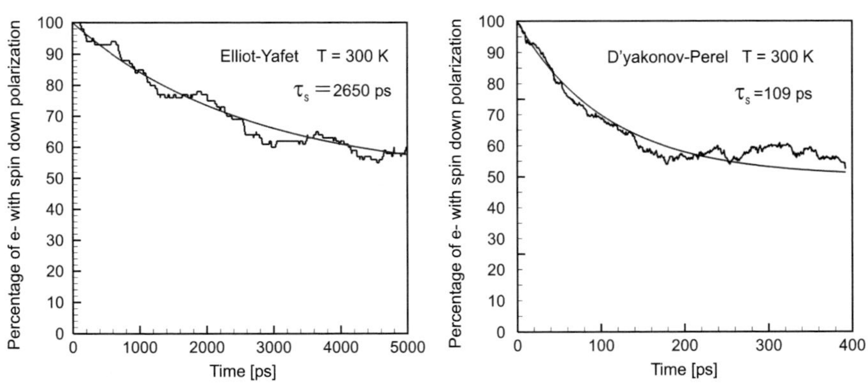

Fig. 3. Simulated time evolution of the spin polarization in bulk GaAs at 300 K, with either the DP or EY mechanism activated. An exponential decay model and a least mean squares approximation method were used to interpolate the data and derive the spin relaxation time.

Simulation results of bulk GaAs 300 K in a population of electrons initially polarized with the same spin are presented in Figure 2. No electric field was applied to the system in order to compare our results with published relaxation times for the EY and DP mechanisms, respectively. These results are in excellent agreement with the predicted spin relaxation time in *n*-type GaAs [6].

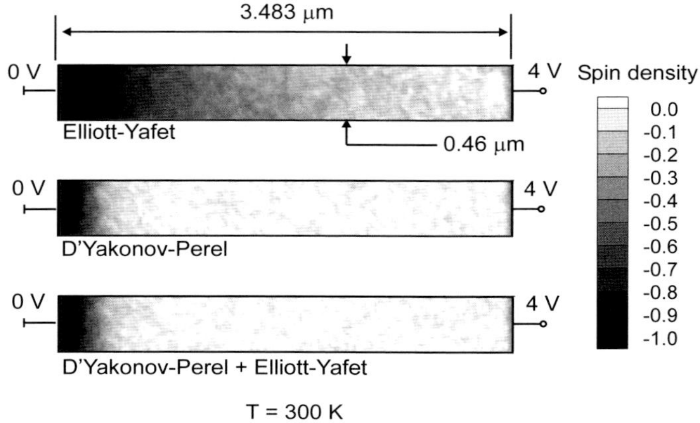

Fig. 4. Spin density within a simulated GaAs resistor, obtained by selectively activating the Elliott-Yafet scattering mechanism (top), the D'Yakonov-Perel scattering mechanism (middle), and both (bottom).

The study of the relaxation length L_s in non-bulk systems is also of particular interest for the design of spin devices where one is interested to know how far the carriers can travel and retain their original spin polarization. We modeled and simulated a 3.5 μm x 0.5 μm GaAs resistor with a homogeneous impurity concentration of 10^{14} cm^{-3}, a lattice temperature of 300 K, and a 4 V bias. The device was discretized using a homogeneous grid of 129x17 cells, while 10,000 super-particles were used to represent the electron population. Carriers were polarized with spin down and injected from the left electrode when needed in order to maintain charge neutrality at the contacts. A set of three simulations were ran, by selectively activating the EY mechanism, the DP mechanism, and both. The spin spatial density is shown in Figure 4 for the three simulations, the numerical value (+1) is used for carriers polarized with a spin up while (-1) is used for the spin down polarization. In the case where only the EY mechanism is activated, *Ls* corresponds to about the length of the sample, while it decreases to 500 nm under the influence of the DP mechanism. In analogy to the results of the bulk simulations, the DP mechanism is therefore dominant, resulting in a relaxation length which is one order of magnitude shorter than the one obtained with the EY mechanism.

4 Conclusion and Future Work

We have presented a simple procedure to extract the band structure from the Bloch spectral function as well as the implementation of major spin flip mechanisms within a full-band framework. Simulation results have been presented as well, showing the relaxation times of two scattering mechanisms and the spin mean free path of spin-polarized electrons in bulk GaAs and in a simple resistor, respectively.

Acknowledgements

This work has been partially supported by HPTi. Results for $Ga_{0.95}Mn_{0.05}N$ were obtained using AkaiKKR (http://sham.phys.sci.osaka-u.ac.jp/~kkr/).

References

1. G.M. Stocks and H. Winter: *'The Electronic Structure of Complex Systems', Plenum Press, NATO ASI Series B: Physics*, **113**, 463-579.
2. M. Saraniti and S. M. Goodnick: *'Hybrid full-band cellular automaton/Monte Carlo approach for fast simulation of charge transport in semiconductors', IEEE Trans. Electron Devices*, **47**,1909-1915, 2000.
3. M. Saraniti, J. Tang, S. M. Goodnick and S. J. Wigger: *'Numerical challenges in particle-based approaches for the simulation of semiconductor devices', Math. Comput. Simul.*, **62**, 501, 2003.
4. H. Akai: *'Fast Korringa-Kohn-rostoker coherent potential approximation and its application to FCC Ni-Fe systems', J. Phys. Conde. Matter*, **1**, No. 43, 8045-8064, 1989.
5. Official Web Site of AkaiKKR: http://sham.phys.sci.osaka-u.ac.jp/~kkr/
6. A. G. Aronov, G. E. Pikus and A. N. Titkov: *'Spin relaxation of conduction electrons in p-type III-V compounds', Sov. Phys.-JETP*, **57**, No. 3, 680-687, 1983.
7. I. Zutić, J. Fabian and S. Das Sarma: *'Spintronics: Fundamentals and applications', Rev. Mod. Phys.*, **76**, No. 2, 323-409, 2004.
8. R. J. Elliott: *'Theory of the effect of spin-orbit coupling on magnetic resonance in some semiconductors', Phys.Rev.*, **96**, No. 2, 266-279, 1954.
9. Y. Yafet: *'Solid State Physics'*, **14**, edited by F. Seitz and D. Turnbull (Academic, New York), p. 2.
10. I. D'Yakonov and V. I. Perel': *'Spin orientation of electrons associated with the interband absorption of light in semiconductors', Sov. Phys.-JETP*, **33**, No. 5, 1971.

Cellular Monte Carlo Modeling of $Al_xIn_{1-x}Sb$/InSb Quantum Well Transistors

J. Branlard, N. Faralli, T. Dutta-Roy, S.M. Goodnick*, D.K. Ferry*, S.J. Aboud[+] and M. Saraniti

ECE Dept., Illinois Institute of Technology, Chicago, IL, USA
* EE Dept., Arizona State University, Tempe, AZ, USA
[+] ECE Dept., Worcester Polytechnic Institute, Worcester, MA, USA

Summary. In this work, an Indium Antimonide (InSb) quantum well transistor is investigated using full-band Monte Carlo simulations. The steady-state characteristic of the device is first analyzed, showing particle transport along the two-dimensional electron gas (2DEG). The small-signal behavior of the device is also investigated. Finally, the noise analysis is performed, allowing for a two-dimensional mapping of the noise within the device.

1 Introduction

Due to its low band gap and high electron mobility, Indium Antimonide (InSb) is a promising material for extremely high frequency devices operating at low voltages [1]. These desired properties are not easily accessible at room temperature because of the high intrinsic carrier density, leading to parasitic leakage currents. However, techniques based on minority-carrier exclusion and extraction [2] have been used to raise the device operating temperature, allowing for room temperature operation of InSb MISFETs [3]. More recently, AlInSb/InSb quantum well transistors have been developed [4]. The purpose of this work is to investigate the static and dynamic behaviour of these devices using full-band Monte Carlo simulations.

2 Cellular Monte Carlo Simulations

The full-band particle-based simulation tool used for modeling the devices is based on the Cellular Monte Carlo (CMC) [5] method. Bulk simulations of InSb have been performed at 77K for validation purposes. The results are compared to published data [6,7] in Fig.1 (a) and (b), showing the electron drift velocity and energy versus the ‹100› electric field, respectively. Good agreement is observed.

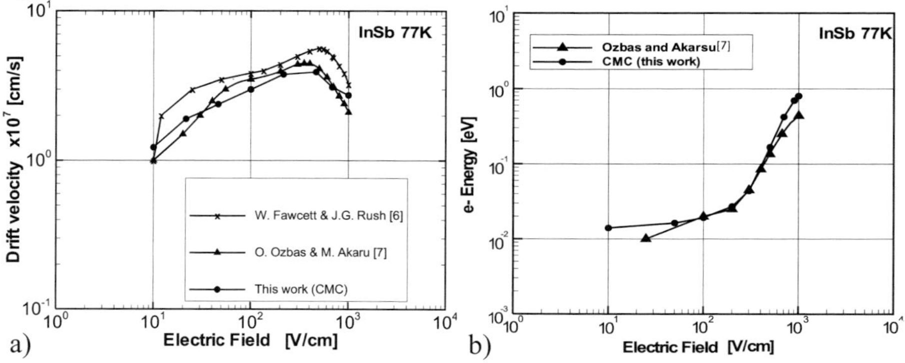

Fig. 1. Electron drift velocity (a) and energy (b) obtained with CMC simulations and compared to other published results [6,7].

3 AlInSb/InSb Quantum Well Transistor

The cross section of the simulated AlInSb/InSb quantum well transistor is shown in Fig.2(a). A 15 nm layer of doped $Al_{0.3}In_{0.7}Sb$ is separated from a 20 nm conducting InSb channel by a 5 nm $Al_{0.3}In_{0.7}Sb$ spacer. The 200 nm $Al_{0.3}$-$In_{0.7}Sb$ substrate underneath is unintentionally doped. The shaded areas indicate the average concentration of electrons. To account for the quantization of the electron motion in the vicinity of the heterojunction, the effective potential approach [8] has been used.

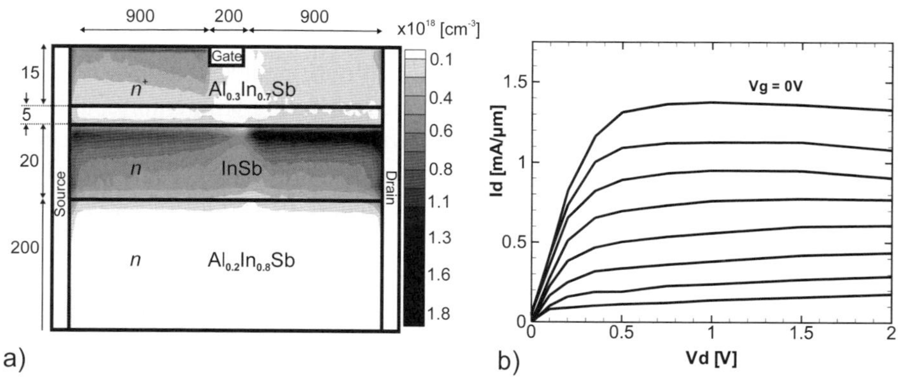

Fig. 2. (a) 2D schematic layout of the simulated device. (b) Drain current-voltage characteristics. The gate biases are applied every 0.1 V.

Fig.2 (b) shows the drain current versus the drain voltage for gate biases ranging from -0.7 V to 0 V. The knee voltage is observed at 0.5 V. The device exhibits good saturation characteristics, and a slight decrease in the drain current can be observed for drain voltage greater than 1 V. This negative differential resistance has already been observed in InSb structures and reported in literature [6].

4 Results

The frequency analysis is performed using both a Fourier decomposition technique [9] and monochromatic sinusoidal excitation [10]. An average of 200,000 super-particles has been simulated to model the electron population. Typical simulated times range from a few picoseconds to over a hundred picoseconds, averaging to 77 minutes of CPU time per simulated picosecond on a Pentium IV 2.4 GHz processor.

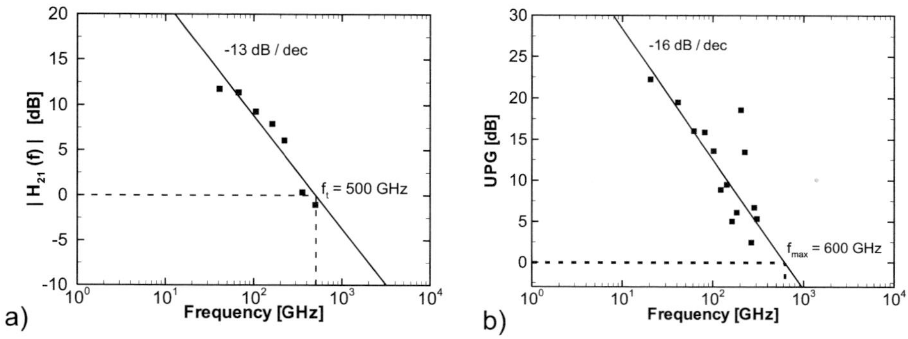

Fig. 3. Short circuit current gain (a) and unilateral power gain (b) as a function of frequency.

The short circuit current gain h_{21} and the unilateral power gain UPG are shown as a function of frequency on Fig.3 (a) and (b), respectively. A linear fit is indicated with a solid line, showing a slope of 13 and 16 dB per decade, respectively. A cutoff frequency f_T = 500 GHz and a maximum frequency of oscillation f_{max} = 600 GHz are extrapolated, and the f_T/f_{max} ratio of 1.2 is in good agreement with published results [4]. The frequencies reported here are larger than the ones obtained by the experiment; however, they correspond to internal small-signal values without consideration of any contact and connector resistance and capacitance, which typically have a negative impact on the device performance. Moreover, the magnitudes of the voltage perturbations used here are on the border of small-signal analysis, to minimize the impact of noise on the results. Longer simulations with a larger number of particles would reduce the noise, allowing for smaller perturbation amplitudes and resulting in a more accurate estimation of the frequencies. Finally, the gate delay CV/I = 0.87 ps is found to be close to the experimental value of 1.06 ps reported in [4].

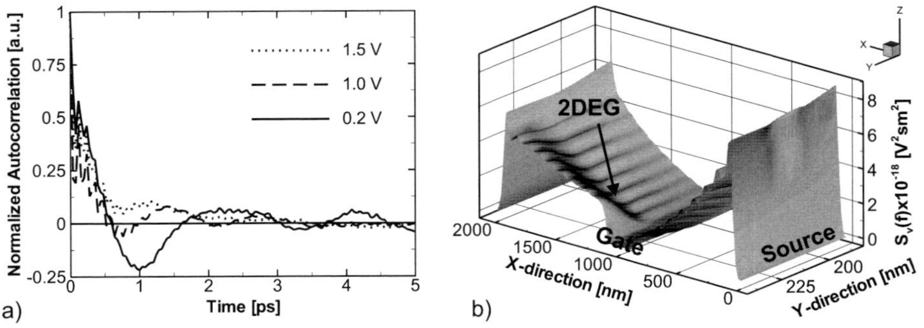

Fig. 4. (a) Voltage autocorrelation function, along a section below the gate. (b) Voltage PSD at 100 GHz as a function of position in the device.

In order to perform noise analysis, the voltage fluctuations about steady state are analyzed in the time domain through calculation of the respective autocorrelation functions, and in the frequency domain through their Fourier transforms yielding their Power Spectral Density (PSD) [11]. The autocorrelation function of the voltage fluctuations is shown in Fig.4(a), along a 2D slice underneath the gate region, for three drain voltages. After an exponential decay corresponding to the dielectric relaxation time, diminishing oscillations are observed and attributed to the concurrent contributions of plasma frequency oscillations and dielectric relaxation [11]. These oscillations tend to disappear as the applied bias is increased, and the negative tail directly following the initial decay is also reduced, resulting in a higher negative differential mobility cutoff frequency [11].

To complete the noise characterization, a plot of the PSD spatial distribution within the device is shown in Fig.4(b) at 100 GHz. As can be seen, the regions where the level of noise is the lowest are the regions where the carrier concentration is low (i.e. in the depleted region under the gate), or in the 2DEG where transport is almost free of scattering. The two peaks on the side coincide with the intersection of the electrode and the different type of semiconductor at the heterojunction. The n^+n regions created at this intersection generate an accumulation of charges responsible for the observed peaks. These peaks would disappear for a fully planar transistor design.

5 Conclusion

Based on recently published data [4], an AlInSb/InSb quantum well transistor has been modeled and simulated using full-band Monte Carlo simulations. A

DC characterization was performed to extract the device current-voltage curves, showing negative differential resistance behaviour for drain voltages above 1 V. A frequency analysis predicted a cutoff frequency and a maximum frequency of oscillation of $f_T = 500$ GHz and $f_{max} = 600$ GHz, respectively. The authors believe these results are higher than what could be measured experimentally and offered explanations for this over estimated values. The noise analysis was performed also based on Monte Carlo simulations and allowed for the mapping of the noise sources within the device. Future work includes improving the accuracy of the frequency analysis, and investigating the impact of scaling the device geometry on its small-signal representation.

6 Acknowledgements

The authors wish to thank M. Fischetti for useful discussions and guidance.

References

[1] T. Ashely, A.B. Dean, C.T. Elliott, R. Jefferies, F. Khaleque and T.J. Philipps, "High-Speed, Low-Power InSb Transistors", Elec. Dev. Meeting, 1997 Technical Digest International, p. 751-754, 7-10 Dec. 1997

[2] E. Sijerčič, K. Mueller and B. Pejčinović, "Simulation of the exclusion/extraction InSb MOSFETs", IEEE W. on Microelec. and Elec. Dev., 2004

[3] T. Ashley, A.B. Dean, C.T. Elliott, G.J. Pryce, A.D. Johnson and H. Willis, "Uncooled high-speed InSb field-effect transistors", Appl. Phys. Lett., **66**, (4), Jan. 1995.

[4] T. Ashley *et al.* "Novel InSb-based Quantum Well Transistors for Ultra-High Speed, Low Power Logic Applications, IEEE, 2004

[5] M. Saraniti and S.M. Goodnick, "Hybrid Fullband Cellular Automaton/Monte Carlo Approach for Fast Simulation of Charge Transport in Semiconductors", IEEE Trans. on Elec. Dev., **47**, (10), Oct. 2000

[6] W. Fawcett, J.G. Ruch, "NDM in InSb", Appl. Phys. Lett., **15**, (11), 1969

[7] O. Ozbas, M. Akarsu, "MC Simulation of Electron Transport in InSb", Turk. Jour. Phys., **26**, p. 283-287, Nov. 2001.

[8] D.K. Ferry, "The onset of quantization in ultra-submicron semiconductor devices" Superlattices and Microstructures, **28**, (5/6), p. 419-423, 2000

[9] S.L. Laux, "Techniques for Small-Signal Analysis of Semiconductor Devices", IEEE Trans. on Elec. Dev., **ED-32**, (10), Oct. 1985.

[10] J. Branlard, S. Aboud, P. Osuch, S. Goodnick and M. Saraniti, "Frequency Analysis of Semiconductor Devices Using Full-Band Cellular Monte Carlo Simulations.", Monte Carlo Methods and Appl., **10**, (3-4), 2004

[11]L. Varani and L. Reggiani, "Microscopic Theory of Electronic Noise in Semiconductor Unipolar Structures", La Rivista del Nuovo Cimento, 1994

Non-Parabolic Model for the Solution of 2-D Quantum Transverse States Applied to Narrow Conduction Channel Simulation

Z. Yang[1], A. Godoy[2], U. Ravaioli[1] and F. Gámiz[2]

[1]Beckman Institute, University of Illinois, 405 N Mathews Ave, Urbana, IL 61801 USA, [2]University of Granada, 18071 Granada, Spain

Summary. Because of the strong confinement in nano-scale semiconductor devices, electrons occupy higher energy states and a simple parabolic band may not be sufficient. A more accurate treatment of Schrödinger equation requires the consideration of non-parabolicity effects [1]. In this work, we extend to 2D a modified version of Schrödinger equation accounting for non-parabolicity [1]. In this work we extend this model to self-consistent solution of the coupled Schrödinger/Poisson problem to obtain transverse energy states and quantum electron density in the cross-section of a prototype quantum channel formed with a T-gate MOS structure [2]. The model can be applied to improve quantum correction in a particle Monte Carlo model, or can be coupled to existing Non Equilibrium Greens Function (NEGF) solvers for transport along the conduction channel.

1 Introduction

Standard quantum models based on Schrödinger equation use a simple parabolic band approximation. This simplification may be adequate for low energy states and low temperature situations. However, for the simulation of realistic silicon devices at normal temperatures, it is desirable to include in the model a more detailed band structure. This is particularly necessary to treat accurately size quantization in narrow nanoscale silicon channels where relevant quantum states tend to be pushed to higher energy levels. In silicon, non-parabolicity effects start to become relevant already at conduction band energy above 0.2 eV. In this paper, we apply a modified 2-D Schrödinger equation with non-parabolicity [3] to a full self-consistent solution of the coupled 2D Schrödinger/Poisson problem for a prototype wire channel in a MOS structure shown in Fig.1 where a T-gate metal structure is used to define the confinement region electrostatically [2]. Results show significant deviation of the quantized energy states when parabolic or non-parabolic bands are considered.

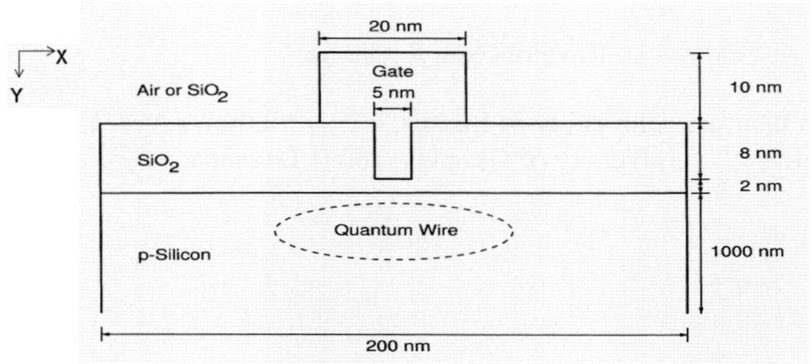

Fig. 1. 2D T-Gate MOS Structure

2 Model and Results

Various approximations can be found in the literature to deal with nonparabolicity effects [1, 3] but commonly a simple truncated series of the dispersion relationship is used to capture the main deviation from parabolic bands in a certain energy range, using the form

$$\frac{\hbar^2 k^2}{2m^*} = E(k)(1+\alpha E(k)) \tag{1}$$

where the non-parabolicity coefficient α has the dimension of an inverse energy. A direct numerical solution of Schrödinger equation incorporating (1) is cumbersome. We extend to 2D an approach presented in [1] expressing first the Schrödinger equation as an infinite series [3] as

$$\frac{1+2\alpha E_x}{2\alpha}\sum_{l=1}^{\infty}\begin{bmatrix}\frac{1}{2}\\ l\end{bmatrix}\left[-4\alpha\frac{\hbar^2}{2(1+2\alpha E_x)^2}\right]^l \times \sum_{q=0}^{l}\left[\frac{1}{m_y^{l-q}}\frac{\partial^{2(l-q)}\psi(y)}{\partial y^{2(l-q)}}\times\frac{1}{m_z^q}\frac{\partial^{2q}\psi(z)}{\partial z^{2q}}\right] + U(y,z)\psi(y)\psi(z) = E_{yz}\psi(y)\psi(z) \tag{4}$$

Rather than attempting to solve this equation by truncating the series, a set of test wave functions are defined in the solution domain and a corresponding non-parabolic dispersion relation is obtained. A shooting method is used to solve the problem, selecting the energy states that correspond to physical wave functions compatible with the non-parabolic dispersion [3]. This methodology is extended here to a self-consistent solution including Poisson equation for the model structure in Fig. 1. The overall convergence for this problem is rapid for the coupled equations, as shown in Fig.2.

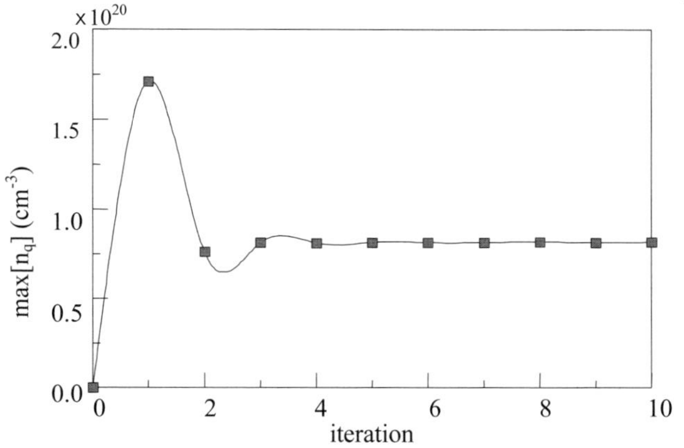

Fig. 2. Maximum of quantum electron density n_q as a function of iteration number

In the non-parabolic case, the energy eigenvalues are expected to be lower than in the parabolic case, because of the higher density of states. A comparisonis given in Fig. 3 for a typical calculation considering the three ladders originated by the three pairs of ellipsoids projected on the [100] plane of the cross-section assumed in this calculation.

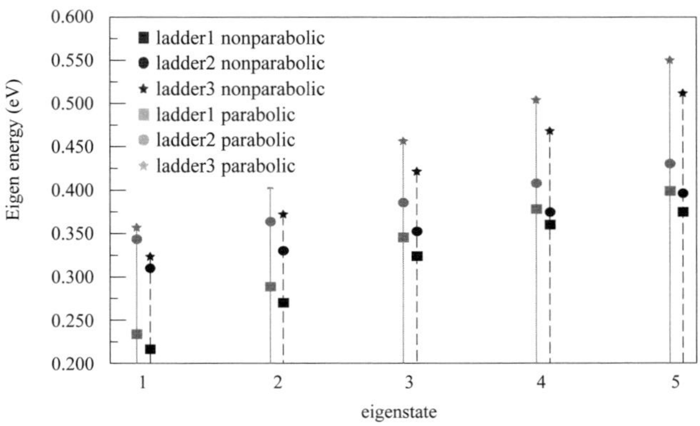

Fig. 3. Comparison of energy eigenvalues for parabolic and non-parabolic model

The quantum electron density for non-parabolic case is slightly higher than the parabolic case as shown in the example of Fig.4.

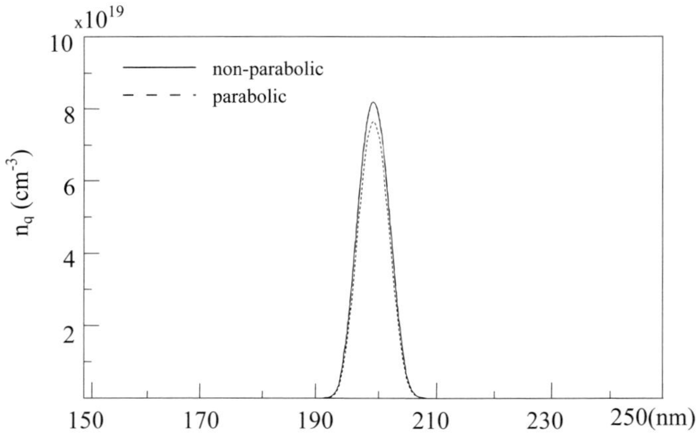

Fig. 4. Quantum electron density on a line parallel to the oxide interface in correspondence of maximum density.

4 Conclusions

We have demonstrated a self-consistent non-parabolic model for 2D solution of quantized states in a conduction channel. Non-parabolicity introduces significant deviations in the energy levels and leads to somewhat higher quantum density. This model is computationally efficient and work is in progress to obtain 2D quantum potential corrections in a 3D particle Monte Carlo simulation.

4 Acknowledgements

US Army Research Office DURINT Program Contract No. SIT52786-08, Semiconductor Research Corporation Contract No. NJ-1044, and Spanish Ministry of Education and Culture Grant No. PR2003-0235.

References

1. J. A. Lopez-Villanueva, I. Melchor, P. Cartujo, and J. E. Carceller, *Phys. Rev B*. **48**, 1626, 1992.
2. A. Trellakis and U. Ravaioli, *J. A. P*hys. **86**, 3911, 1999.
3. A. Godoy, Z. Yang, U. Ravaioli, and F. Gámiz *J. A. P*hys. **98**, 13702, 2005.

Self-Consistent Quantum Transport Theory of Carrier Capture in Heterostructures

T. Kubis, A. Trellakis and P. Vogl

Walter Schottky Institute and Physics Department, Technische Universitaet Muenchen, 85748 Garching, Germany

Summary. We present fully self-consistent NEGF calculations for GaAs/InGaAs heterostructures in a regime where multiple scattering, interference, carrier confinement, and carrier capture must be treated on an equal footing. We include the coupling between $G^<$ and G^R within the device self-consistently and take into account scattering within the leads.

1 Introduction

One of the most widely used methods for a realistic prediction of carrier transport in nanodevices is the non-equilibrium Green's function method (NEGF) [1-3]. The seminal paper by Lake *et al.* provided its first detailed numerical implementation for semiconductor devices [4] and provided the basis for the NEMO package [4]. More recently, this method has been applied to several other devices [5-6].

Nevertheless, it has been notoriously difficult to apply this method to resistive nanodevices where scattering is not weak. In Ref. 4, self-consistent coupling between scattering states (represented by the retarded Green's function G^R) and their occupation (represented by the lesser function $G^<$) has been neglected. This effectively violates Pauli blocking. In Ref. 5, these effects have been taken into account, but all scattering self-energies have been taken as momentum-independent scattering. In addition, the leads were treated ballistically in Ref. 4, whereas Ref. 6 considered periodic boundary conditions for the device.

In this paper, we have implemented the NEGF formalism fully self-consistently for resistive open devices were multiple scattering is significant. We take into account acoustic and polar-optical phonon scattering, including their full momentum and energy dependence, and the electron-electron scattering in Hartree approximation. The coupling between $G^<$ and

G^R is completely taken into account, thereby preserving Pauli blocking effects. In addition, we include the scattering within the leads self-consistently with the device regions close to the Ohmic contacts. The structures specifically considered in this paper are assumed to be homogeneous in the lateral (x, y) direction, and to be in contact with two reservoirs at $z = 0$ and $z = L$. The latter are in thermal equilibrium and we assume room temperature throughout. The electrons are described by a single conduction band with effective mass m^*.

2 Theory and Results

We represent all Green's functions in terms of a discrete real space basis along the device direction and in a plane wave basis in the lateral direction. The only remaining major approximation is the neglect of the off-diagonal elements in the inelastic self-energy in real space. For polar-optical phonons, these off-diagonal elements decay as $\Sigma_{po}(z,z') \propto |z-z'|^{-2}$.

In order to illustrate the influence of scattering in a short device where quantum mechanical effects such as interference, carrier confinement, bound state formation, and carrier capture play a significant role, we have performed systematic NEGF calculations for GaAs n-i-n structures with an InGaAs quantum well within the intrinsic region. The total length of the device is 50 nm with 16 nm intrinsic region embedded in between the two 17 nm n-regions with $n = 1\times10^{18} cm^{-3}$ each. Within the i-region, there is a 12 nm $In_{.14}Ga_{.86}As$ quantum well of 150 meV depth. The polar optical phonon energy is $\hbar\omega = 35\,\text{meV}$.

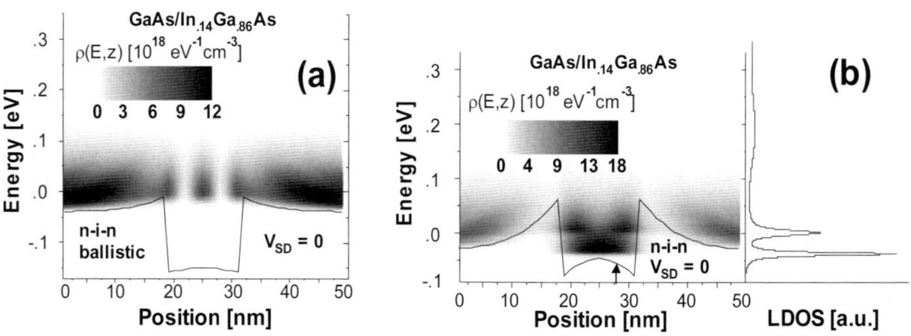

Fig. 1. (a) Contour plot of energy-resolved electron density as a function of energy and position for the GaAs n-i-n with InGaAs quantum well at zero bias, as calculated for zero scattering (ballistic device). The full line shows the self-consistent potential. (b): Same as (a), but with scattering included fully self-consistently.

In Fig. 1(a), we show a ballistic NEGF calculation for the n-i-n structure *in the absence of any scattering* and for zero applied bias. Only the Hartree Coulomb interaction is taken into account. The figure displays a contour plot of $\rho(E,z) = -2\,\mathrm{Im}\int d^2k G^<(z,z,k,E)/(2\pi)^3$, i.e. the local, energy-resolved density, as a function of energy and position within the device. The self-consistent total potential is indicated by a full line. When $\rho(E,z)$ is integrated over the energy, we obtain the density. The zero of energy is always put at the Fermi level of the left contact. Due to the absence of inelastic scattering, the carriers cannot fill the quantum well but occupy a resonance that is derived from the third quantum well state. This explains the interference pattern in the well region at room temperature.

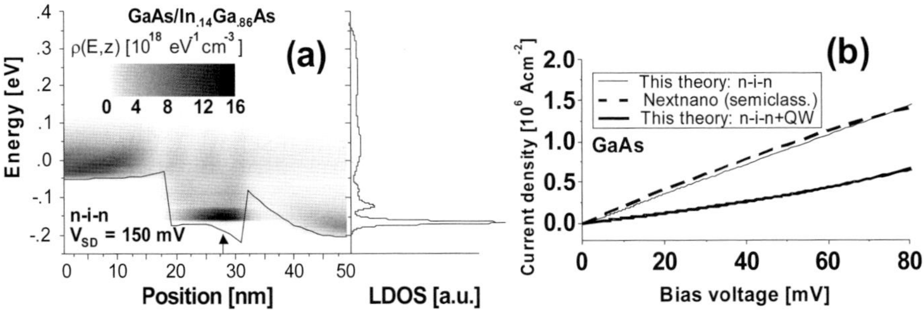

Fig. 2. (a) Same as Fig. 1(b), but with a bias of effectively 150 mV applied. **(b)** Current-voltage characteristics of the GaAs n-i-n structure. The thin line is calculated without a quantum well and compared to a semiclassical calculation carried out with the simulator nextnano3 [7]. The thick line is a full NEGF calculation for the n-i-n structure with a GaInAs quantum well.

The effect of scattering can be seen from the contour graph in Fig. 1(b) where all phonon scattering mechanisms are fully included. To the right of the figure, we display the local density of states near the center of the well region (marked by an arrow). The lowest well state is a true bound state whereas the second one is a resonance state an electron can tunnel out from into the leads. The bottom of the second state lies 59.4 meV below the top edge of the well which is significantly larger than an optical phonon energy. The lowest bound state lies 37.6 meV below the second level which is also a bit larger than $\hbar\omega$. Because we treat scattering within the self-consistent Born approximation, the combination of elastic and inelastic scattering can fill both states in the stationary state.

In Fig. 2(a), we show the same device under an applied bias of effectively 150 mV. At the start of the self-consistent calculation, we apply a bias of 200 mV and allow the Fermi levels to adjust to obey charge neutrality within the device and obtain flat band conditions at the contacts. This requires to include the self energies Σ^R near the contacts also within the leads so that the transitions from lead to device are smooth. As one can clearly see, the resonance corresponding to the second well state has been largely emptied by tunneling towards the right lead so that only the lowest well state remains significantly occupied.

Fig. 2(b) displays the I-V characteristics of this diode for small bias. The NEGF calculation agrees very well with a standard semiclassical calculation [7] for a n-i-n structure without a quantum well. The mobility is smaller in the case with the quantum well due to the additional electron charge accumulation within the well that builds up due to the inelastic scattering.

Acknowledgement

Financial support by Deutsche Forschungsgemeinschaft (SFB 631) is gratefully acknowledged.

References

1. Schäfer, W.; Wegener, M.: *Semiconductor Optics and Transport Phenomena*, Springer, 2002.
2. Ferry, D. K.; Goodnick, S. M.: *Transport in Nanostructures*, Cambridge University Press, 1997.
3. Datta. S.: *Electronic Transport in Mesoscopic Systems*, Cambridge University Press, 1995.
4. Lake, R.; Klimeck, G.; Bowen, R. C.; Jovanovic, D.: *J. Appl. Phys.*, 81, 7845, 1997.; see also http://dynamo.ecn.purdue.edu/~gekco/nemo1D/index.html
5. Wacker, A.: *Physics Reports*, 357, 1-111, 2002.
6. Wacker, A.; Lee, S.-C.: *Phys. Rev. B*, 66, 245314, 2002.
7. Hackenbuchner, S.; Sabathil, M.; Majewski, J.A.; Zandler, G.; Vogl, P.; Beham, E.; Zrenner, A.; Lugli, P.: *Physica B*, 314, 145 (2002); see also http://www.wsi.tum.de/nextnano3

SPRINGER PROCEEDINGS IN PHYSICS

64 **Superconducting Devices and Their Applications**
Editors: H. Koch and H. Lübbing

65 **Present and Future of High-Energy Physics**
Editors: K.-I. Aoki and M. Kobayashi

66 **The Structure and Conformation of Amphiphilic Membranes**
Editors: R. Lipowsky, D. Richter, and K. Kremer

67 **Nonlinearity with Disorder**
Editors: F. Abdullaev, A.R. Bishop, and S. Pnevmatikos

68 **Time-Resolved Vibrational Spectroscopy V**
Editor: H. Takahashi

69 **Evolution of Dynamical Structures in Complex Systems**
Editors: R. Friedrich and A. Wunderlin

70 **Computational Approaches in Condensed-Matter Physics**
Editors: S. Miyashita, M. Imada, and H. Takayama

71 **Amorphous and Crystalline Silicon Carbide IV**
Editors: C.Y. Yang, M.M. Rahman, and G.L. Harris

72 **Computer Simulation Studies in Condensed-Matter Physics IV**
Editors: D.P. Landau, K.K. Mon, and H.-B. Schüttler

73 **Surface Science**
Principles and Applications
Editors: R.F. Howe, R.N: Lamb, and K. Wandelt

74 **Time-Resolved Vibrational Spectroscopy VI**
Editors: A. Lau, F. Siebert, and W. Werncke

75 **Computer Simulation Studies in Condensed-Matter Physics V**
Editors: D.P. Landau, K.K. Mon, and H.-B. Schüttler

76 **Computer Simulation Studies in Condensed-Matter Physics VI**
Editors: D.P. Landau, K.K. Mon, and H.-B. Schüttler

77 **Quantum Optics VI**
Editors: D.F. Walls and J.D. Harvey

78 **Computer Simulation Studies in Condensed-Matter Physics VII**
Editors: D.P. Landau, K.K. Mon, and H.-B. Schüttler

79 **Nonlinear Dynamics and Pattern Formation in Semiconductors and Devices**
Editor: F.-J. Niedernostheide

80 **Computer Simulation Studies in Condensed-Matter Physics VIII**
Editors: D.P. Landau, K.K. Mon, and H.-B. Schüttler

81 **Materials and Measurements in Molecular Electronics**
Editors: K. Kajimura and S. Kuroda

82 **Computer Simulation Studies in Condensed-Matter Physics IX**
Editors: D.P. Landau, K.K. Mon, and H.-B. Schüttler

83 **Computer Simulation Studies in Condensed-Matter Physics X**
Editors: D.P. Landau, K.K. Mon, and H.-B. Schüttler

84 **Computer Simulation Studies in Condensed-Matter Physics XI**
Editors: D.P. Landau and H.-B. Schüttler

85 **Computer Simulation Studies in Condensed-Matter Physics XII**
Editors: D.P. Landau, S.P. Lewis, and H.-B. Schüttler

86 **Computer Simulation Studies in Condensed-Matter Physics XIII**
Editors: D.P. Landau, S.P. Lewis, and H.-B. Schüttler

87 **Proceedings of the 25th International Conference on the Physics of Semiconductors**
Editors: N. Miura and T. Ando

88 **Starburst Galaxies**
Near and Far
Editors: L. Tacconi and D. Lutz

89 **Computer Simulation Studies in Condensed-Matter Physics XIV**
Editors: D.P. Landau, S.P. Lewis, and H.-B. Schüttler